AF520587

Evolution of Geographical Thought

Evolution of Geographical Thought

SIXTH EDITION
(Revised and Enlarged)

MAJID HUSAIN

RAWAT PUBLICATIONS
Jaipur • New Delhi • Bangalore • Guwahati • Kolkata

ISBN 978-81-316-0715-2 (HB)
ISBN 978-81-316-0716-9 (PB)

First Edition, 1984
Second Edition, 1988
Third Edition, 1995
Fourth Edition, 2001
Fifth Edition, 2004
Sixth Edition, 2015

Reprinted, 2021

Published by
Prem Rawat for **Rawat Publications**
Satyam Apts, Sector 3, Jawahar Nagar, Jaipur 302 004 (India)
Phone: 0141 265 1748 / 7006 Fax: 0141 265 1748
E-mail: info@rawatbooks.com
Website: www.rawatbooks.com

New Delhi Office
4858/24, Ansari Road, Daryaganj, New Delhi 110 002
Phone: 011 2326 3290

Also at *Bangalore, Guwahati* and *Kolkata*

Typeset by Rawat Computers, Jaipur
Printed at Nice Printing Press, New Delhi

To
my grandfather
who visualized geographical variations
as evidence of the handiwork and
almighty wisdom of God

Contents

PART III
CONTEMPORARY MOVEMENTS IN GEOGRAPHY

Foreword

Dr. Majid Husain has written an interesting book on the *Evolution of Geographical Thought*. He has dealt with the philosophy, history and methodology of geography. The book is divided into three sections: pre-history, modern geographical thought and contemporary movements in geography from antiquity to the 1980s.

The physical aspects of the earth and the elements it contains have part of man's search for knowledge for a long time. This vast knowledge has been carefully set out in this book which is indispensable for instant reference as well as for reading at leisure. The book gives quick access to fascinating and wide-ranging subjects covered in it. It is a work of great scholarship and is an absorbing reference book. Written in a clear and interesting style, it reveals the profound knowledge of the author on the subject of geographical thought. I have no doubt, the book will be of immense help to those interested in the subject and will also encourage them to search further and learn more.

M. Hidayatullah
Former Vice President of India

Preface

I have carefully gone through Dr. Majid Husain's work on the *Evolution of Geographical Thought* and have been impressed by the wide canvas covered by him. The scholar has attempted to build up the story of geographical thoughts, ideas and knowledge right from the early Greek period to modern contemporary geography.

The book has twelve chapters in all which have been divided into three parts. Each chapter is a landmark in the evolution of geographical thought. The first two chapters are devoted to the development of scientific geography during the Greek and early Roman periods and its setback with the rise of Christianity. The ancient Indian geographical concepts have been systematically presented in the third chapter. The role of Arab geographers in advancing geographical ideas and geographical knowledge has been treated at length. The influence of the Greeks on Arab geographers, and the transition to renaissance geography generated by Arab geographical ideas, have been rightly emphasized in the fourth chapter.

The impact of explorations on geographical concepts has been discussed in the fifth chapter. It also gives an account of the re-emergence of scientific geography through the works of Kant and Varenius. This resurgence of scientific trend in the development of geography was reinforced by the scientific works of scholars of the eminence such as Alexander von Humboldt and Carl Ritter. Their comprehensive, empirical and scholarly works firmly established geography as a scientific discipline in Europe. In the seventh chapter, various schools of geography have been described.

The rise of any discipline to maturity is always marked by philosophical conflicts of ideas and ideologies. Chapters eighth and ninth discuss these controversies in detail. Dr. Husain rightly points out that geographical concepts such as determinism and possibilism and dichotomies in geography ought not to be conceived as contradictions. They rather represent the diversity of human response to environmental situation and variety of approaches to the study of geography.

In order to be treated at par with contemporary scientific disciplines, geography could not escape their influence and had to move the realm of quantitative revolution and consequently took to the model building, paradigms and systems analysis. No explanation of geography is complete without these trappings of contemporary science. These have been discussed in the tenth and eleventh chapters.

That modern geography is also value oriented is stressed by the author in last chapter of the book.

The greatest merit of this book is that it brings together in one volume the entire gamut of geographical thinking and development from the ancient Greek to the contemporary quantitative explanatory phase. It underscores the point that geography has not cast off its descriptive gazetteer role to blossom into a mature modern scientific discipline.

S. Manzoor Alam
Former Vice Chancellor
University of Kashmir, Srinagar

Note on the Sixth Edition

It is almost ten years the book was revised in the form of the fifth edition. During this period, I received numerous letters from the graduate and post-graduate students of geography and the concerned faculty members in appreciation of the book, its utility and simple style of presentation of the difficult and complex concepts. They also gave a number of suggestions to make the book more useful and readable. Looking at the wonderfully generous reception and the useful suggestions given by the eminent scholars, the book has been thoroughly revised and updated, making it a comprehensive textbook for those who are interested in knowing more about the philosophy and methodology of geography. The main objective of the sixth edition is to incorporate a number of relevant new topics and concepts to streamline the text which may go a long way in expanding the geographical base of the interested students about the philosophy and methodology of geography. Moreover, the present edition is an attempt to overcome the shortcomings of the previous editions. This new edition is presented in a larger format with all improved maps and illustrations. Biographical details of geographers have been updated and assisted with photographs, wherever possible. Besides removing number of errors from the previous editions, the entire text has been revised and updated for better clarity and readability.

The chapters dealing with the contributions of Greek, Roman, Indian, Arab and Chinese geographers, and also the sections on Modern Geographical Thought and Contemporary Movements in Geography have been restructured to make them more easy to understand. The book now introduces to the students a contemporary sub-discipline of Feminist Geography. This chapter seeks to explore how gender relations differ throughout the world and have important bearings on the social structure and geography of an area. In the concluding part of the book, a description and explanation of some of the important technical geographical terms, having their origins from different languages, has been given. A new comprehensive list of biographic notes on selected geographers is also appended at the back.

Written in a cogent and highly readable style and updated with the latest developments in geography, it is hoped that this book will play a creative role in preparing students for their graduate and post-graduate examinations. It will also serve the need of the candidates who prepare themselves for the Indian Administrative Service examination and for the National Eligibility Test (NET). The book, I am sure, will be of immense help for the students and teachers alike.

Majid Husain

List of Figures

Introduction

Geography is all around us, it is an inextricable part of our everyday life. The book attempts to capture the spirit of geography. It provides a full-length survey of major theories and debates, key thinkers, models and schools of thought in geography. Presenting an introduction to the philosophy, history and methodology of geography, the book maps the intellectual changes in the discipline over the years. These are the aspects of the discipline which have been increasingly discussed in books and research papers in recent years. In the present work, an attempt has been made to define and explore complex ideas and difficult concepts in an easily intelligible and lucid manner. Moreover, developments in the subjects have been viewed in the light of influential thought in many other disciplines.

The book incorporates three parts, namely, prehistory of geographical ideas, modern geographical thought, and contemporary movements in geography. The first part deals with the historical development of geographical ideas, discoveries and tools which helped in the preparation of an outline of the world. The second part of the book gives an account of the discoveries and explorations made by Columbus, Vasco da Gama, Magellan and Captain Cook. It also highlights the works of Kant, Varenius, Humboldt and Ritter, and also various German, French, American, British and Soviet geographers who helped in the re-emergence of geography as a scientific discipline. Besides, concepts like dualism and dichotomy, quantitative revolution in relation to the concept of paradigm in geography, models in geography, the new critical revolution emphasizing the influence of values on research, the search for geographical synthesis as it developed from simple description to system analysis have been analysed. The third part of the book also puts together the diverse philosophies relevant to geography. The basic themes of major contemporary philosophies (positivism, pragmatism, functionalism, existentialism, idealism, behaviouralism, humanism and Marxism), their impact on geography and their potential as a forum for geographical research have also been investigated. A whole new chapter on feminist geography is also included here.

The history of geographical thought is largely a history of dichotomies and abandoned theories. Throughout history, geographers have developed successive images concerning the arrangement of things on the face of the earth. The mental images and explanations of one generation were seldom satisfactory to succeeding generations. There was a continuous search for new and more perfect images and for explanations that conform to contemporary scientific analysis. These changing images and theoretical structures, set up to give them plausibility, make up the content of the history of geographical ideas.

In the initial phase, geography was considered as a description of the earth surface and explorations. In the 18th century, it gradually emerged as a discipline of the earth science dealing with man-nature interaction. For examining this man-nature interrelationship, ideographic and monothetic approaches were adopted by geographers. Recent years have, however, seen a more active discussion on philosophical themes in geography.

The book opens with a discussion on the prehistory of geography. The first chapter deals with the contributions made by Hecataeus, Homer, Herodotus, Eratosthenes and Hipparchus. The Greek contribution reveals that geography has a longer genealogy than any other science. During the Greek period, geography was considered to be a branch of knowledge of cosmology which concerns the study of stars, determination of latitudes and longitudes, description of places, mountains, rivers, tribes and nations. This was the period when the habitable world was extremely limited and therefore well within the comprehension of scholars. Consequently, the Greek philosophers were well-versed in history, philosophy, astronomy, politics, medicine, mathematics, and geography. The world maps prepared by Hecataeus, Herodotus and Eratosthenes have also been examined at length in this chapter.

The second chapter reveals the contribution made by the Romans in the various branches of geography. The Romans, with the help of their well-disciplined armies, conquered distant countries and they added many of the unknown lands of Europe including Britain to the world map. Polybius and Posidonius contributed to physical geography, while Strabo and Ptolemy were the well-known regional and mathematical geographers, respectively. Strabo was a prolific writer who wrote seventeen volumes on world geography, while Ptolemy laid the foundation of mathematical geography. For centuries, Ptolemy's *Almagast* and *Geography* dominated and guided the geographical works of Arab scholars and cartographers of the Great Age of Discoveries. Ptolemy developed the concept of Hipparchus that a map should be based upon points of which the latitudes and longitudes are known. This led to the concept of 'graticule' which helped in the development of map projection and the science of cartography. The third chapter presents a concise account of the ancient Indian geographical concepts which hinged on Hindu mythology, philosophy, epics, history, sacred laws, and travelogues. The contribution of Chinese to cosmography, geography and cartography has also been briefly described.

The fourth chapter gives an account of the progress made in geographical knowledge by the Arabs. The human progress had been like the confused movements of a crowd, whose advance is only to be clearly seen after many swayings and stoppages rather than the orderly forward march of an army along a military road. Early Arabian geography underscores this fact. For about two hundred years, there was little contribution of the Arabs to geography. Geography and other sciences, however, started with great vigour under the first four of the 'Abbasid Caliphs'. They collected geographical ideas from the Greeks and the Romans in the west and from China, India and Persia in the east. They added to these the results of their own observations, researches, explorations and discoveries. They discovered new lands and diffused geography from the Atlantic to the Pacific Ocean. The major contributions made by Ibn-Hawqal, Al-Masudi, Al-Biruni, Al-Idrisi, Ibn-Battutah, and Ibn-Khaldun have been examined in this chapter. The Arabs, in fact, remained masters of geography and allied sciences for several centuries.

The 'Renaissance' in Europe brought about a sudden and complete emancipation from the old modes of thought. In the period of Great Age of Discoveries, scholars like Sabastain, Carpenter, Varenius and Kant laid the foundations of the classical period. Varenius was the first scholar who distinguished between general geography (geography generalis) and special geography (regional geography). Moreover, he imbibed or assimilated the most recent ideas of Copernicus, Kepler and Galileo and accepted the heliocentric universe. Kant, who lectured in physical geography at the University of Koningsberg, stressed the point that knowledge about the earth as home of man was necessary support for philosophical studies. He was able to give geography a respectable place in the field of learning.

In the second part, the first two chapters (6 and 7) deal with the development of modern geographical thought, the foundation of which was mainly laid by Alexander von Humboldt and Carl Ritter. These two great German geographers loom large across the pages of history of science. They brought the period of classical geography to an end and produced massive synthesis of the voyages and explorations which helped in the development of new concepts and methods of study of man-nature interaction.

The seventh chapter examines the development of geography and geographical thought in Germany, France, Britain, the United States and the Soviet Union. A major innovation in the world scholarship took place in the 19th century. The university as an institution appeared in Medieval Europe when charters were issued by religious and secular authorities giving certain facilities and the right to teach social and natural sciences to universities. Geography, as a field of advanced study, first appeared in Germany in 1874 which was taught by professionally qualified individuals. Within a few decades, geography departments were established not only in Germany, France and Britain but also in many other parts of the world. Thus, geography as a discipline came into existence that could establish the paradigm of geographical study. The development of various philosophies and methodologies adopted by geographers of various schools of thought has also been analysed in this chapter.

In the eighth chapter, the dichotomy between determinism and possibilism has been examined in historical perspective. Geographical research after Darwin was primarily concerned with recognizing the laws of nature. In human geography, this approach led to a rather deterministic view of struggle and survival. Nature was studied with open eyes, seeking as objectively as possible, to identify the forces of processes which governed the formation of valleys, uplands and coastlines. A more restricted view was taken of human activity and the relationship between nature and man was considered to be of primary interest. Man's achievements were to be explained as consequences of natural conditions. The British philosopher, Herbert Spencer, enunciated 'Social Darwinism'. He believed that human societies closely resembled animal organisms and that human societies must struggle in order to survive in a particular environment as much as plants and animal organisms do. The role of Ratzel and his pupil Semple, as staunch supporters of the philosophy of determinism, has been highlighted.

The deterministic approach was, however, criticized in the later decades of the 19th century. The French scholar, Febvre, founded the school of 'possibilism'. He and his followers believed that there are no necessities, but only possibilities. This viewpoint was stressed by Paul Vidal de la Blache and Jean Brunhes in France and later on by Bowman and Carl Sauer in USA. The conceptual and philosophical controversy between determinism and possibilism has been elucidated with examples from the different regions of India and the world.

Chapter ninth examines the concept of dualism and dichotomies in geography at length. In the delineation of the sphere of geography, its concepts and methodology, dichotomies have always been there and they exist as of now too. Some of the major dichotomies are general geography vs. regional geography, physical geography vs. human geography and historical geography vs. contemporary geography. Most of the dichotomies have been discussed in historical perspectives and it has been established that in the final analysis they do not oppose but support one another.

Theory building for generalization has been an area of major concern for geographers. This feeling became stronger after the Second World War. Now more stress is being laid on abstract models. The tenth chapter covers quantitative revolution, paradigms, system analysis and regional concept. It also examines Kuhn's paradigms, who postulated that science is not a well-regulated activity where each generation automatically builds upon the results achieved by earlier researchers. The concepts of ideographic, monothetic, regional synthesis and system analysis have been elaborated. A detailed discussion has also been given on region as a concept, its attributes and relevance in geographical studies.

In the second part, a separate chapter (eleventh) has been added to discuss the types, attributes, utility and shortcomings of geographical models.

The third part of the book has two chapters - modern themes in geographical thought and feminist geography. Geographical thought, at any point of time, is a manifestation of the interaction between the prevailing philosophical viewpoints and the major methodological approaches in vogue. Because of the extreme diversity of viewpoints with regard to both philosophy and methodology, there has been a constant extension and even a shift in the focus of discipline. This chapter examines the diverse philosophies prevalent in geography with special stress on positivism, pragmatism, functionalism, existentialism, idealism, realism, Marxism, radicalism, behaviouralism, humanism and postmodernism. Stress has also been laid on geography and public policy. The role of geographers in building a society of well-being has also been highlighted which shows that in the spatial organization of social amenities and other socio-economic phenomena the role of geographers is indispensable. The last chapter looks at feminist thought in geography in the context of broader economic, political, philosophical and social-theoretic tendencies. Feminist geography bears the marks of women's exclusion from the upper echelons of the organized discipline and from a research agenda which long ignored their existence.

The author is profoundly grateful to Justice M. Hidayatullah, former Vice President of India, who inspired him to embark upon this venture. The generous 'Foreword' written by him to the book is a great moral booster to the author. Prof. Shah Manzoor Alam, former Vice Chancellor, University of Kashmir, Srinagar, was a continuous source of suggestions, comments and assistance. He also encouraged the author by writing 'Preface' to the book, for which he deserves author's deep gratitude.

In the writing of the present book, the author is aware of the debt of gratitude which he owes to late Prof. S.M. Tahir Rizvi, late Prof. S.M. Ali, late Prof. Mohammed Shafi, late Prof. Enayat Ahmad, late Prof. Moonis Raza, late Prof. S.M. Rafiullah, late Prof. Mohammad Anas, late Prof. D.D. Deshpande, late Prof. F.A. Ahmad, late Prof. N. Ahmad, late Prof. G.S. Gosal, Late Prof. A. Ahmad and Late Prof. R.C. Sharma. The synopsis of the book was discussed with them at different occasions and the author is obliged to them for their constructive suggestions and modifications. My colleagues at the Aligarh Muslim University, Aligarh, J.N.U., New Delhi, University of Kashmir, Srinagar, North-Eastern Hill University, Shillong, and Jamia Millia Islamia, provided encouragement, for which I would like to express my thanks. Valuable assistance was rendered by Ms. Ghadir Fatima, Mr. Khurshid Anwar, Mr. Ali Mohammad Mir, Mr. U. Singh Bali, Mr. Shokat Ali Bhat, Dr. Haseena Hashia and the technical staff of the department who assisted in the preparation of maps and diagrams. Ms. Ulfat and Mr. G.M. Shaikh helped in the correspondence and secretarial aspect of the work for which the author expresses thanks. The author is indebted to Rawat Publications, Jaipur which published the first edition of the book in 1984 and improved the subsequent editions, making it indispensable for the graduate and post-graduate students of geography as well as those appearing in the IAS and other civil services examinations with geography as their optional subject. Rawat Publications has a dedicated, committed and skilled team who not only improved the quality of book but also popularized it in all parts of the country, the neighbouring countries and other English-speaking countries the world over. The publication of the sixth edition is evident of its popularity.

The author wishes to express special thanks to his wife for her forbearance over the last several years. Thanks are also due to Javed Alam, Dr. Shagufta Bano Husain, Abdul Rehman Mehdi, Mansoor Majid, Majid Abdul Rahman, Madeeha Javed, Jibran Mansoor, Rashad A. Rahman, Fareeha Javed, Areeb Mansoor, Hiba-A-Rahman, Uzair-Alam and other family members who helped in this endeavour.

In spite of the action taken with regard to the valuable suggestions and comments received from learned scholars and colleagues, imperfections are bound to creep in. In order to overcome the shortcomings of the present edition, comments and suggestions from teachers, students and experts are gladly invited so that subsequent editions could be improved.

Majid Husain

Part I

Prehistory of Geographical Ideas

1

Greeks

The Pioneers of Prehistorical Ideas

Geography has a longer genealogy than any other science. The earliest records of man's interest in the nature of the physical world around him contain observations and speculations of geographical type. In the ancient period, geography grew out of explorations, mapping of the areas known and the speculation about the material collected. It is difficult to trace the development, diffusion and spread of geographical knowledge during the prehistoric periods. The archaeological evidences obtained from the different centres of civilization, however, reveal that in the initial stage geography marched at a snail's pace. There is a consensus that all civilizations contributed to the development of geographical concepts. In fact, geographical knowledge had been a concomitant of civilization, which had been pursued by different civilizations according to the circumstances of their physical and geographical surroundings. Nevertheless, it is believed that astronomy flourished in Chaldaea, Assyria, and Babylonia, where the skies were mostly clear; geometry developed in the fertile, arable lands of Nile Valley; and physical geography in Greece, characterized by diverse relief features and indented coastline. Almost all the great civilizations originated in river valleys, were nourished by trade, and came to maturity in cities. The progress of a civilization is marked by man's increasing control over nature through applied mathematics and science, the evolution of writings, and political and religious organizations. The union of the peoples of the upper and the lower Nile some 5,000 years ago heralded the first major civilization in history. During its development, mathematics made possible the building of Gizeh pyramids. A most vigorous civilization developed in the fertile valleys of Mesopotamia. Here, the Babylonians and Assyrians had adopted the cuneiform writing. Sumerians were the first to develop the city states in the Tigris and Euphrates valleys.

Before the rise of Hellenic culture, geography was regarded as the knowledge of topographical features, mountains, rivers and places of one's own country and its boundaries. Later on, maritime trade and commercial relations provided a store of geographical information. In Asia Minor, along the coast of the Mediterranean Sea, Egyptians and Jews were the early influential rulers. *The Old Testament* – the religious book of the Jews – contains numerous geographical details concerning Palestine and neighbouring districts.

PHOENICIANS

After the Jews and Egyptians, it was the Phoenicians who contributed to the advancement of geographical knowledge. Phoenicians occupied Asia Minor (coastal Turkey, Lebanon, Syria, Israel) while Tyre (Tyr) and Sidon (Saida) were their major ports and towns. The Phoenicians period is about 3,000 years before Christ. Gadeira (Gadis) was established by the Phoenicians as early as 1,100 BC, while Carthage (near the present Tunis) and Utica cities were established in about 813 BC along the northern coast of Libya (Africa) as the colonies of Phoenicians. The Phoenicians, too, developed the world's first phonetic alphabet. It was made up entirely of consonants, like the modern Semitic alphabet. The Greeks added the short vowels to the phoenician alphabet. Thus, Phoenicians were the first repositories of geographical knowledge. But their narrow, selfish and secretive policies prevented them from communicating to others the information they had obtained about the distant nations and trading centres.

GREEKS

The Greeks were pioneers in many branches of knowledge. Their period is known as the 'Golden Age of Greece'. They borrowed many of the concepts of astronomy, geometry and mathematics from the Egyptians, Chaldaeans, Sumerians and Assyrians. The Egyptian and Sumerian priests collected large number of observations on the position and movement of celestial bodies under the clear skies and fertile levelled plains. The geographical knowledge in the earliest ages was, however, limited to the countries and islands adjacent to Greece, and to the group of nations, surrounding the Aegean Sea.[1] All beyond was vague and indefinite: derived from hearsay reports which would be mixed up with mythological fancies and fables.[2]

The Greeks possessed philosophical and scientific aptitude, versatility of intellect, inquisitive nature and comprehensiveness of mind. These qualities of character enabled them to make acute observations to record the peculiarities of their country and parts of the world they visited. Moreover, by temperament, the Greeks were not secretive unlike Phoenicians who were given to obsessive secrecy.

The location of Greece, situated on both sides of the Aegean Sea, was also conducive to geographical study.[3] The great diversity in its topography and physical features provided great impetus to the growth and development of geography. In fact, Greece is a hilly and undulating country. Many of its peaks become snow-clad in winter. Many of the rivers are mainly torrents which debouch from the hills into the coastal plains. There are numerous straits which penetrate in the land making the coastline highly indented. Such a coastline helps in the development of harbours and ports. The capes project into the sea inspiring traders to go to the neighbouring islands and nations. The Greeks, living in such a physical setting, were able to make tremendous advancements in the knowledge of geomorphology, climatology and oceanography. Moreover, in the limestone topography of the mainland, many of the rivers would disappear emerging once again from the subterranean course. Greece, which lies in the weak zone of the earth, records tremors and earthquakes. These phenomena, the Greek scholars tried to study and explain. There are numerous hot springs and volcanoes which were considered by some of them as supernatural things while others tried to explain them with scientific reasonings.

Between the 5th and 3rd century BC, the Greek colonies (Figure 1.1) were established in different parts of the Mediterranean Sea and Euxine (Black Sea).[4] In the 5th century BC, Miletus, owing to its location and the colonies on the Euxine, became the main centre of geographical enquiry. The early expedition of Hanno along the western coast of Libya (Africa) and that of Alexander towards the east, provided factual knowledge to the Greeks about distant lands and their people. The establishment of the famous Library Museum at Alexandria provided an impetus to Greek scholars to know more about the phenomena of nature, places and people. The scholars exchanged their views and itineraries with the traders and navigators. It was at this Library Museum that Eratosthenes and Hipparchus made their scientific observations about the size, shape and circumference of the earth.

Homer

Homer was a very well-known Greek. He was probably the greatest Greek poet. His work was published in the form of *Illiad* and *Odyssey*. These long epic poems describe the episodes of Trojan war (1280 and 1180 BC) and contain valuable information about historical geography of the then known world. Homer believed the earth to be a place of circular form, surrounded on all sides by the Ocean river. The sky (vault of heaven), which he conceived as a solid concave surface, equals in extent to the earth, resting on tall pillars. These pillars are in the charge of Atlas mountains. It is repeatedly stated in the Homeric poems that the Sun rose out of the Ocean Stream, and again sank into the same at its setting.[5] How it was carried back to the point from which it was to start afresh on its course no one in his days ever troubled himself to

Homer

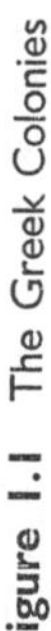

Figure 1.1 The Greek Colonies

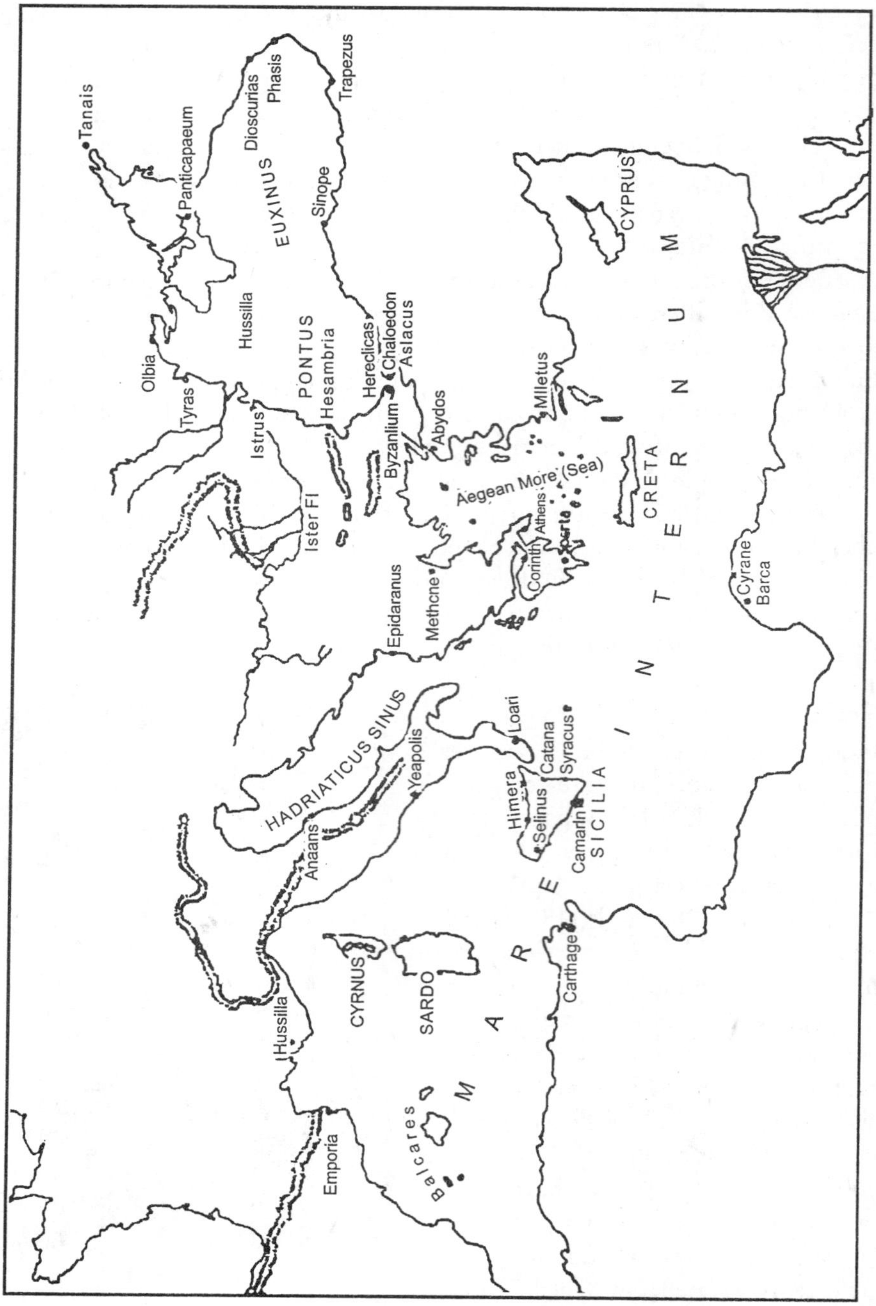

inquire.[6] The stars also are represented as following the same course and bathing everyday in the waters of the Ocean.

Homer described the four winds coming from the four different directions. *Bores* was the north wind, strong and cool with clear skies; *eurus* was the east wind, warm and gentle; *notus* was the south wind on the front of an advancing storm, wet and sometimes violent; and *zephyrus* was the west wind, dreaded, balmy with gale force.[7]

Homer was not familiar with the terms 'Europe' and 'Asia', but at some later time the name 'Europe' was applied to the shore of the Aegean Sea towards the setting sun, and 'Asia' was applied to the shore towards the rising sun. The origin of these names, however, is not certain.[8]

Thales of Miletus

Thales of Miletus

Thales was a resident of Miletus. Miletus in Ionia (in present-day Turkey) is situated on the eastern side of the Aegean Sea near the mouth of Meander River (now the Meanders). Miletus became a major centre of trade and commerce and attracted Phoenician and Greek ships from all around the Mediterranean and the Black Sea. The sailors and merchants brought to Miletus a wealth of information about Europe, Asia and the land north of the Black Sea. Thales was the first Greek thinker, philosopher and traveller who originated several basic theorems of geometry. He was also the first person to initiate measurement of the earth and to try to locate things on the face of the earth.[9] He lived in the 7th century BC. Thales was a practical businessman. On a trip of Egypt he observed that some priests at work were measuring angles, baselines and computing areas. Thales was inspired by the Egyptian monks and he solved many problems of geometry and trigonometry. There are six geometric propositions that are credited to him: (1) that the circle is divided into two equal parts by its diameter; (2) that the angles at either end of the base of an isosceles triangle are equal; (3) that when two parallel lines are crossed diagonally by a straight line, the opposite angles are equal; (4) that the angle in a semicircle is a right angle; (5) that the sides of similar triangles are proportional; and (6) that two triangles are congruent if they have two equal angles and a side respectively. He visualized the earth as a disc floating in water.[10] Thales is remembered for his cosmology based on water as the essence of all matter and for his prediction of an eclipse of the Sun, generally agreed to be that of 28 May 585 BC.

Anaximander (610–546 BC)

Anaximander

Anaximander was a disciple of Thales in Miletus (modern Turkey). Anaximander was a great philosopher, astronomer, mathematician and geographer. Pythagoras and Anaximenes were his pupils. He is often called the 'Father of Cosmology' and founder of astronomy. Anaximander explained how the four elements of ancient physics (air, earth, water and fire) are formed and how earth landforms are formed through their interactions. He attributed thunder and lightning to the intervention of elements, rather than the divine causes. In his system, thunder results from the shock of clouds hitting each other. He is credited with the introduction into the Greek world of a Babylonian instrument known as 'gnomon'.[11] Gnomon is a pole set vertically above a flat surface on which the varying position of the Sun could be measured by the length and direction of the shadow cast by the vertical pole. This was just like a sun dial.[12] With the help of gnomon, noon

Figure 1.2 Anaximander's World Map

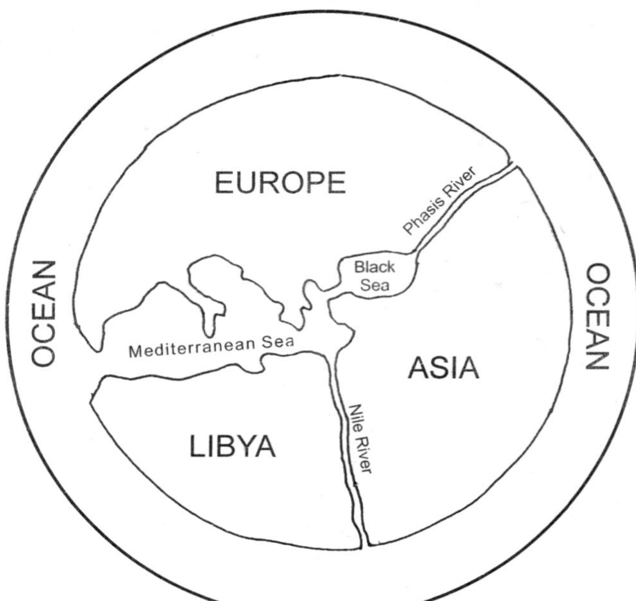

could be established by noting when the shadow was the shortest; the noon shadow provided an exact north-south line, or meridian (from merides, meaning noon).[13]

Anaximander was a pioneer. He was the first who prepared a world map to scale, though the Sumarians before him had drawn pictorial maps of some of their cities as early as 2700 BC (Figure 1.2). In this map, Europe has been shown in the centre of the world. The map was circular and was bounded on all sides by the Ocean river. Thales and Anaximander are generally recognized as the founders of mathematical geography.

Hecataeus (550–476 BC)

Hecataeus was a resident of Miletus. Very little is known about his date of birth and early life but he was from a wealthy family a great statesman, historian, and pioneer geographer. He was the first writer of Greek prose. Hecataeus collected and classified information of the known Greek world and the unknown distant areas.[14] His main book is *Ges-periodos* (description of the earth) which was published most probably before the end of the 6th century. It is the first systemic description of the world and because of this fact Hecataeus is known as the 'Father of Geography'. It also gives a detailed account of the Mediterranean Sea, islands, straits, and describes the general outline of all the countries of the world. Unfortunately, most of this work is not in existence.

Hecataeus travelled the then known world extensively. According to Herodotus, Hecataeus had visited Thabes in Egypt (Figure 1.3) but with regard to other countries there is not much direct evidence to prove that he visited the distant lands of Spain, Ister (Danube), Colchis (Armenia) and the coastal lands of Euxine (Black Sea). He, however, did not write anything about Italy and Sicily which shows that his westward journey was highly unlikely. Hecataeus used to consult merchants and travellers who had themselves visited the more remote regions. The extensive commercial relations of his native city of Miletus could afford him most favourable opportunities in this respect.

In his work, Hecataeus gives a general survey of the world. It describes the places and districts situated in the vicinity of the Mediterranean Sea, especially those near the Aegean Sea. This work could be called as *Periplus*, which means, in the Greek language, the coastal survey.[15]

The work of Hecataeus was divided into two books: the first book contains geographical information of Europe and the second deals with Libya (Africa plus Asia) (Figure 1.3). In the first book dealing with Europe, he gave a vivid account of the shores of Greece and the European coast of Aegean. After describing the geographical facts of Greece, he proceeds to the geography of Adriatic, Italy and Spain, after which he narrates Thrace, Scythia (northern parts of the Black Sea).[16] In his second book, he described Hellespont, southern coast of Euxine up to Caucasia, Asia Minor, Syria, Egypt and Libya. He described Medes, Parthians, Persians and even Indians but the source of information is not given and the descriptions are vague.[17]

Figure 1.3 World after Hecataeus

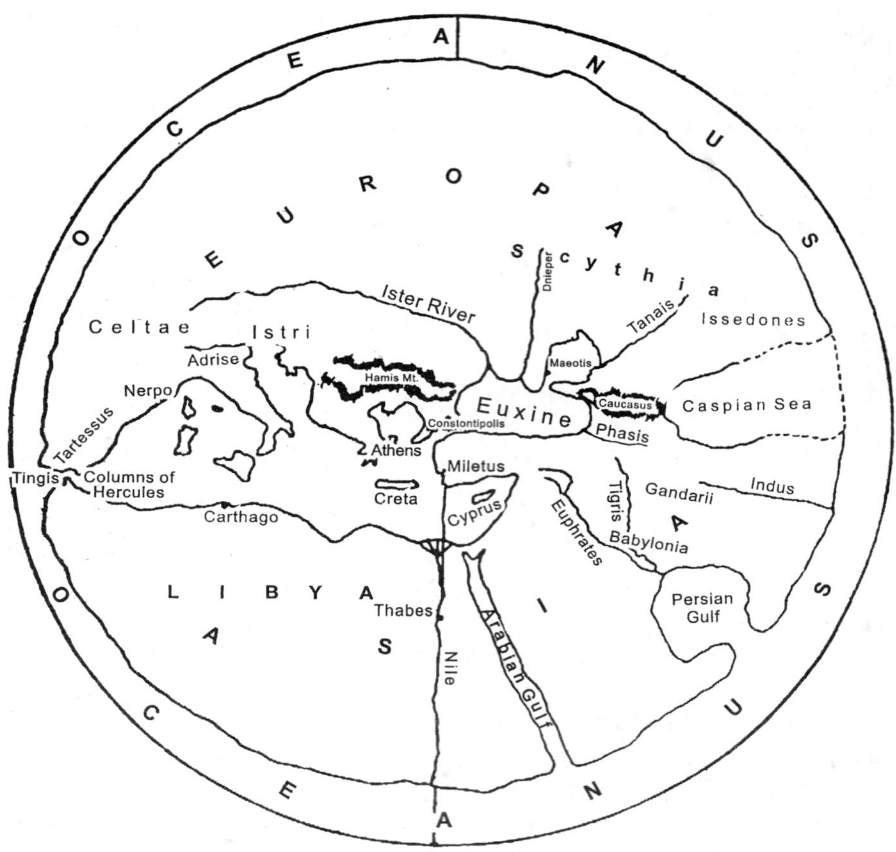

So far as the shape and size of the earth is concerned, Hecataeus maintained the traditional view of the Ionian philosophers that the earth is a circular plane and Greece is in the centre of the world. The world was assumed as surrounded by Ocean river. The landmass was divided into two equal parts, i.e. the continent of Europe towards the north and that of Libya (Africa and Asia) to the south.[18] The division between the two continents was made by a line along the Hellespont, the Euxine, the Caucasus mountains and the Caspian. Hecataeus treated the Caspian as an inland sea or a bay of the Eastern Ocean. The world map prepared by Hecataeus was based on the map of Anaximander – the great mathematical geographer – who made the first map of the world.

So far as Hecataeus' knowledge about Europe is concerned, it was mainly confined to the Greek colonies in the Mediterranean Sea. He was well acquainted not only with the shores of the Aegean and Ionian seas, and with those of the southern Italy and Sicily, but also with the eastern coast of the Adriatic, when he enumerates various obscure tribes with which the Greek colonies had connections. Near the head of the Adriatic Sea he placed Istri tribe. He describes a river of northern Italy, also called by the name of *Istri* (Po) where the fertility of land is extremely high.[19] Though he was familiar with Corsica and Sardinia, no reference is found to any place or city of Italy, and the entire western coast of Italy was also omitted. He mentions a number of place names in Spain, especially those in the neighbourhood of the Pillars of Hercules. But there is no evidence to suggest that he had knowledge of the Atlantic shore of Spain. Even the important place Gadeira (Gadis) has not been mentioned by him. Hecataeus had no knowledge of the western and northern limits of the continent of Europe. He assumed the ocean occupied the western and northern parts of this continent. This was merely a conjecture drawn from the then established notion of the Greeks that the whole world was surrounded by a circumfluent ocean.

The land situated to the north of Euxine (present Ukraine) had many Greek colonies. Hecataeus was quite familiar with the coastal land lying between Ister (Danube) and Boresthenes (Dnieper) rivers to the north of Euxine. In this region were many colonies of Milesians. He referred to the barbarian tribes like Melachlaeni (Dandarii) of Scythia. From these descriptions it may be inferred that his knowledge was not confined to the coastal tract of Euxine Sea but that he was familiar with the primitive tribes living in the remote areas of Scythia.[20] He plotted Tanais (Don) river also and has referred to the Issedonean tribes which lived in the northern parts of the Caspian Sea.

The second book deals with Libya (Asia and Africa). In Asia, Hecataeus was quite familiar with Asia Minor (the present coastal Turkey, Lebanon and coastal Israel). He has described the Phasis river and its neighbouring areas occupied by the Colchis (Figure 1.3). But his knowledge of the Colchis, Chalybes, Moschi and other barbarian tribes of the south of the Euxine Sea was inadequate. His account of the Matieni (Armenians) tribe was, however, fairly reliable. Caspian Sea, known as Hyrcanian during the time of Hecataeus, was regarded as an arm of ocean, bounded to the west by lofty snow-covered mountains (Caucasus).

The land between the Caspian Sea and the Gulf of Persia was plotted by him as 'Media'. But Babylon and its major cities were strangely omitted which shows that he did not travel through the land of Babylonia. It is, however, commendable that he also collected information concerning India. India and the Indus river were plotted by Hecataeus and he has referred to several tribes and cities of India, especially the Gandarii people who occupied the country between upper Indus and the valley of Cabul (Kabul). Caspatyrus, the capital city of the Gandarii, situated on the banks of the Indus, was known to Hecataeus. His information about India was, however, confined only to the western parts of the country, lying to the west of the Indus river.

Egypt, especially the Nile Valley, is a part of the world which was extensively toured by Hecataeus and therefore he gave adequate space in his writings to the geography of this country. It is reliably established by the subsequent geographers and historians that he ascended Nile up to Thebes (Figure 1.3) and met the priest of the temple of Ammon.[21] The account of phoenix, hippopotamus and crocodiles as given by Hecataeus was literally copied later by Herodotus. Hecataeus regarded the Nile river as having its source in the Southern Ocean while its upper reaches passed through the land of Negroes and Pygmies. The coast of Libya and the strait of Hercules are described. He also described the coasts and many islands along the northern coast of Libya.

Herodotus (484/485–425 BC)

Herodotus

Herodotus was the first and foremost Greek historian, and is regarded as the 'Father of History'. Little is known of his personal history. The *Histories* – his masterpiece and the only work he is known to have produced – is a record of his 'inquiry' being an investigation of the origins of the Greco-Roman Wars and including a wealth of geographical and ethnographical information. He placed historical events in a geographical setting; some of his writings are truly geographical in character. He not only described geographical phenomena as, for example, the annual flow of the Nile but also attempted to explain them. He was also one of the pioneer geographers. He was a strong supporter of the idea that *all history must be treated geographically and all geography must be treated historically*.[22] His work is an excellent example of historical geography. While describing the surface of the earth, he gave an interesting account of the then existing tribes and their lifestyles. Anthropologists consider him as the foremost enthnographer.

Herodotus was born at Halicarnassus in the 5th century BC. He stayed at Athens – the main centre of Hellenic culture; afterwards from Athens in 433 BC, he went to Thurii – a town in South Italy.[23] Herodotus wrote most of his works at Athens before leaving for Thurii, but these were completed at Thurii only.

Herodotus was a great traveller and his contribution to geography is highly remarkable as he wrote after making personal observation during many years of travel. Westward he visited Italy and through the strait of Marmara and Bosphorus (Turkey) reached the Euxine Sea. Traversing the Euxine he reached the Persian Empire, and stayed at Susa and Babylon.[24] Moreover, he was quite familiar with the coasts of Asia Minor, as also with the island of the

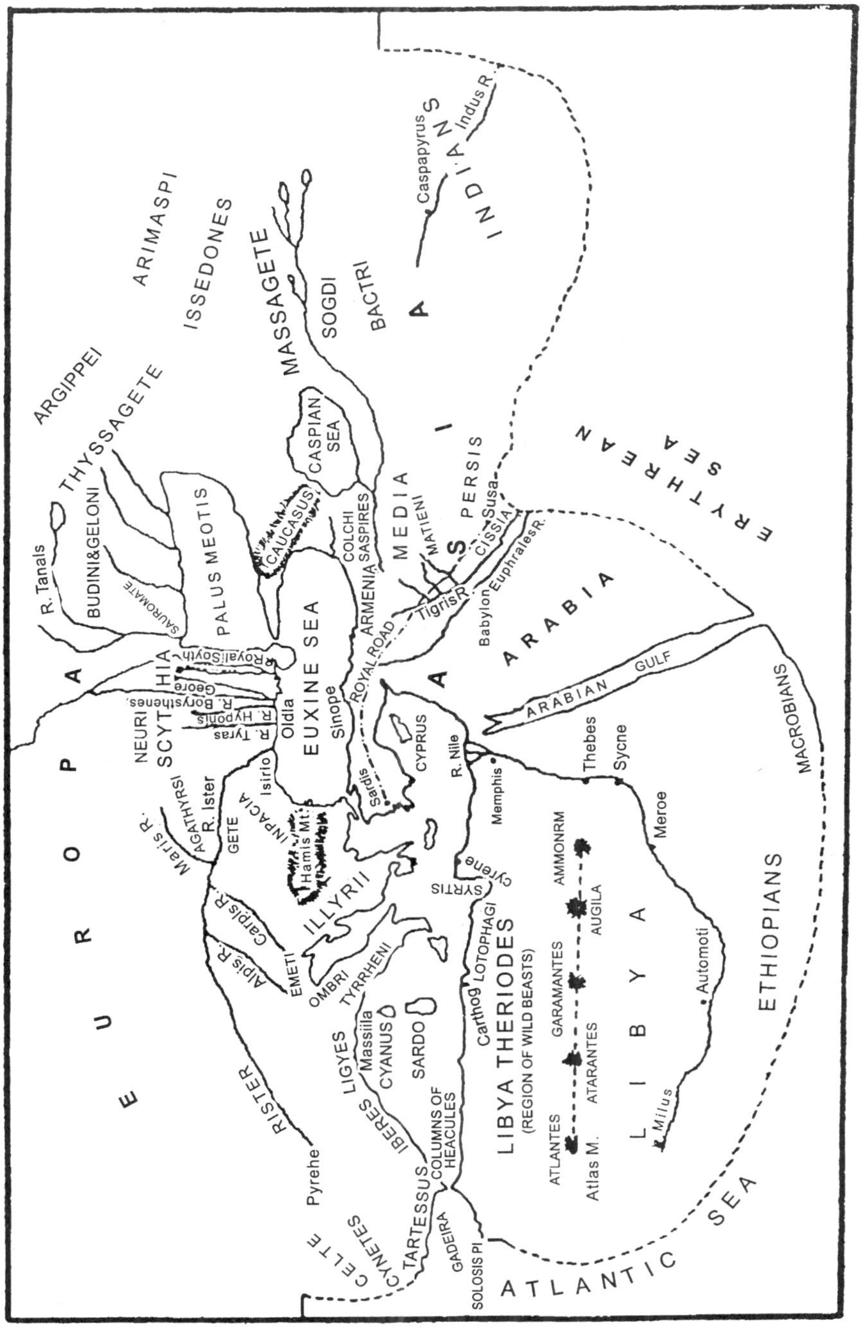

Figure 1.4 World after Herodotus

Aegean, the mainland of Greece, Byzantium, and the neighbouring shores of Thrace. He made voyage to Tyre. The land of Colchis was also visited by him. Towards the south he travelled several times to Egypt and ascended most probably up to the Elephantine cataract (present Aswan). In Libya, he reached Cyrene to see the statue there of which he gave an eyewitness description.

Herodotus' views about the shape of the earth were not in conformity with those of Hecataeus that the earth, as a circular plane, is surrounded by an ocean stream. He accepted the Homeric view of the earth as a flat disc over which the Sun travelled in an arc from east to west. He belonged to the Pythagorean School of Philosophy and thus tried to establish a symmetrical correspondence in the distribution of land, and in the source, direction and course of the Ister (Danube) and Nile rivers. His knowledge of the source and upper course of the Nile was highly erroneous.[25] He was convinced that when there were inhabitants on the back of the northerly winds (Bores), there should also be tribes on the back of the southern winds. The land, according to him, is divided into two equal parts, one lying to the north of the line passing through Hellespont, the Euxine, on the Caucasus mountains and the Caspian Sea. Thus, Europe was taken to be equivalent to Asia and Libya (Africa) combined. About the Nile river, he stated that it flowed in a direction from west to east, dividing Libya through the middle into two parts. The source of the Nile was in the west of Libya, almost the same distance as that of Ister in Cells near the city of Pyrene.[26] Ister, according to him, flowed through the whole of Europe and before discharging its water into the Euxine Sea, adopted a north-south course like the Nile river. Egypt, according to him, lay almost exactly opposite the mountain portion of Cilicia (Turkey) and the peninsula of Sin̆ope. Cilicia (Turkey) lay opposite to the place where the Ister fell into the Black Sea. It was a crude manner of drawing a meridian from the mouth of Nile to that of Ister (Figure 1.4). The primitive cosmic beliefs of Herodotus were highly erroneous. He believed that the Sun was driven southward out of its regular course by the winds at the approach of winter. In spite of all such unscientific beliefs, he was the first scholar who tried to draw a meridian on the world map. The meridian was drawn from Egypt to Cilicia (south coast of Turkey), peninsula of Sinope and the mouth of Ister (Danube). In his opinion, all these fell in a north-south line, which could be taken as the meridian.

So far as the spread of continents is concerned, Herodotus did not have a clear idea and could not fix the northern limit of Europe. He also did not have any idea of the existence of the north-eastern seas. On the southern side, he felt that the ocean sprawled continuously from the coast of India to that of Spain (Figure 1.4). In support of his conviction, he asserted that Scylax had sailed from the mouth of Indus to the Red Sea, and Necho voyaged from Egypt to explore the coast of Africa and succeeded in reaching the Pillars of Hercules by the southern side. He was familiar with the Arabian Sea, the Indian Ocean (Arythraean) and the Atlantic Ocean and believed in only two inland seas, one stretching northward and the second eastward from the Indian Ocean and the

Atlantic Ocean respectively, i.e. Red Sea (Arabian Gulf) and the Mediterranean Sea. He was ignorant of the Persian Gulf (Figure 1.4).

So far as the Euxine (Black Sea) is concerned, he had himself navigated in it. In fact, the Greek traders were quite active in the Black Sea. His knowledge of the Euxine and the land and tribes lying to the north of it was more correct as compared to his predecessors. He wrote that the Euxine, 'the most wonderful of all seas', is 1,100 stadia (110 miles) in length and at the widest portion it is 1,300 stadia (130 miles). His idea of the size and dimensions of Palus-Maeotis, 'the mother of the Euxine' (Sea of Azov), was, however, erroneous. In reality, the Sea of Azov (Palus-Maeotis) is not much more than one-twelfth of the size of Black Sea (Euxine). The supporters of Herodotus feel that the Sea of Azov is a shallow one and might have shrunk in the last two and a half thousand years or so. According to geologists, the Don river is pushing its delta into the Sea of Azov. Even making allowance for the silting of the Palus-Maeotis, its sprawl is greatly exaggerated. Herodotus showed the Palus-Maeotis in a north-south direction drawing a boundary between Europe and Scythia which is also not correct.

Herodotus is the first geographer who regarded the Caspian as an inland sea, whereas Hecataeus and his contemporaries as well as all the geographers of Alexandrian era considered it as an arm of the Northern Ocean (Figure 1.4). In respect of the Caspian, it is remarkable that Herodotus was in advance of almost all his successors.[27]

Herodotus was well aware of some of the physical processes that transform the surface of the earth. He insisted that the Nile Valley, especially its delta, has been built by silt and mud brought down by the river from Ethiopia. This mud is black in colour which can be ploughed easily. Furthermore, he supports the hypothesis that the Nile mud deposited into the Mediterranean Sea had built the delta. He reconstructed the ancient shoreline and showed that many sea ports were now far inland. He explained the process of delta making and stressed the point that the delta of Meander River (West Turkey) was also the result of river deposition.[28] Similarly, he tried to establish a relationship between temperature and the movement of winds.

Herodotus was also the first scholar who divided the world landmass into *three continents,* namely, Europe, Asia and Libya (Africa). The size of Europe has, however, been taken as equivalent to Asia and Libya (Africa) combined. It is surprising that he took the western frontier of Egypt as the boundary between Asia and Libya. Asia and Europe were divided by the Strait of Bosforus, the Tanais (Don) river, the Caspian Sea and Araxes (Amu). It would be worthwhile to examine in brief the knowledge which he possessed about the physical features and tribes of these three continents.

Of the greater parts of Europe his knowledge was scanty. He described Ister (Danube) as the greatest river of the world and the Carpis (Carpathian mountains) and Alpis (Alps mountains) as its two major tributaries.[29] The description of Central and Western Europe was scanty. He plotted the Ombri and Eneti nations in the northern parts of Italy but was not familiar with

Great Alps which separate these from the tribes of the north. Scythia, the land to the west of Palus-Maeotis and to the north of the Euxine, was best known to him as this land was occupied by the Greek traders and he himself also travelled that part of the world. He probably stayed for some time in the land lying between Olbia and Borysthenes (Dniester river) (Figure 1.4).

The Borysthenes is considered by Herodotus as the largest river of Scythia after the Danube. Its plain was regarded as the most productive in the world excepting the Nile Valley.[30] The inhabitants of this region were called 'Olbia' (the prosperous). Tanais (Don) is another river mentioned by Herodotus. Thus, the knowledge of Herodotus about the Scythian rivers was appreciably good. But, as he recedes from the coast, his information becomes vague and untrustworthy.

The Scythian people as conceived by Herodotus were divided into several tribes. Characterized by some difference in their modes of life and habits, these tribes were spread in different geographical locations. He held that the tribes dependent on agriculture dwelt in the valley of Borysthenes; moving eastward the area was occupied by nomads and along the coast of Palus-Maeotis lived the royal tribe. Among the other tribes Agysthrsi, Neuri, Androphagi, Melanchaeni, Geloni, Budini and Sauromatae were prominent.[31] All these tribes had their own separate rulers, and were, in the opinion of Herodotus, distinct from the Scythians. Agythrsi people were considered the most refined among them; they wore gold ornaments. The Neuris resembled Scythians in manners but were said to have the peculiar power of transforming themselves for a few days every year into wolves. Beyond the Neuris were the Androphagis (Cannibals). Their manners in all respects were most rude and savage and they spoke a language different from Scythians. To the east of these was the homeland Melanchalaenis (Black Cloaks) about whom much information is not given. The Budinis have been considered as blue-eyed, with red hair, a well-built powerful tribe. They were nomads, like their neighbours on both sides, but their land was thickly forested. The Geloni were settled farmers. According to Herodotus, their origin was from Greece, having migrated to Scythia. The Argippaens were the last people towards the north. They were the people who lived to the east of Urals. The people living to the east of the Caspian Sea were Issedones and Massagataes.

Herodotus' knowledge of Asia was confined mainly to the Persian Empire which sprawled over the whole of Western Asia (with the exception of the Arabian Peninsula) from the Erythraean Sea to the Caucasus and the Caspian, and from the coast of the Mediterranean Sea and Black Sea to the Indus. Beyond these regions his knowledge was vague.

The Persian Empire for the purpose of administration and revenue collection was divided into twenty *satrapies* (provinces). He was acquainted with these *satrapies* and their principal tribes. From the Erythraean Sea (Arabian Sea) towards the Caspian Sea he placed four major tribes, i.e. Persians, Medes, Saspirians and Colchians.[32] His knowledge about the Peninsula of Anatolia (Turkey) and Asia Minor which was surrounded by the

Greek colonies was very inaccurate. He was not aware of any of the great mountain chains of Asia like Tarus, Elburz, Zagros, Hindu-Kush and the Himalayas. Nevertheless, he was conversant with the courses of Tigris and Euphrates and their sources in the highlands of Armenia.

Herodotus gave a good account of the Royal Road, joining the city of Sardis to Susa (Figure 1.4). This road, according to him, was marked by royal stations at regular intervals. At each station, there were 'carvan-sarais' (rest houses). This road was about 13,500 stadia (1,350 miles) which is very close to the actual distance between Sardis and Susa.[33]

Herodotus alluded to the Erythraean Sea as situated to the south of Asia, extending from the Arabian Gulf (Red Sea) to the mouth of the Indus. The account of India and its inhabitants by Herodotus is interesting and instructive. He was not familiar with the fertile Gangetic plain and considered the Indus as flowing in a west-east direction. To the east of the Indus there was no tribe and the area as described by Herodotus was a big sandy desert. He was not sure of the eastern limit of Asia and the existence of a sea to its east. He had, in his opinion, the only river, except the Nile, in which they were found. Curiously enough, he has nowhere referred to the elephants of India. Albeit he knew that the Indians used or grew cotton which resembled wool. They also grew a large kind of reed (bamboo), used for making bows. Caspatyrus was the only city known to him.

Herodotus also possessed enormous information about the continent of Africa. His knowledge of the southern coast of the Mediterranean Sea was as accurate as that of its northern coast. In fact, he himself visited Cyrene (Aswan), which was at that time an important centre of Greek life and culture.[34] Up to the point of Carthage, his knowledge was pretty correct. With regard to the interior of the continent his knowledge was confined to the course of the Nile river. He himself ascended up to the Elephantine just below the First Cataract. The Cataract he described as that owing to the rising of the ground 'it is necessary to attach a rope to the boat on each side, as men harness an ox, and so proceed on the journey'. The next important station along the bank of the Nile was Maroes – the capital of Aethiopians (Ethiopia). Meroes city was to the north of Khartoum. The only people of whom he had heard as situated beyond Meroes were a race called Asmach (or deserters). These people occupied the territory to the south of Khartoum between the two branches of the Nile, i.e. the Blue and the White Nile. According to Egyptians, they had left Egypt during the period of Psammitichus owing to his brutality and hard service conditions.

The Macrobian Ethiopians were regarded by Herodotus as the remotest dwellers, occupying the southern most parts of Africa. They, he presumed, were the tallest and the most handsome race in the world and lived for an average of 120 years. Gold was so abundant in their country that it was used even for the chains and fetters of prisoners. They enclosed the dead in pillars of transparent crystal instead of coffins, their food consisted solely of meat and milk and this was the main cause of their high longevity.[35] These people were none but the inhabitants of Somalia, opposite the Red Sea.

With the western coast of Africa, Herodotus was less acquainted. The northern coast of Africa was divided into the eastern and western portions. The eastern part was sandy and its people were nomads. The land was barren; the city of Syrene was situated in this tract. The western tract was characterized by hills, fertile valleys and dense forests. The western coast of the coastal Africa was occupied by primitive tribes with whom the Carthagians practise 'dumb commerce'. Still there are certain primitive tribes in the tropical world who practise this type of barter. The dumb commerce has been described by Herodotus in the following words:

> There is a country in Libya, and a nation beyond the Pillars of Hercules, which the Carthagians used to visit, where they arrived and unloaded their wares, and, having disposed them in an orderly fashion along the beach, leave them, and returning aboard their ships, raise a great smoke. The natives, when they see the smoke, come down to the shore, and such gold as they think the worth of the wares, withdraw to a distance. The Carthagians then come ashore and look. If they think the gold is enough, they take it and go their way, but if it does not seem to them sufficient, they go aboard ship once more, and wait patiently.[36] Then the others approach and add to their gold, till the Carthagians are content. Neither party deals unfairly with the other, for they themselves never touch the gold till it comes up to the worth of their goods, nor do the natives carry off the goods till the gold is taken away.

No indication is furnished by Herodotus of the locality where this dumb commerce was carried on but the fact of gold being the object of the trade leads to the inference that it was at a considerable distance towards the south, as there is very little gold found to the north of Sahara.

Herodotus divided the interior parts of Africa into three latitudinal zones. The first zone is the Mediterranean coast from Atlas mountains to the delta of Nile. It is partly occupied by nomads and partly by cultivators. The second zone to the south of it is the area of 'wild beasts'. By the Arabs this was called the 'land of dates'. The third zone which lies further south is the true Sahara desert. Herodotus has mentioned that in the desert of Sahara there are five oases, namely, Ammonium (Siwah), Augila, Garamantes, Atarantes and Atlantes (Figure 1.4). They lie at a distance of ten-day journey from one another.

Alexander the Great (20/21 July 356 – 10/11 June 323 BC)

The multifarious works of the great philosopher – Aristotle – created a keen desire in his disciples to test theory by direct observation. He taught his pupils to 'go and see' for themselves whether a particular theory could or could not be accepted.[37] One of his outstanding and adventurous pupils was Alexander who marched towards east with a

Alexander the Great

well-trained and disciplined army to see what lies in reality in the far remote areas beyond the Persian Empire about which there were many fables and fallacies. Alexander was born in Pella in 356 BC and studied with Aristotle for a period of three years (343–340 BC), i.e. between the age of thirteen and sixteen. Alexander succeeded his father, Philip II to the throne of Mecedonia at the age of twenty (336 BC).

He spent most of his ruling years on an unprecedented military campaign through Asia and northeast Africa. At the age of thirty he had created one of the largest empire of the ancient world. His empire stretched from Greece, Egypt, Central Asia and present Pakistan. He was an accomplished military commander who encouraged soldiers by his own heroic efforts. His main objective was to establish Greek rule all over the world and to reorganize the world administrative system. His treatment to conquered people was exemplary. All people, said Alexander in the 4th century BC, are brothers and should be treated as brothers. The second objective of Alexander was to determine the outer limits of the world and thus his expedition may be termed as an armed exploration.

Before the eastward expedition of Alexander, the knowledge of Greeks about Asia, especially Central Asia and India, was imperfect and vague. Before entering into Asia in 334 BC at Hellespont (Turkey), he conquered the barbarian tribes of the Ister (Danube).[38] Then, he invaded the Central Turkey (the Plateau of Anatolia) and parts of the Persian Empire. Thence, he continued southward along the eastern side of the Mediterranean Sea and reached Egypt (Figure 1.5). In 332 BC, he founded the city of Alexandria, which, in later years, became a great intellectual and commercial centre of the classical period. In the Egyptian capital, he stayed for nine months and then ascended the Nile river up to Marcotis and the island of Pharos.[39] To have blessings, he visited the temple of Zeus Ammon in the desert of Libya before quitting Egypt in 331 BC to make an eastward march. He defeated Darius – the emperor of Persia – in the battle of Arbela (Mesopotamia). Chasing the King Darius, Alexander reached Elburz and the coast of Caspian Sea, where the Greeks for the first time had a view of the Caspian Sea. Returning from his expedition, Alexander marched eastwards through the north of Parthna (Figure 1.5) and reached Artacoana (Herat). Crossing the terrain of Drangiana (Seistan) he advanced towards Archosia (Candhar). Moving northward, Alexander passed through Alexandria and Caucasus and finally reached Alexandria-Ultima (Figure 1.5). In the spring of 327 BC, he crossed Hindu-Kush and passed the Khyber Pass. Crossing the Indus he entered the plains of Punjab or the 'land of five rivers'. Punjab is traversed by the five great tributaries of the Indus, namely, Hydaspes (Jhelum), Acesipes (Chinab), Hydraotes (Ravi), Hyphasis (Bias), and Zaradrus (Sutlej).[40] Believing that he was only a short distance from the eastern limit of the habitable world, he wanted to march farther eastward, but his troops mutinied and insisted on returning to Greece.[41]

Figure 1.5 Alexander's Eastern Expedition

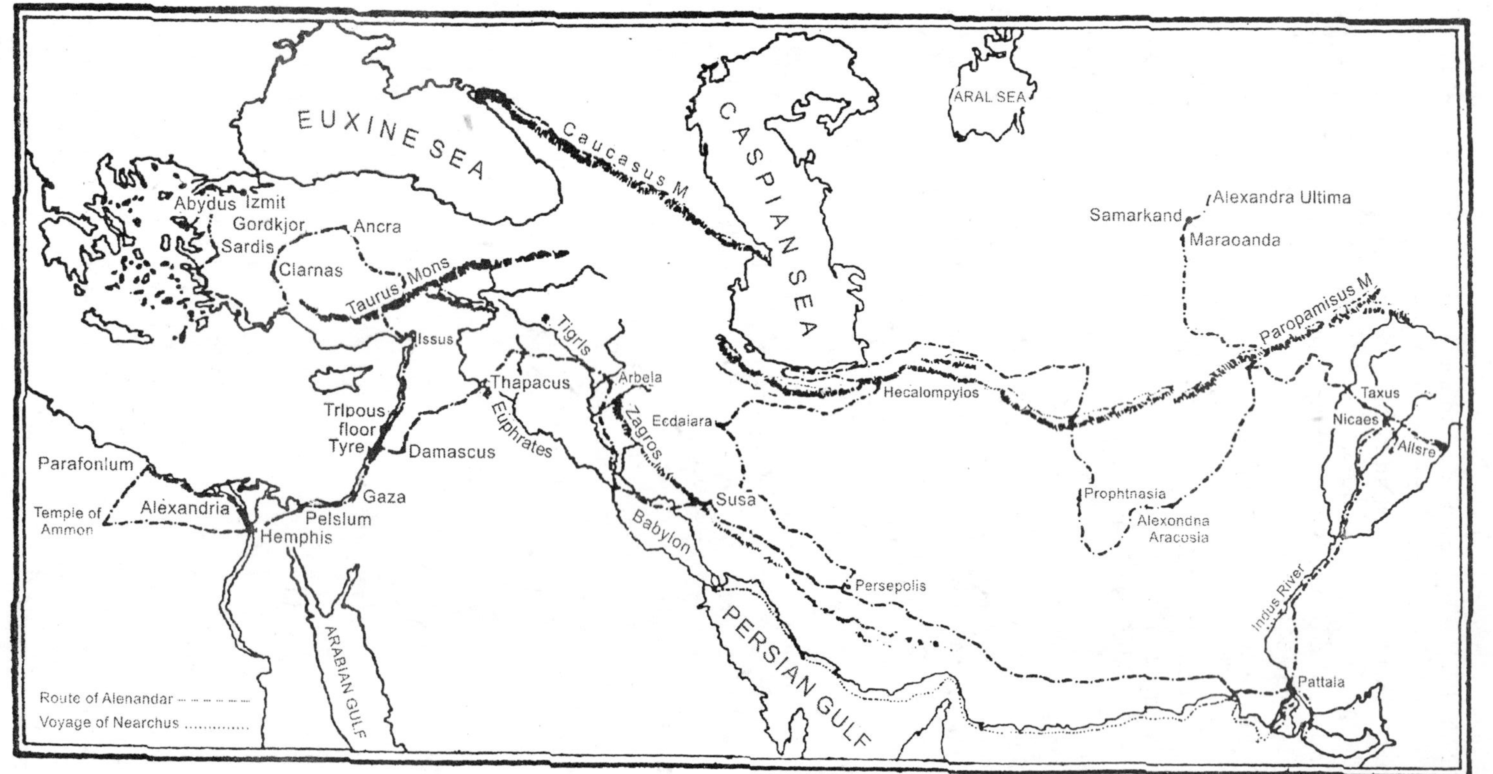

Alexander, for his return, adopted the land and the sea routes. One of the armies was sent under the command of Nearchus to navigate the Arabian Sea (Indian Ocean) from the mouth of the Indus to the head of the Persian Gulf. The remainder of the army he led by an unexplored route from Pattala (Hyderabad-Sindh) westward through the southern parts of Mekran, Baluchistan and Iran. He reached Mesopotamia and planned to circumnavigate the Peninsula of Arabia and Libya but could not do so as he expired in 323 BC at Babylon the city he planned to establish as his capital.

Alexander's staff included writers to describe the lands they crossed and astronomers to take observations of the height of the bright star Canopus in order to fix latitude, or distance to north of the equator. There were trained pacers whose duty was to measure distances on the march. As a result he gathered enormous geographical information.

Records and itineraries hold that Alexander contributed substantially in the fields of historical and regional geography. The Greeks became familiar with the natural features of the Persian Empire, Central Asia, Afghanistan, India and the coastal parts of Iran. The Greek armies crossed deserts, plains, salt-steppes, fertile valleys, snow-covered peaks and high mountain passes which added enormously to the then existing knowledge of the Greeks about the terrain, tribal people, flora and fauna of Asia. He founded some twenty cities that bore his name, most notably Alexandria in Egypt. Thus, Alexander gathered a wealth of new observations about the region which was beyond the Greek horizons.

Eratosthenes (276–194 BC)

Eratosthenes was a Greek mathematician, geographer, astronomer, poet and music theorist. Eratosthenes may be regarded as the first scientific geographer who ascertained the length of equator with remarkable precision. He also developed the systems of co-ordinates for the world, i.e. latitudes and longitudes. This made it possible for him to draw the first passably accurate map on sound principles. He was well acquainted with the astronomical knowledge of his period.

Eratosthenes

Eratosthenes was born in the Greek colony, Cyrene (Libya), in 276 BC. He got his education at Cyrene and then at Athens. From Athens he was invited by the ruler of Egypt, Ptolemy Euegetes, and appointed chief librarian of the Library of Alexandria. This post was considered to be the highest academic honour of that period.[42] He served in that capacity for about forty years until his death in about 194 BC at the age of eighty. During his tenure as the chief librarian, he wrote several scientific and literary books. All his works have unfortunately perished. Nevertheless, it is a known fact that Eratosthenes attempted to give the subject of geography a systematic form and tried to construct the world map on more scientific principles. However, the materials at his command were very imperfect and the observations made with the help of the available apparatus (gnomon) about the heavenly bodies and latitudes and longitudes were erroneous.

About the problems of shape, size, position and spherocity of the earth, Eratosthenes adopted the views of Aristotle and Euclid and regarded the earth as a sphere, placed in the centre of the universe, round which the celestial spheres revolved every twenty-four hours. Further, the Sun and the Moon had independent motions of their own.[43] Thus, his idea of the shape of the earth is in conformity with the modern geographers.

Eratosthenes tried to measure the circumference of the earth. Apart from his other contributions in the fields of astronomy and geography, he is famous for the correct measurement of the length of equator with the help of an indigenous apparatus known as ‘gnomon’. He made two separate observations of the position of the Sun.[44] One observation was done at Syene (Aswan). At this place, there was a deep well, and at the bottom of the well, at the summer solstice, the image of the Sun was reflected in the water. The existence of this well had been known for a long time, and tourists in ancient times travelled up to Syene to witness this strange occurrence each year. This meant that on that date the Sun was directly overhead. The second observation was made outside the museum in Alexandria, where there was a tall obelisk. Using the obelisk as a gnomon, Eratosthenes measured the length of the shadow at the solstice. He was thus able to measure the angle between the vertical obelisk and the rays of the Sun. With these data in mind, Eratosthenes made use of the well-known theorem of Thales, which states that when a diagonal line crosses two parallel lines, the opposite angles are equal. The parallel lines were given by parallel rays of the Sun (Figure 1.6). The rays of the Sun at Syene, which were vertical, could be extended to the centre of the earth (SC). Also the obelisk, which was vertical at Alexandria, could be extended to the centre of the earth (OC). Then, the angle between the Sun’s rays and the vertical obelisk, which was vertical at Alexandria (BOC) must be the same as the opposite angle at the centre of the earth (OCS). The next question was: How much of the whole circumference of a circle to be subtended by the angle OCS? Eratosthenes measured this as one-fiftieth of the whole circumference. What remained was to fill in the distance between Syene and Alexandria, which the Egyptians said was the equivalent of about 500 miles, and then to multiply this distance by 50. Erotosthenes, therefore, concluded that the whole earth was about 25,000 miles in circumference (actually the circumference measured through the poles in 24,860 miles).[45]

The only theoretical error in the measurement of the circumference of the earth was that Eratosthenes took the earth as a perfect sphere instead of oblate spheroid. As such the length of the meridian great circle was taken to be equal

Figure 1.6 Measurement of the Circumference of the Earth by Eratosthenes

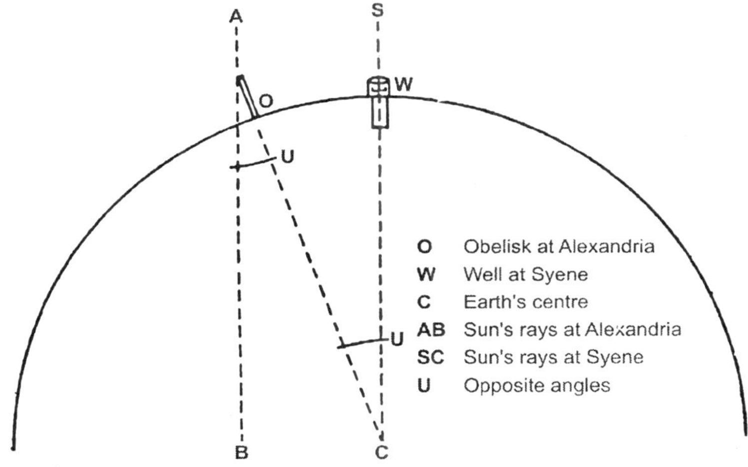

Distance OW subtended by angle OCS equals 1/50 of the circumference of a circle (After Eratosthenes).

to the length of equator, i.e. 25,000 miles, though, in fact, it is only 24,860 miles. But, this type of error is to be considered insignificant as the equipment of measurement (gnomon) was not very precise and accurate. Moreover, the linear measurements along the roads were also far from exact. Eratosthenes assumed that Alexandria was due north of Syene, whereas, in fact, it is about longitude 3°W of Syene. The length of the road between Syene and Alexandria, which the Egyptians said was the equivalent of 500 miles, is actually 453 miles and Syene is actually at latitude 24°5'N, a little to the north of Tropic of Cancer. But all these errors canceled out so that the resulting calculation was amazingly close to the correct figure.

Eratosthenes also attempted to determine the distance of the Sun and the Moon from the Earth. He computed the distance of the Moon at 7,80,000 stadia (78,000 miles) and that of the Sun 40,00,000 stadia (4,00,000 miles). No account is preserved of the process by which he arrived at these conclusions. But these results are far from accurate.[46]

Eratosthenes tried to determine different latitudes and longitudes. With the help of gnomon, he determined the latitude of Rhode Island. He continued this latitude through the Strait of Gibraltar in the west and through Thapsacus on the Euphrates and the Himalayas as far as the eastern ocean. The Arctic circle was placed in Thule.

Determination of longitudes was even a more difficult task as the sophisticated instruments of measurement were not available. The longitudes were thus determined on the basis of the calculations made with the help of unscientific instruments. Eratosthenes drew a meridian southward through

Alexandria, Syene, Meroe and passing through Rhodes, Troad (Traos) and Byzantium and the mouth of Boresthenes (Dniester) on the northern shore of the Black Sea.[47] Figure 1.7 shows the meridian drawn. Another meridian Eratosthenes drew through Carthage, and Straits of Messinia and Rome. This was also erroneous. In spite of all these shortcomings, Eratosthenes has been called the founder of 'geodesy'.

Figure 1.7 Meridian drawn by Eratosthenes

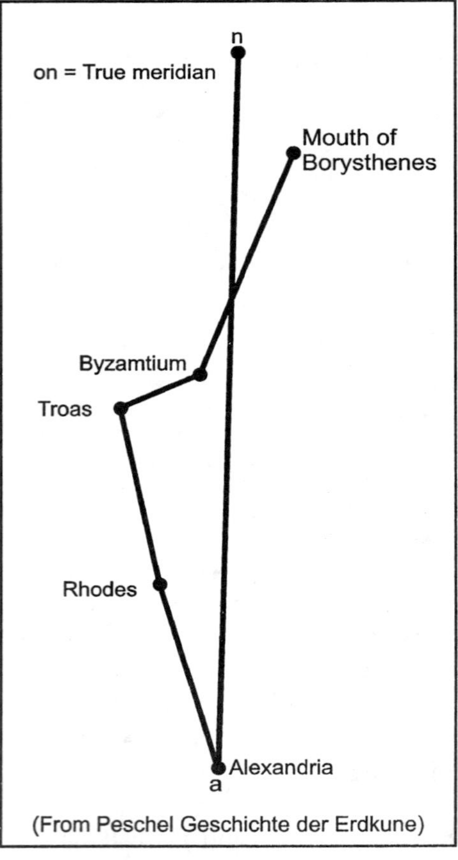

Eratosthenes was also of the opinion that the spread of the world from west to east is more pronounced than from north to south. According to his calculations, the length of the known world from the Atlantic to the Eastern Ocean amounted to 78,000 stadia (7,800 miles), while its breadth from the parallel of Cinnamon land to that of Thule did not exceed 38,000 stadia (3,800 miles).[48] He extended the habitable world from Thule to Taprabone (Ceylon) and from the Atlantic Ocean to Bay of Bengal.

Eratosthenes possessed fairly accurate knowledge of the shores of Spain and Gaul that border the Mediterranean but his information of the rest of the north-west European countries was very imperfect. His knowledge of the western regions of Europe was based on the accounts of Pytheas (the famous navigator). He appeared to have had no accurate knowledge of Scythia – the land lying to the north of Euxine (Black Sea). He was equally unacquainted with the northern shores of Germany.

Regarding Asia he relied on the itineraries of Alexander and his record keepers. He was familiar with the fact that the Mt. Taurus was connected with Armenia, Koordistan, and Elburz to the Himalayas. He believed that the Ganges flowed from west to east and merged into the Eastern Ocean. His ideas of the geographical position and configuration of India was greatly erroneous. He conceived the sub-continent of India to be of rhomboidal form. And supposed the range of Imaus (Himalayas) which bounded the country to the north to have its direction from west to east, while the Indus flowed from the north to the south. Moreover, he conceived the peninsula projecting towards the south-east, instead of the south.

He was familiar also with the name of Taprabone (Ceylon) which was known to the Greeks since the days of Alexander, but he placed Taprabone south of Coniaci (Cape Comorin), at a distance of seven-days journey from the mainland of India.[49]

Eratosthenes was well acquainted with the extent and dimension of the Red Sea, which he described as extending 9,000 stadia (900 miles) from the head of gulf (Gulf of Suez) to the station of Ptolemais Epithera. This is very fair estimate. The knowledge possessed by him about the Nile was superior to that of his predecessors.[50] His knowledge of the lower Nile (Egypt) was perfect and he was the first to mention the name of Nubians who occupied the land on the west of the Nile. Of the rest of Africa, Eratosthenes knew little, but he conceived Africa to be surrounded by ocean. Caspian Sea was shown by him as an arm of the Northern Ocean.

The book written by Eratosthenes describes the ekumene – the inhabited earth – in which he accepted both the major divisions of Europe, Asia and Libya (Africa), and five climatic zones, i.e. a torrid zone, two temperate and two frigid zones. The torrid zone he thought was 48 degree of the whole circumference (24 degree north and south was calculated as the location of tropics). The frigid zone extended 24 degree from each pole.[51] The temperate zone was between the tropics and the polar circle.

Hipparchus (190–120 BC)

Hipparchus

Hipparchus was a Greek astronomer, mathematician and geographer. He is considered the founder of trigonometry. He is most famous for his discovery of equinoxes. He calculated the length of the year to within six and half minutes. He compiled the first known star catalogue. Hipparcus was born in Nicaea (now Iznik, Turkey) and probably died on the island of Rhodes. After the death of Eratosthenes, Hipparchus was appointed as the librarian of Alexandria (162–127 BC). In his work, he stressed the importance of parallels of latitudes and meridians of longitudes. He made a marked development in the field of scientific cartography. Hipparchus' catalogue, completed in 129 BC, listed about 850 stars, the apparent brightness of which were specified by a system of six magnitudes similar to that used today. For its time, the catalogue was a monumental achievement. He applied rigorous mathematical principles to the determination of places on the earth's surface and he was the first to do so by specifying their longitudes and latitudes – the method used today. On the basis of latitudes, he divided the world into *climata* (bells of latitudes).

Hipparchus was the first to divide the circle into 360 degree, based on Assyrian arithmetic. The equator, he pointed out, was a great circle (one that divides the earth into two equal parts) and the meridians that were drawn converging on the poles were also great circles. The parallels, on the other hand, became shorter and shorter as they approached the poles. Since the earth makes one complete revolution in twenty-four hours and there are 360 meridians drawn from poles to equator, each hour the earth turns through 15 degree of longitudes.[52]

For the determination of latitudes and longitudes, Hipparchus invented an instrument known as *astrolabe*. This apparatus was easier to handle and more accurate than gnomon. Moreover, it was very useful in the open seas where it could be used by hanging on the rigging of a ship. The astrolabe made possible the measurement of latitude at sea by observing the angle of the Polar Star.

The credit for the conversion of a three-dimensional sphere into a two-dimensional plane also goes to Hipparchus. He devised two kinds of projections so that the curved sphere of the earth might be converted into plane surface on sound mathematical principles.[53] He told how to make a stereographic projection by laying a flat parchment (plane) tangent to earth and extending the latitude and longitude lines from a point opposite the point of tangency. The orthographic projection is similarly produced, but by projecting the lines from a point in infinity. The orthographic and the stereographic are the two important projections designed by him. These projections show only a hemisphere, not the whole earth.

Posidonius (135–51 BC)

Posidonius was an important Greek historian, politician, astronomer, geographer and teacher, who lived shortly before the time of Christ. He was a native of Syria and was acclaimed as the greatest polymath of his age. His vast body of work exists today only in fragments. His two important contributions to geography are well known; out of which, the one, a wrong idea persisted for centuries, and the other, a correct idea that was overlooked.

Posidonius

He calculated the circumference of the earth and arrived at a much smaller figure than that of Eratosthenes. He observed the height above the horizon of Conopus (a star of the first magnitude) at Rhodes and

Alexandria, which he assumed to be on the same meridian. He then estimated the distance between them based on average sailing time for ships. The figure he arrived at for the circumference of the earth was 18,000 miles. He also greatly overestimated the west to east distance from the western-most part of Europe to the eastern end of the ekumene (habitable world) then thought to be occupied by India. He declared, therefore, that a ship sailing westward across the Atlantic from western Europe would reach the east coast of India after a voyage of only 7,000 miles. It was this estimate on the basis of which Columbus argued before the scholars at Spanish court that he could sail west to India.

Posidonius, however, was right about another matter. He refused to follow Aristotle in believing that the equatorial part of the torrid zone was uninhabitable because of heat. The highest temperature and the driest deserts, he insisted, were located in the sub-tropical zone near the tropics, and the temperatures near the equator were much less extreme. He arrived at the conclusion – amazing in the 1st century BC – on purely theoretical grounds, for he had no access to creditable reports from anyone who had crossed the Sahara, including Hanno's voyage along the African west coast. The Sun, he pointed out, pauses longest near the tropics and is overhead for a much shorter time at the equator. The interesting point is that Posidonius' incorrect estimate of the circumference of the earth was widely accepted by those who followed him, while his correct belief concerning the habitability of the equatorial regions was overlooked.

GREEKS' CONTRIBUTION TO MATHEMATICAL AND PHYSICAL GEOGRAPHY

The Greeks not only extended the horizon of geography from the Aegean Sea to Spain and Gaul, the Russian steppes in Central Asia, the Indus river in the east and Ethiopia in the south, but also put the subject on a sound footing by making remarkable contributions in the field of mathematical, physical, historical and regional geography. Mathematical geography developed by Thales (580 BC), Anaximander (610–546 BC) and Aristotle (384–322 BC) reached its zenith with Eratosthenes (276–194 BC). The sphericity of the earth was proved, its size computed accurately. The latitudes and longitudes of many places were calculated and the practice of plotting world maps on a grid was initiated. The physical geography progressed less rapidly. Various scholars speculated about the phenomena of weather, tides and volcanism. Causes of Nile's floods and the formation of deltas were studied. How streams slowly carved their valleys was examined and understood. Posidonius (135–51 BC) investigated the cause of tides at Gades and measured the depth of sea at Sardinia. The Greeks also attempted to write historical and biographical account of the known world. In the following paragraphs, their major contributions in the various branches of geography have been briefly described.

Mathematical Geography

During the Golden Age of the Greeks – 5th century BC to 2nd century BC – there was a bunch of scholars who were engaged in determining the shape, size and climatic zones of the earth, and to ascertain the Ocean River, encompassing the habitable world. Moreover, philosophers like Plato and Aristotle encouraged their pupils to make efforts to determine the distances and latitudes with the help of astronomical observations. Plato was the first philosopher to announce the concept of a round earth located in the centre of the universe with the celestial bodies in circular motion around it. Whether it was Plato's original concept or whether it was suggested to Plato by Socrates, whom he quotes, cannot be determined. It was, however, Aristotle who first looked for support of the concept.

Heracleides Ponticus, the renowned associate of Plato, established the rotation of the earth on its axis, though still regarding it as the centre of the universe.[54] Anaximander introduced the Babylonian gnomon to the Greek world. He measured the latitudes of important places and prepared the first map of the world to scale.[55] Thales and Anaximander are considered as the founders of mathematical geography. Thales and Aristotle established the spherical shape of the earth. Aristotle, by philosophical reasoning and astronomical observations, arrived at a conclusion that the earth was a sphere. His speculations about the shape of the earth were seconded by Eratosthenes based on a limited, measured arc of longitudes. The astronomer Hipparchus of Rhodes (2nd century BC) developed a location system of lines on the surface of the earth – the forerunner of latitudes and longitudes. Hecataeus and Herodotus were, however, not the followers of the idea of the sphericity of the earth. Eudoxus of Cnidus – a contemporary of Plato – developed the theory of zones of climate based on increasing slope (*klima*) away from the Sun on a spherical surface. All these formulations were deductions from pure theory that all observable things were created in perfect form and that the most perfect form was a sphere.[56] Aristotle was the first philosopher who wrote with definite arguments about the spherical shape of the earth. He gave two arguments in support of his statement about the sphericity of the earth. First, he deduced it from the law of gravitation, or, as he expressed it, the tendency of all things to move towards the centre. Through the operation of this principle, when the earth was in the course of formation, and the component elements were coming together equally from every quarter, the mass thus formed by acceleration was so constituted that its entire circumference must be equidistant from its centre. Secondly, he infers it from what is seen to take place in lunar eclipses; for, when the earth is interposed between the Sun and the Moon, the special form of the obscured part of Moon's surface shows that the body which causes the obstruction is also spherical.[57] Archimedes (250 BC) deduced from the interference that the surface of the sea must be convex – a conclusion which would naturally involve the gradual revelation of objects approaching upon it.[58]

Plato stressed on the theory of intuition and reasoned from the general to the particular (deductive approach), while Aristotle built the theory by reasoning from the particular to the general. These two ways of thinking about things/phenomena are known respectively as deductive approach and inductive approach (scientific approach).

In addition to this Aristotle postulated the existence of four basic substances: earth, water, fire and air. All material objects on earth are made up of these basic elements in varying proportions. Aristotle also added a fifth substance, aether, which did not occur on the earth but was the material from which celestial bodies were made.

Aristotle was the first teleologist to believe that everything is changing in accordance with a preexisting pattern or plan. Aristotle said, all things are not deteriorating from an ideal state, but are rather developing toward an ideal state.

Archytas measured the total length of the land and sea. Aristotle agreed with the calculations of his predecessors stating that the circumference of the earth is 40,00,000 stadia (40,000 miles). Eratosthenes who made observations at Syene (Aswan) and Alexandria calculated the circumference of the earth as 250,000 stadia (25,000 miles). Looking at the indigenous gnomon he used, it can be said that he was very near to the truth.

Herodotus attempted to determine the meridian of longitude. He drew a meridian from Meroe, Syene (Aswan), Alexandria, Troad, Byzantium and the mouth of Borysthenes[59] (Figure 1.7). Hipparchus pointed out that the true method of determining longitudes was by the comparative observation of eclipse. However, we have no evidence to show that any investigations were made. The Greek scholars, especially Herodotus, Anaximander, Hipparchus and Eratosthenes drew the parallels of latitudes also. The first parallel drawn by Eratosthenes passed from the Pillars of Hercules to the extremity of India. Other parallels of latitudes were drawn through the Cinnamon region (Ethiopia), Meroe, Syene (Aswan), Alexandria, Rhodes, Troad, the mouth of Borysthenes, Pytheas and Thule. Similarly, the meridians of longitudes, drawn by Eratosthenes, passed through the Pillars of Hercules, Carthage, Alexandria, Thapsacus (on the Euphrates), the Caspian Gates, the mouth of the Indus and the mouth of the Ganges.

So far as the shape of the earth is concerned, the Ionians (Thales, Anaximander and Hecataeus) considered the earth as a circular plane, surrounded on all sides by Ocean River. The Ionians divided the habitable world into two continents, namely, Europe and Asia including Libya (Africa). Herodotus, however, did not agree with this idea of the Ionian School. He conceived the earth as a plane and was of the view that there is symmetrical arrangement of land and water and inhabited parts on the earth's surface. He compared not only the lands to the north and south of the Mediterranean Sea but also the rivers (Nile and Ister), their directions and deltas. Eratosthenes treated the inhabited world as an island, and made it in shape of irregular oblong, the extremities of which tapered off to a point both east and west. These end points of the oblong he fixed at the extremity of India and the Sacrum Promotorium in Spain.[60] He divided the world by the Mediterranean Sea and the Tarus Mountains. To the north of these he called the land as Europa and to the south of it as Asia and Libya (Africa). The subsequent Roman scholars like Strabo also followed the same theory of the shape of the earth.

One of the major contributions of the Greek scholars was in the realm of theory building and hypothesis testing. Plato mostly built theories by intuition and reasoned from the general to the particular which is called the deductive approach. Contrary to this, his disciple Aristotle built theory by reasoning from the particular to the general.[61] This is known as the inductive approach. Aristotle recognized that observations made through the senses can never provide explanations. Our senses, he said, can tell us that fire is hot but cannot tell us why it is hot.[62] Aristotle was also the first teleologist as he believed everything was changing in accordance with the preexisting pattern or plan, just as a mason/carpenter building a house knows in advance what the house will be like when it is finished. Aristotle reasoned that the parts of the earth close to the equator (the torrid zone) were not suitable for human habitation. He also opined that all things are not deteriorating from an ideal state, but are rather developing towards an ideal state.

Physical Geography

The Greeks also made remarkable development in the field of physical geography. Greece was in many ways a suggestive country having diverse topographic and physical features. As has already been stated at the outset, Greece was a land of mountains, many of which were of great height to be snow-clad in winters. The rivers for the most parts were torrents which flowed with a rushing current in winter, and were dry in summer. There were perennial streams also like Acheolous and Alpheius.[63] The capes which projected into the Aegean Sea and the straits which penetrated into the mainland inspired the Greek sailors to make observations and to explore the neighbouring islands. There were some peculiar features also like subterranean streams in limestone areas, volcanoes and occurrence of earthquakes which provided incentives to the Greeks to find the causes of their occurrence. The effect of these features on the mind of Aristotle is especially traceable.[64] The destruction of the cities of Helice and Bura on the coast of Achaia by an earthquake took place during the early life of Aristotle. This event seems to have greatly touched his imagination, for he refers to it more than once in his *Meterologica.*[65]

The works of the Greeks contain numerous references to mountains, delta building, winds, change of weather, rain, earthquakes and their causes, volcanoes and transformation in the topographic features. Aristotle explained the phenomena of expansion of land in the shallow seas and the formation of delta. He correctly pointed out the process of alluvial deposition through which in so many places the land was gaining on the sea, especially in the

Palus-Maeotis (Sea of Azov), which he affirmed, was continually becoming shallower, and would be one day entirely filled up and converted into dry land.[66] The delta formation of the Nile was also attributed to the enormous silt carried by the river from its upper reaches (Ethiopia).

The Greeks were of the opinion that all the perennial rivers had their sources in great mountain ranges. Agatharchides has described the occurrence of gold ore in the Ethiopian gold mines visited by him and has narrated the process of its extraction from the veins of rock strata. Plato has discussed some of the barren lands of Attica (Greece) and has explained that such waste tracts in the past were full of vegetative covers and fertile soils. Under the impact of the external forces, the forests had been depleted and soils leached resulting into barren topography. Such wastelands, he said, were like the skeleton of a sickman, all the fat and soft earth having been wasted away, and only the bare framework of the land being left.[67] Plato considered man an active agent who changed the face of the earth.

From the Aegean Sea, the Greeks expanded the horizon of knowledge in the study of seas and oceans and distinguished the varying properties of their coastlines, salinity, waves, tides and winds. Posidonius wrote a book – *The Ocean*. On oceanography he was considered an authority. Herodotus observed the phenomenon of tides in the Red Sea and the Matiac Gulf. Aristotle observed the tidal movements in his book – *Meteorologica*. But the cause of tidal waves he attributed to the winds. Later on, Nearchus observed the tides in the Arabian Sea and Pytheas in the Atlantic Ocean. Pytheas, who was a scientist, made careful observations on the regular recurrence of tides, with the aim of determining the causes which produced them. He established the correspondence between their diurnal recurrence and movement of the moon.[68] It was Posidonius who pointed out that at the new moon when the sun and the moon were in conjunction, and also at the full moon, the tides were the highest (spring tides), whereas at the first and last quarters they were the lowest (neap tides).[69]

The Greeks, right from the Homeric period, recognized four major winds, having different properties and directions. These winds were called bores (north wind), eurus (east wind), notus (south wind), zephyrus (west wind). In the second century BC, the Athenians built a tower identifying eight wind directions with sculpture illustrating the weather types associated with each.[70] The tower still stands in the midst of a Roman market at the base of the Acropolis.

The Greeks divided the world into torrid, temperate and frigid zones. They were familiar with the excessively high temperatures to be experienced in Libya along the southern side of Mediterranean. The Greeks believed that the Libyans are black because they had burned black by exposure to the sun and deduced that further southward (near equator) life must be impossible. Aristotle reasoned that the parts of the earth close to the equator (the torrid zone) were uninhabitable. The parts away from the equator being very cold and frigid were also uninhabitable. The Greeks established a relationship between temperatures and ecumenic regions of the world.

The mountains of Greece belong to the Alpine tectonic activity. These are the young folded mountains in which occurrence of earthquakes is a frequent phenomenon. Earthquakes attracted the attention of Greek thinkers. Herodotus expressed his opinion with regard to the disruption of Olympus and Ossa. Anaximander described earthquakes as fractures of crust of the earth, which were produced by its passing through a process of drying, after having previously been saturated with moisture.[71] According to Aristotle, earthquakes and volcanoes were caused by winds (gases) which were confined beneath the surface of the earth, and were trying to find a vent. The Greeks made very careful observations about volcanoes and felt that they were vents in the earth crust, by means of which the frequency and violence of the earthquake movements were lessened.[72]

The Greek scholars also recognized differences in the fauna and flora of the different parts of the world. Some of the pupils of Aristotle, particularly Theopharastus, described the habitat of different trees and shrubs and established a correlation between climate and natural vegetation.[73] As regards the fauna, the Greeks described a great number of strange animals of different nations and regions.

Notes

1. Casson, L., 1959, *The Ancient Mariners*, New York, p. 37.
2. Bunbury, E.H., *A History of Ancient Geography*, Vol. I, p. 4.
3. Ibid., p. 35.
4. Tozer, H.F., 1897, *A History of Ancient Geography*, Trans. J.D. Akhtar, New Delhi, p. 14.
5. *Illiad*, vii, pp. 421–423.
6. Bunbury, E.H., op. cit., p. 34.
7. Ibid., p. 35.
8. Nick, M., 1945, *Die Entdeckung von Europa durch die Griechen*, Basel, Benno Schwabe.
9. James, P.E. and Martin, G.J., 1972, *All Possible Worlds*, New York, p. 7.
10. Ibid., p. 18.
11. Heidel, W.A., 1937, *The Frame of Ancient Greek Maps*, New York.
12. James, P.E., 1972, op. cit., p. 18.
13. Ibid., p. 18.
14. Eratosthenes, ap. Strab, i, p. 7.
15. Bunbury, E.H., op. cit., p. 137.
16. Thomson, J.O., 1965, *A History of Ancient Geography*, New York, p. 35.
17. Tozer, H.F., op. cit., pp. 15–30.
18. Ibid., p. 65.
19. Bunbury, E.H., op. cit., p. 148.

20. Tozer, H.F., op. cit., p. 148.
21. Bunbury, E.H., op. cit., p. 150.
22. Wright, K.J., 1925, 'The History of Geography: A Point of View', *Annals of the Association of American Geographers,* 15, pp. 200–201.
23. Sarton, G.,1952, *A History of Science, Ancient Science Through the Golden Age of Greece* (Reprinted) New York, John Wiley and Sons, 1964.
24. Bunbury, E.H., op. cit., p. 89.
25. Tozer, H.F., op. cit., p. 89.
26. Ibid., p. 91.
27. Heidel, W.A., op. cit., p. 45.
28. James, P.E. and Martin, G.J., op. cit., p. 21.
29. Bunbury, E.H., op. cit., p. 165.
30. Ibid., p. 170.
31. Ibid., p. 191.
32. Strabo, *The Geography of Strabo*, Trans. H.L. Jones, 1917, New York, G.P. Putman's Sons.
33. Bunbury, op. cit., p. 254.
34. Ibid., p. 262.
35. Ibid., p. 272.
36. Tozer, H.F., op. cit., p. 79.
37. James, P.E. and Martin, G.J., op. cit., p. 28.
38. Sarton, G., op. cit., p. 60.
39. Tozer, H.F., op. cit., p. 138.
40. James, P.E. and Martin, G.J., op. cit., p. 28.
41. Tozer, H.F., op. cit., p. 134.
42. Bunbury, E.H., p. 615.
43. Ibid., p. 620.
44. James, P.E. and Martin, G.J., op. cit., p. 32.
45. Ibid., p. 621.
46. Thomson, J.O., op. cit., p. 61.
47. Bunbury, E.H., op. cit., p. 625.
48. Tozer, H.F., op. cit., p. 140.
49. Bunbury, op. cit., p. 640.
50. Ibid., p. 645.
51. Ibid., p. 649.
52. Thomson, J.O., op. cit., p. 33.
53. James, P.E. and Martin, G.J., op. cit., p. 26.
54. Lewis, C., op. cit., p. 33.
55. Heidel, W.A., op. cit.

56. James, P.E. and Martin, G.J., op. cit., p. 26.
57. Aristotle, De Caelo, 2,14, pp. 8–13.
58. Archimedes, Trans. 1,2.
59. Tozer, H.F., op. cit., p. 178.
60. Ibid., p. 181.
61. Sarton, G., op. cit., pp. 10–60.
62. James, P.E., and Martin, G.J., op. cit., p. 26.
63. Tozer, H.F., op. cit., p. 10.
64. Ibid., p. 185.
65. Aristotle, *Meteorologica*, Vol. I, pp. 6–8.
66. Bunbury, op. cit., p. 398.
67. Glacken, C.J., op. cit., pp. 70–92.
68. Tozer, H.F., op. cit., p. 193.
69. Ibid., p. 193.
70. James, P.E. and Martin, G.J., op. cit., p. 16.
71. Aristotle, op. cit., p. 6.
72. Tozer, op. cit., p. 199.
73. Theoph, *History of Plants*, pp. 4–10.

2

Romans

Their Contribution to Geography

ROMAN PERIOD

After the Greeks, the political power passed into the hands of the Romans. The Roman armies made advances in different directions, especially in Central Europe, France, Britain and Asia Minor. They, with the help of their well-disciplined armies, were able to conquer distant countries. Pompey in Albania and Asia Minor made significant conquests while Gaul and Britain were overwhelmed by Caesar. During the Augustan Age, the knowledge of the world which the ancients possessed reached its farthest limits. The Romans' major contribution was to historical and regional geography. Whereas Polybius and Posidonius contributed to the field of physical geography, it was Strabo who compiled the regional and historical geography of the world.

A very large part of what scholars thought they knew about ancient geography came from Strabo. Most of the books written by earlier scholars have disappeared entirely or survived only in fragments. But Strabo's monumental work on geography is almost intact, with only a very few minor parts missing.[1] Thus, Strabo's *Geographica,* whatever its defects, is our great repertory of information concerning knowledge of the subject which the ancients possessed.

Strabo

Strabo (64 BC – 24 AD)

Strabo was a Greek geographer, born in Amesia, about 75 km to the south of Black Sea coast in Turkey in 64 BC. Amesia is in the interior of Asia Minor (Turkey) and still retains its ancient name (Figure 2.1). It was the capital of barbarian kings and had a large Greek population. It is certain that Strabo got a good Greek education which made him one of the leading scholars of his period.[2] Strabo is considered as the 'Father of Regional Geography', because he substituted divisions based on natural boundaries

Figure 2.1 World after Strabo

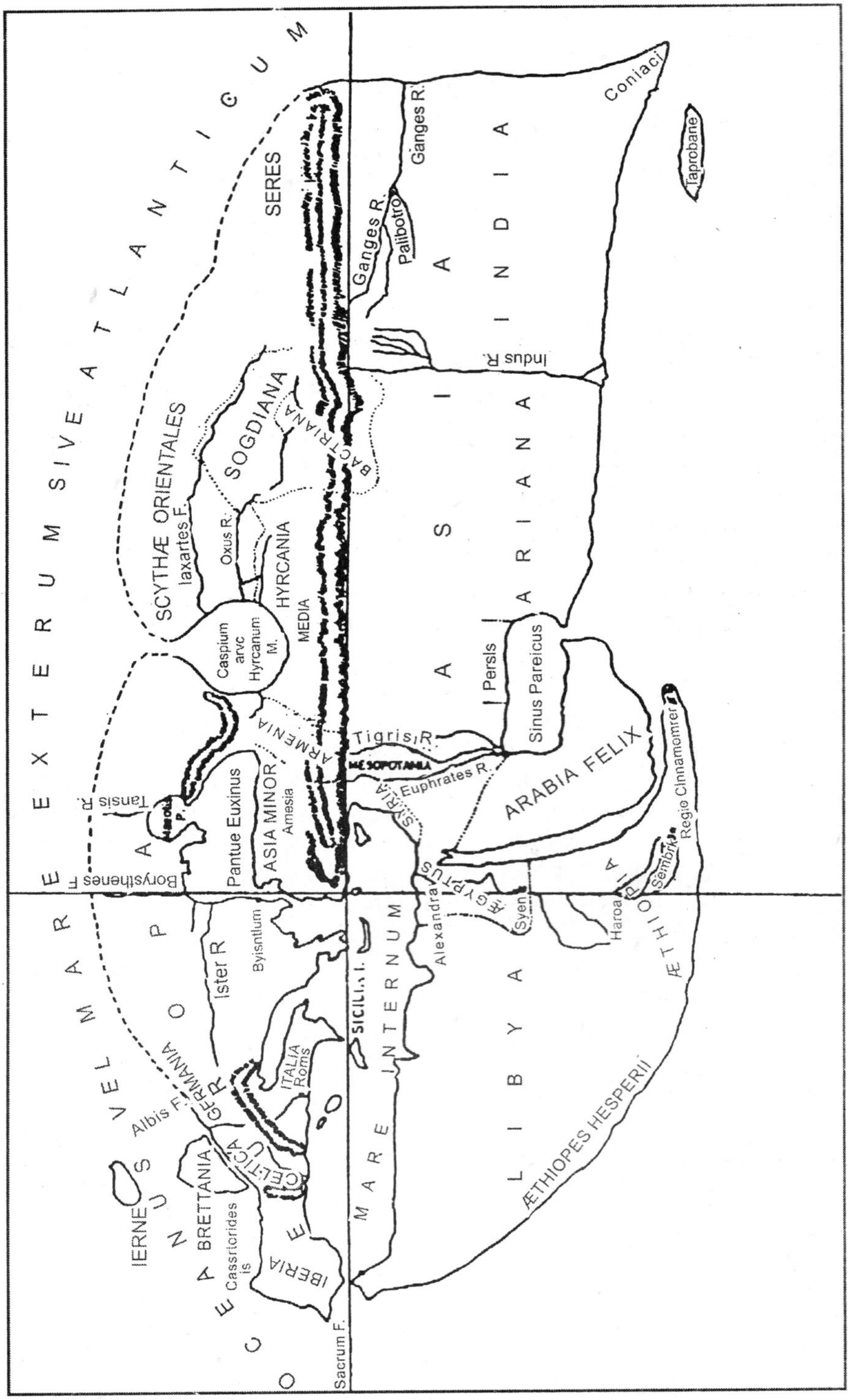

(such as mountains, rivers etc.) for the less permanent and artificially drawn political units. He was the first to declare geography as a *chorological science.* About Strabo, Humboldt has justly remarked that he 'surpasses all the geographical writings of antiquity, both in grandeur of plan and in the abundance and variety of its materials'.[3] Later on he stayed at Cnossus in Crete to which he adverts in his description of that island.[4]

Very little is known about his early life and the exact date of his birth. From his writings, it can be ascertained that Strabo got his early education at Nysa under the supervision of Arstodamus who was a great grammarian.[5] Strabo's life was characterized by extensive travels. He journeyed Egypt, Ethiopia, and Asia Minor. He visited Corinth (in Greece) at the time of Augustus and in about 29 BC went to Rome where he stayed for several years. From Rome he came to Alexandria and with Gallus (the Roman Governor) made a voyage on the Nile up to Syene in 24 BC.[6] He travelled from the frontiers of Armenia on the east to the shores of Tyrrhenian Sea on the west, and from the Euxine (Black Sea) to the borders of Ethiopia. It is, however, doubtful if he visited all the countries and lands between these limits. In fact, he saw very little of Italy and Greece. In Greece, his visits were confined to Corinthus, Athens, Megara and Argos. The Adriatic coast of Italy was also a Terra-Incognita (unknown world) to him. He was better acquainted with Asia Minor. His accounts of Armenia and Colchis are rather vague and superficial. Of the land lying to the north of Caucasus mountains and the Black Sea, his knowledge was highly imperfect. Strabo, who died at the age of about 84 (24 AD), wrote most of his works after his return to his native city.

The geographical treatise written by Strabo is not only the most important geographical work that has come down to us from the classical period, but also unquestionably one of the most important works ever produced by the scholars of antiquity. The main feature of Strabo's geographical account lies in the fact that it was the first attempt at bringing together all the then known geographical knowledge in the form of *general treatise.* He laid the foundation of chorological writing in geography, and he was the first who codified the term '*chorology*' most elegantly. The criticism that Strabo's geographical treatise is just an improvement over the work of Eratosthenes (a Greek geographer) does not carry much credibility. The work of Eratosthenes was based on just three volumes while Strabo wrote as many as 43 volumes under the title *Historical Memoir*. Moreover, he wrote 17 volumes of his geographical treatise. Strabo is the first scholar who conceived the idea of a complete geographical treatise, comprising all four branches of the discipline, namely, mathematical, physical, political, and historical geography. In estimating its importance from a modern point of view, we have to take into consideration not only its intrinsic merit, but also the enormity of the loss which we would have suffered had it perished.[7] It is the one complete treatise on geography and acquaints us with the writings of his predecessors whose works are entirely lost. Their passages are found in the form of quotations in Strabo's works.

Strabo's geographical treatise, namely, *Geographica* was designed not for geographers but for politicians and statesmen. It also included attempts to explain cultural distinctiveness, types of governments and customs in particular places. In other words, it was meant for general reader and not for mere geographers. The author thus endeavoured to present a general picture of each country, its character, physical properties, surface configuration and natural productions. It was Strabo who stressed on the division of the world into natural and not into political boundaries. In his opinion, divisions of an area can be established only by boundaries and geography should divide the world by the 'natural boundaries of lands' rather than the 'political boundaries of states'.

In the field of mathematical geography, Strabo's contribution cannot be termed outstanding when compared with that of his predecessors (Eratosthenes and Posidonius). His work was not designed for astronomers and mathematical geographers. Nor was it meant to help them to determine the shape and size of the earth, its relation with heavenly bodies and the important latitudes (Equator, Tropic of Cancer, and Tropic of Capricorn). Nevertheless, he quotes with approval the assertion of Hipparchus (a leading mathematical geographer) that it was impossible to make any real progress in geography without the determination of latitudes and longitudes.[8] He was of the opinion that for the astronomical and mathematical part of the subject a geographer may content himself with taking for granted the conclusion of physical philosophers and mathematicians. Thus, he assumes that the earth is spherical in shape and situated in the centre of the universe. He also assumes the division of the earth into five zones and the circles upon the sphere derived from the motion of the celestial bodies, i.e. the equator, the zodiac, the tropics and the Arctic circle. He saw the earth as an *oblong*. He regarded Ireland as the most northerly of all the known lands.

In the field of physical geography also, his work cannot be regarded as outstanding but there is no denying the fact that it was a great improvement over his predecessors' works. Unfortunately, Strabo gave little attention to topographic features, mountains, rivers and their courses while giving geographical accounts of different regions. Strabo's remarks on physical geography are of great value. He has brought together a large amount of material to throw light on the changes which have occurred over the face of the earth owing to the transgression and regression of the sea, and due to earthquakes and volcanic eruptions. He also discusses the causes which have brought about these changes. The two main principles which he enunciates as his own are mentioned with high praise by Sir Charles Lyell, as being anticipations of the latest conclusions of modern science. These are: (i) the importance of drawing inference with regard to the more extensive physical changes from those which take place on a lesser extent under our own eyes; and (ii) the theory of alternate elevation and depression of extensive areas.[9]

Strabo's work is mainly historical. Not only does he everywhere introduce the history of a country side by side with its geography, but he also illustrates the one by the other, and endeavours to point out the intimate connection that exists between geography and history. He also attempted to trace the influence of the physical features of an area on the character and history of its inhabitants. To illustrate this point, he wrote that Italy was in a peculiarly protected geographical location and owing to this fact the people of this country are more advanced and developed. The physical location of Italy contributed to the development of power of Rome. He dilates upon the advantages Italy derived from its natural geographical situation. It offered her protection against attacks from outside; its natural harbours gave a boost to its commerce and business activity. Further, Italy's physical location was responsible for its varied and temperate climate as also for the influence of elevation in different parts which caused it to enjoy the products both of a mountainous country and of the plains. It had a beneficial effect upon her water supply, and, above all, on her central position among the great races of the world. Moreover, Strabo gave an artistic treatment to his geographical writings which are not a dry account of facts and places. The main objective of Strabo in his geographical treatise was to present a general survey of the entire habitable world known during that period. Spain, Gaul (France), the coast of the Atlantic, south-eastern parts of Britain – all these areas were fairly known and thus the Romans opened out all the western parts of Europe up to the river Albis (Elbe) and the region beyond the Danube and the river Tyras. The tracts on the north of Euxine (Black Sea) and along its eastern coast to the borders of Colchis were plotted in the world map of Strabo (Figure 2.1). In fact, Mithridates and his army generals did enough exploration in this part of the world. Unfortunately, Strabo did not consult the Greek historian and geographer, Herodotus, who had given a vivid account of the region and tribes situated to the north and east of the Euxine Sea. Herodotus, in the opinion of Strabo, was a *retailer of fiction*.[10] It is because of this attitude of Strabo towards Herodotus that his knowledge of the Scythian races is quite meagre and erroneous.

Of the Caspian Sea, Herodotus has given a correct account describing it as a closed sea but Strabo believed that it communicated with the Northern Ocean, and beyond it the Jaxartes remained, as it was in the days of Alexander, the limit of discovery. With regard to India, the Peninsula of Hindustan continued to be unknown, and the Ganges was regarded as flowing into the Eastern Ocean. Regarding Africa, the upper course of Nile (*Cinnamon Land*) was the southern-most limit, as far as Strabo was concerned. He did not describe Mauretania and the western coast of Africa though a good account of these regions was given by the Greeks and his own contemporary Juba. He compared the deeds of the Roman army with those of Alexander's eastern expedition by saying that the Romans had opened out all the western parts of Europe in the same manner as the conquest of Alexander had done shortly before the time of Eratosthenes. It is worthwhile to give a brief account of the contents of the different volumes of Strabo's geographical treatise.

The first two volumes of his *Geographica* are devoted to an introduction of the subject in which he discusses the aims and objectives of his treatise and the fundamental principles on which he conceives the general features which characterize the entire area of the world and the then known continents. These two volumes can be regarded as the most difficult and unsatisfactory part of his work. These volumes comprise a historical review of the progress of geography from the earliest days but the approach is not methodical. In these works, he reviewed the work of Eratosthenes and his other predecessors but most of the time he criticizes the previous attempts of geographers. He, however, appreciates the work of the great Greek poet, Homer, and considers him as the founder of all geographical knowledge.

In the second volume, he examines in detail the work of Eratosthenes and discusses the various changes introduced by him in the map of the world. He appreciates the work of Eratosthenes in which he gave an account of Asia. In fact, in regard to the whole of Asia, Strabo adopted the map of Eratosthenes with hardly any alteration. It was only about the land lying between the Euxine and the Caspian Sea that Strabo had acquired more information than his predecessors, and even this knowledge was of such imperfect character that he believed the Caspian to communicate with the Northern Ocean. Little or no change was made in the details of Africa, but in the map of Europe, especially its north-western parts, he inserted many new details. About the shape of the inhabited world he followed the view of Eratosthenes who had described it as forming an irregular oblong with tapering extremities towards the east and the west (Figure 2.1).

The third volume gives an account of Europe with stress on the geography of Spain, Gaul (France) and Britain. For the description of these areas, Strabo mainly relied on Polibius and Posidonius who had travelled to Spain. He also gathered information about these countries from Caesar. While describing Spain he refers to the Pyrenees mountains as forming a continuous chain from Gaulis Gulf (Bay of Biscay) to the Mediterranean Sea in a north-south direction which is not correct. Moreover, he considered the Sacred Promontory (Cape St. Vincent) to be the most westerly point of Europe. In the last section of the third book, Strabo treats the islands adjacent to Spain and describes at length Gadeira (Gadis) which was one of the important commercial centres of that period.

The fourth volume is devoted to Gaul, Britain and the Alps. His description of the Gaulis Gulf (Bay of Biscay) 'as looking towards the north and towards Britain' is also erroneous. He conceived the northern coasts of Gaul as maintaining the same direction from the Pyrenees to the mouth of the Rhine. He held that the four great rivers – the Garumna (Garonne), the Liger (Loire), the Sequana (Siene), and the Rhine, flowed from south to north. The mouths of all these rivers he considered as being opposite to Britain. Considering the rivers of Gaul as the most perfect drainage system he felt they provided easy means of trade routes and transportation. Of considerable interest are the paras in which he describes the primitive tribes of Iberia (Spain) and the civilized and developed societies of Gaul. Of Britain he had very little knowledge except what he had derived from Caesar. He conceived Ireland to be situated to the north of Britain. Its length he reckoned to be more than its breadth. About its inhabitants he writes that they were savages, and cannibals. Regarding the Alps he writes that it forms a great curve having its concave side turned towards the plains of Italy.[11]

The fifth and sixth volumes are devoted to Italy and Sicily. The major source for the description of these countries was Posidonius. He described Italy according to the popular belief of its north-south direction, but in his map he has shown Italy stretching from the west in the eastern direction (Figure 2.1). He considered the Alps mountains as the northern boundary of Italy. The Apennines are described by Strabo as extending directly across the whole breadth of Italy. He vividly describes the volcanic eruptions on the island of Pithecusa (Ischia) and Mount Vesuvius. Vesuvius has been referred to as a burning mountain. Of the streams of lava he gives an accurate account, pointing out how the burning matter that overflows from the crater in a liquid form gradually hardens into a compact and hard rock-like millstone. He notices also the great fertility of the soil produced by the volcanic ashes for the growth of vines. He devoted very little space to the description of Corsica and Sardinia as such. This description is very brief and imperfect.

In the seventh volume he gave a brief and general account of the countries extending to the east of Rhine and to the north of Danube. This geographical account is highly defective. In fact, his knowledge of Central Europe and the land lying to the north of the Euxine was so imperfect that he did not write anything about the sources of Tanais river. This area was inhabited by barbarians and the Greeks had very little commercial relations with the interior. Consequently, Strabo did not have reliable information about this region.

The eighth, ninth and tenth volumes are devoted to the geography of Greece and the neighbouring islands. Strabo, for the information on Greece and its adjacent islands, relied upon Homer – the great Greek poet – as a result of which all these three books are 'a desultory and rambling commentary upon the Homeric catalogue rather than a geographical treatise'. He had himself visited only a few points of Greece (Athens, Megara and Corinth) and was therefore compelled to collect his information at second hand for which he relied on poets instead of Greek historians like Herodotus. Thus, he followed the example of his predecessors, especially Hipparchus, Polybius and Posidonius, and not that of Eratosthenes who was opposed to the Homeric concepts of geographical thought. The description of the configuration of the northern part of Greece was even more erroneous. He gives very little information about the physical geography of Greece except the inlets, straits and the subterranean drainage of some of the rivers. In the limestone topography of Greece many of the rivers pursue their courses for some distance underground, and then appear again over the surface.[12]

About the islands situated in the Aegean Sea he gives a meagre account and their geographical locations have not been correctly indicated.

Six volumes – eleventh to sixteenth – are devoted to the geographical descriptions of Asia. In all these books, he relied upon Eratosthenes, especially with reference to the configuration, topography and drainage system. He assumed that Taurus mountains traverse Asia from west to east (Figure 2.1). He took Taurus mountains as the dividing line between the Northern Asia and the Southern Asia.[13] He divided Northern Asia into four divisions: (1) Tanais to Caspian Sea; (2) Caspian to Scythians; (3) Medians and Armenians; and (4) Asia Minor.

The portion south of the Taurus mountains consisted of India, Ariana (Iran), Persia and all the nations that extend from the Persian to the Arabian Gulf (Red Sea), the Nile and the land lying to the east of the Mediterranean Sea, i.e. Assyria, Babylonia, Mesopotamia, Syria and Arabia. He was, however, not aware of the fact that both the Tigris and Euphrates have two sources and flow for a considerable distance in two separate streams.

The eleventh volume is devoted to the border land of Asia and Europe taking river Tanais as the boundary between these two continents. In this volume, he gives an account of the land lying between Euxine and Caspian, and Parthia and Media.

Strabo's next three volumes (12th to 14th) are filled with the description of Cappodocia and Pontus and the northern provinces of Asia Minor along the coast of Euxine (Black Sea). Since he was a native of this region, the regional and historical accounts are reliable and of a high order. Of this region he writes that it is an open upland without forest but not devoid of fertility, producing abundance of corn as well as supporting immense quantity of sheep and an excellent breed of horses. He also notices many mineral produce of the country. The main mineral being red-earth which was called Sinopic-earth because it was brought down from the interior to Sinope for export. He also describes the volcanic activity of Mount Argaeus.

The mainland of Asia, situated to the south of Taurus, comprising the nations of Assyria, Persia, Babylonia, Mesopotamia, Syria, Arabia and India, is discussed in the fifteenth and sixteenth volumes. The fifteenth book deals with India and Persia and the intervening districts. The sixteenth volume discusses the geography of Assyria, Syria, Palestine and Arabia. For the geography of India, he relied on Nearchus, Aristobulus and others who accompanied Alexander in his eastward expeditions. He also consulted the records and treatise of Magasthenes. He felt that India's greatest length was from west to east. Hence, he regards the Promontory of Coniac (Cape Comorin) to project to south-east. His conception of the map of India did not differ from that of Eratosthenes. He cites Artemidorus for the correct statement that the Ganges had its source in Emodi mountains (one of the many names of the Himalayas known to the Greeks). He held that the Ganges flowed at first to the south and then to the east and passed by the side of Polibothra (Patliputra, Patna) and from thence to the Eastern Sea.[14] His knowledge of the tributaries of the Indus and the Ganges was, however, vague. The Peninsula of India was also not described by him. He considered Taprobane (Ceylon) as situated at the southern limit of the known world.

The lands lying between India and Persia and the Taurus and Persian Gulf were considered as Ariana (Iran). In fact, it is the central plateau of Iran extending from Seistan (Drangiana) to those of Yezd and Kerman.[15] It is this region through which Alexander passed on his way back from India. Strabo briefly gives an account of this region which can hardly be accepted as a geographical work.

So far as the description of Persia is concerned, he rightly distinguishes, according to their climates, the three regions into which the country is divided: (1) the Persian Gulf and the Median uplands, characterized by sandy tract, and date-palm as the main crop; (2) the fertile and well-watered tract of the interior plain and lake; and (3) the northern mountains of extreme cold. The elaborate canal irrigation system of Babylonia has also been described. He has also noticed the peculiar feature of the Dead Sea – its salinity.[16] The description of Arabia with which this book concludes is as complete as the knowledge of that age allowed.

The seventeenth and the last volume of Strabo's great work is devoted to Africa. Two-thirds of the book deal with the geography of Egypt. It provides adequate information on Egypt since library at Alexandria had detailed records of the Greeks. Moreover, as stated earlier, Strabo himself had travelled up to the upper reaches of the Nile river (First Cataract). Consequently, he describes with considerable minuteness the geography of the Nile delta and the sources of the Nile. He also gives a graphic description of the inundation of the Nile. With regard to the cause of inundation, which had been a subject of so much discussion and curiosity among the early Greeks, he tells us that it was caused, as generally believed in his time, by the heavy rains that fell in the summer on the mountains of upper Ethiopia.[17]

His account of the voyage of Nile is especially interesting. He saw the river of Thabes. He ascended the river as far as Syene. Most probably he reached up to the lake Moeris (Figure 2.1) and the celebrated Labyrinth. Strabo also describes the oases of Libya, referring to them as the inhabited districts surrounded on all sides by vast deserts just as islands by sea.[18]

With regard to the rest of Africa, Strabo had little knowledge. His knowledge of the shape of the continent was like that of the Greek geographers. He described it as a right-angled triangle, having at its base the Mediterranean Sea coast and the shorter side was formed by the Nile through Ethiopia up to the ocean. He also asserted that all the Libyan tribes resembled one another in their dress and habits. He hinted that in the interior side of Libya are two nations, namely, Pharusians and Nigrates which occupy the land to the west of Ethiopia. The account of coast between Carthage and Cyrenaic is given in considerable detail. The Islands of Fortune were, however, omitted by him.

From the above description, it is clear that Strabo was the only geographer of the ancient period who lucidly wrote about all the branches – historical, political, physical and mathematical – of geography.

Ptolemy (90–168 AD)

Ptolemy

Claudius Ptolemy was a native of Egypt a Roman provine. He lived and wrote at Alexandria about the middle of the 2nd century of the Christian era. He was a Greco-Roman writer of Alexandria, known as a mathematician, astronomer, poet and geographer. Throughout his life he stayed at Alexandria, where he died in around 168 AD.

Ptolemy was one of those geniuses who developed sound principles of mathematical geography. His writings inspired the geographers and explorers of the Great Age of Discovery (14th, 15th century AD) to explore the *Terra-Incognita* (unknown land).

Very little is known about Ptolemy's place of birth and early life and belief that he was born at Pelusium is not acceptable to all.[19] The period when he produced great work is also obscure. But, as De Morgan observes that an astronomer always leaves his date in his works[20], it is certain that he made observations in 139 AD Some critics have considered him a plagiarist who was obliged to Marinus of Tyre and thus only improved upon this pioneer astronomer. This criticism is unjust to Ptolemy. Undoubtedly, he made use of materials of Marinus and Hipparchus but he presented them in a more convenient shape and scientific manner. His concept of the universe coincided with that of Aristotle: the earth was a sphere that remained stationary in the centre while the celestial bodies revolved round it in circular orbits. This geocentric concept remained accepted doctrine until the time of Copernicus in the 17th century.

Ptolemy was the author of several scientific treatises, three of which were of continuing importance to later Islamic and European science. These three treatises include (i) Almagest (Mathematical Treatise), (ii) the Geography, and (iii) Astrology.

His own contribution in the field of geography, especially in mathematical geography, is highly commendable and has been acknowledged throughout the ages. His best known work is *The Syntaxis* (popularly known as *The Almagast* – an Arabic name derived from the Greek title *E-Megiste Syntaxis* – means the Great Synthesis). *The Almagast,* probably the earliest of his works, is a great contribution to classical astronomy. It presents in detail the mathematical theory of the motion of the Sun, the Moon and planets. Ptolemy accepted the solar theory of his predecessor – Hipparchus, improved on the lunar theory, and made his most original contribution by presenting details for the motions of each of the planets. Ptolemy's geometric models, used only to predict the positions of these bodies, employed combinations of circles known as *epicycles*, with the framework of the basic earth-centred system (geocentric).

He believed that the stars were fixed points in a rotating sphere. He stated that the planets are much closer to the earth than the stars, but are farther away than the moon.

He devoted two parts of *The Almagast* to a *catalogue of stars*. He described a mathematical arrangement of the stars and gave the celestial latitudes and longitudes, as well as magnitude (brightness) for each of them. This catalogue includes 1,022 stars grouped into 48 constellations. Ptolemy also discovered the irregularity of the moon in its orbit. *The Almagast* was not superseded until a century after Copernicus presented his Heliocentric Theory in the *De Revolutionibus* of 1543.

His second most important work *The Geography*, also known as *The Guide to Geography*, opens with an excellent theory of map projection. The book is a catalogue of places with their latitudes and longitudes and describes briefly each continent, country and tribe. It also contains a map of the world including Europe, North Africa, and most of Asia as well as 26 maps of specific areas.

Two relatively minor works, *The Optics* and *The Tetrabiblos*, dealt with astrology, and respectively, with reflection and refraction. With the exception of *The Optics*, all the Ptolemy's works were immensely influential.

The book *The Guide to Geography* consisted of a list of all known places tabulated according to latitudes and longitudes - the system - Ptolemy devised himself. In his opinion, the purpose of geography was to provide a view of the whole, analogous to the drawing of the whole head, and this meant that he separated geography from chorography which, so he said, 'has the purpose of describing the parts, as if one were to draw only an ear or an eye'. He also opined: 'Geography is a science which deals with the art of map-making'.[21] This conception dominates the entire book of Ptolemy. The basic objective of Ptolemy was 'to reform the map of the world' on the basis of astronomical principles. Thus, he followed in the steps of Eratosthenes and Hipparchus who described geography as the 'science of map-making'. He was a staunch follower of Hipparchus who stressed that a map of the world could correctly be laid down only by determining the latitudes and longitudes of all the important points on its surface. He was aware of the fact that to achieve satisfactory results, it was necessary that all such positions should be determined by direct astronomical observations. Unfortunately, the number of such observations at his command was very small. Thus, he had to rely upon the distances computed by travellers and navigators. These estimates and itineraries of travellers were invariably inaccurate and many a time were highly exaggerated.

The Guide to Geography consisted of eight volumes. He also promulgated the concept of *Terra-Australis-Incognita* declaring that the Indian Ocean is a closed sea. This idea probably he borrowed from Hipparchus.

The major contribution of Ptolemy to the field of mathematical geography can be studied under the sub-headings: (i) circumference of the earth, (ii) dimensions of the habitable world, (iii) prime meridian, graticule and design of projection, and finally, (iv) the salient features of his map and geographical account of the major features of the different parts of the world (Figure 2.2).

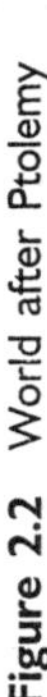

Figure 2.2 World after Ptolemy

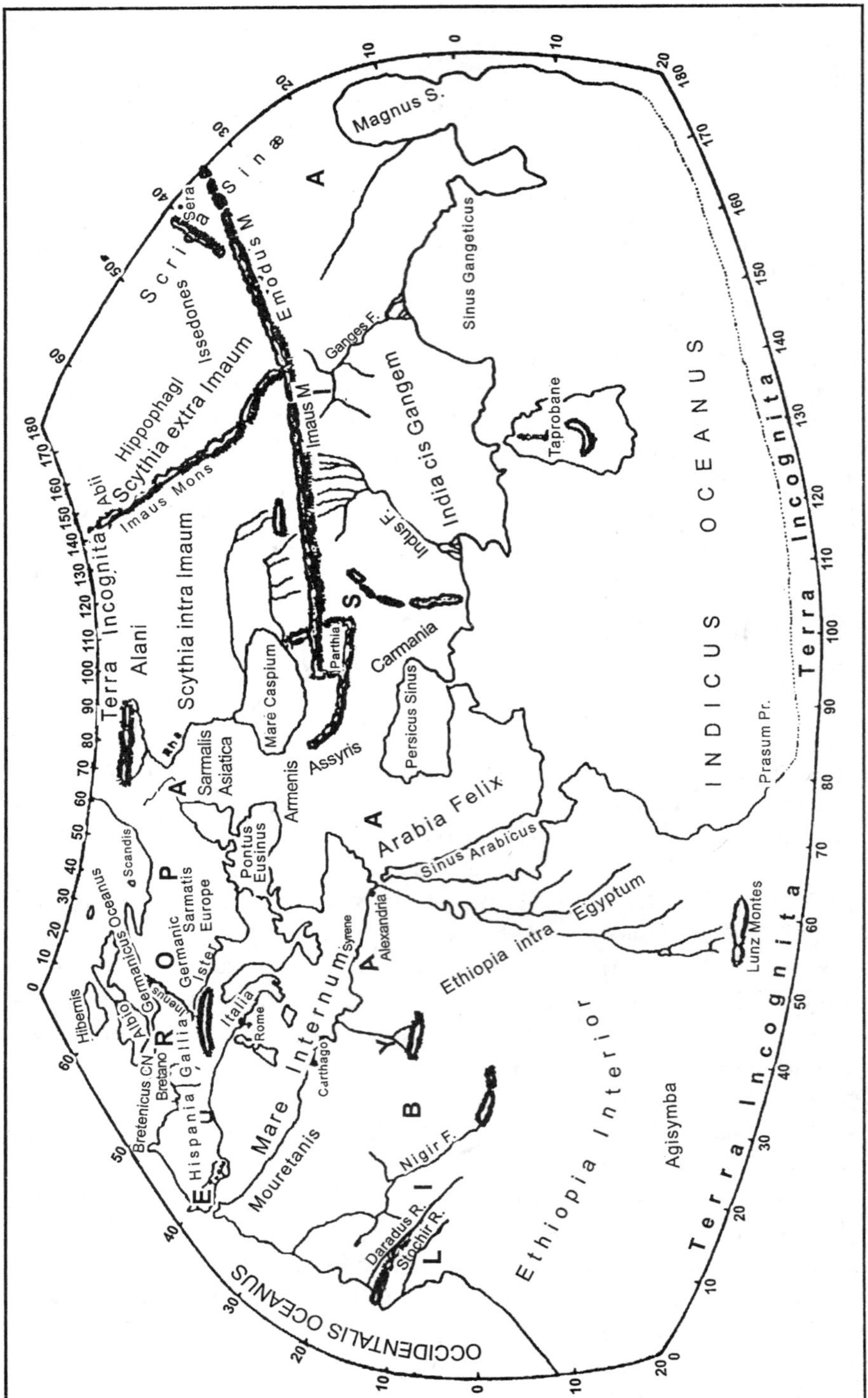

With regard to the measurement of the circumference of the earth, he adopted the division of equator and other great circles from Hipparchus, who for the first time, divided circle into 360 degree. He considered every degree equal to 500 stadia (50 miles) instead of 600 stadia (60 geographical miles).[22] Owing to this wrong notion, the error about the circumference of the earth got multiplied. Apart from this, since the travellers and navigators would exaggerate the distances on their travels between different places and since Ptolemy relied on these accounts and itineraries the world map prepared by him got distorted.

Eratosthenes fundamental parallel extending from Sacred Promontory (Cape St. Vincent) through the Strait of Gibraltar and the Rhode Island to the Gulf of Issus was considered by Ptolemy a latitude to the north of equator corresponding to 36 degree. Ptolemy measured his longitudes along this parallel to which was considered the latitude of Rhode. Ptolemy made a mistake in bringing down the Island of Sardinia fairly south which is really situated at 39° 12'N. This mistake distorted the shape and three sides of Sicily. The latitude of Massilia (43°5') was, however, correctly determined, which was obscure to Strabo.

Ptolemy presumed the Fortunate Islands (the Canaries) as the point through which the prime meridian passed. The westernmost island (Ferro) of the Canaries Islands continued to be treated as the prime meridian, and is so even among some German geographers of the present day. But in Ptolemy's days the position of those islands was not determined, and thus it was only by conjecture that he placed them two and a half degree to the west of the Sacred Promontory (Cape St. Vincent) instead of nine degree which is the true estimate. The total result which he produced for the length of the known world, from the Fortunate Islands in the west of the city of Sera in China towards the east, was 180 degree, whereas in reality it is about 130 degree. In one respect this mistake proved advantageous one in the consequences which proceeded from it at a later period (14th to 15th century). By diminishing the internal distance between the eastern and western extremities of the world, it encouraged the idea that the passage from the one to the other might be accomplished and thus indirectly contributed to the discovery of America by Columbus.[23]

Ptolemy rejected the hypothesis of his predecessors (Hecataeus, Herodotus, Strabo, etc.) of an ocean to the east of Asia. In his opinion, Sera and Sine (in China) were merely the most easterly points in Asia, just as Agisyamba was the most southerly.[24] Beyond Sera, according to him, lies the 'unknown land' (*Terra Incognita*). In regard to the breadth of the known world he assumed the parallel of Thule (Shetland or Orkney) which has been placed by Marinus in 63°N where the longest day is of 20 hours and Prasum (165° lat.) is its southern limit.

In the art of map-making, the great contribution made by Ptolemy lies in the great improvements he made over the previously drawn maps. He was able to do so by adopting projection for the world map showing the graticule of latitudes and longitudes. In fact, in regard to the mathematical construction of the projection of his maps, Ptolemy was far ahead of his predecessors. He represented the equator and the latitudes by parallel curves, and the meridians as straight lines bisecting the equator at right angles which converge at a point (pole) situated beyond the limits of the map (habitable world). Subsequently, he reduced the meridians also to a curved form so as to make them correspond more nearly with the reality.[25] The map on which his network of latitudes and longitudes was drawn was not a perfect hemisphere. The *climata* (climatic zones) of Ptolemy, also marked on his map, were like those spaces on the surface of the globe to which Hipparchus assigned that name.[26] The width of these intervals, however, was not measured in degrees, as was the case with *climata* of Hipparchus, but by the increase in the length of the longest day, proceeding northwards from the equator.[27] From this line as far as 45° N latitude, where the longest day was of fifteen and a half hours, the breadth of a *climata* (climatic zone) was determined by the difference of a quarter of an hour in the length of the longest day; but beyond the 45° N by the difference of half an hour.[28]

As stated at the outset, the main objective of Ptolemy was to elucidate his world map and to make mathematical geography perfect and complete. It would be worthwhile to present a brief account of the information that he was having about the different parts of the world in order to identify the improvements he made over the maps of his predecessors (Figure 2.3).

Ptolemy described the regions and nations of Western Europe at length. In his second book, he devoted two sections to the geography of the British Isles. The Roman armies, under the command of Caesar, achieved tremendous success in the British Isles, especially in England and Wales. Moreover, Agricola (a Roman commander) sailed to the eastern and northern coasts of England and Scotland. The neighbouring island, Hibenia or Ivernia (Ireland),

Figure 2.3 World According to Ptolemy, 150 AD

however, remained untouched by the Romans, except that Eblana (Dublin) part was known to them.[29] The map of the British Isles constructed by Ptolemy with the help of latitudes and longitudes corresponds very closely to the reality in shape (Figure 2.3). The gradual southward slope of the southern coast, the important inlet of the British Channel (Estuary of Thames), the Wash, the Sloway Firth and that of Clyds are all very correctly delineated. The Land's End (Cornwell), St. David's Head (Pembroke-Wales), Cardiganshire (Wales) and the western shore of Scotland are all very correctly plotted. However, the portion of Scotland, north of Firth (Forth), is erroneously pushed farther north, eastward. The Orcas (Duncans by Head), instead of pointing northward, faces the east. This may be because of the design and weakness of the projection or because of the exaggerated information.[30] So far as the Island of Ivernia (Ireland) is concerned, it was plotted in higher latitudes (Figure 2.3). Ptolemy's geographical enumeration of names, physical features, places, tribes, and his understanding of the British Isles is of a very high order but his ideas regarding the north and western coasts are exaggerated and erroneous (Figure 2.4).

Figure 2.4 Coasts of the British Islands after Ptolemy

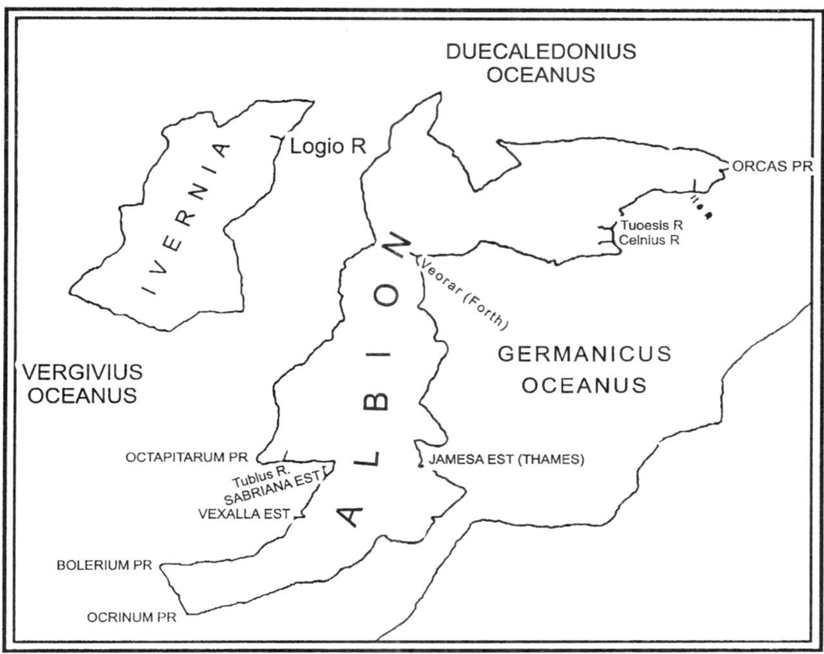

So far as the geographical knowledge of Gaul (France) and Iberia (Spain) is concerned, these parts were well known to the Romans. In fact, they constructed many roads through these countries to improve accessibility and to provide easy means of transport and communications with North-West Europe and the British Isles. Consequently, Ptolemy was better equipped to give a reliable picture in shape of maps and description of these countries. In fact, the regional Roman administrators were under an obligation to provide various socio-economic and administrative information to the centre (Rome). In spite of all these advantages, with regard to Hispania (Spain), Ptolemy committed errors. His main error was in the delineation of Spain's western coast and the undue extension which he gave to its north-western extremity – the Sacred Promontory (Figure 2.2) which he showed westward than the mouth of Tagus (Tejo) river.[31] He had a very imperfect notion of the great projecting headland or the Peninsula of Breatange (north-west Peninsula of France) towards the west. Moreover, he did not take notice of Cotentin in Normandy which constitutes the most marked feature in the British Channel. His conception of the physical features of the interiors of Hisponia (Spain) and Gaul (France) were even more erroneous.

The trans-Rhein (Rhenus) region (Germany) was not captured by the Roman armies. Consequently, Ptolemy gave a distorted and erroneous picture of this region which cannot be taken as an improvement over the works of his predecessors. His knowledge of the supposed islands of the Baltic was still more imperfect. While he mentioned the name of Scandia (Scandinavia), he reduced it to a mere island of ordinary dimensions in the Sarmaticus Oceanus (Baltic Sea). The Scandia Island, according to Ptolemy, was situated opposite to the mouth of Vistula. He, however, gave more precise information about the sources of Weser, Elb and Vistula. According to him, these rivers of North-West Europe originate in the Sormation (Carpathian) Mountains. The land between the Baltic and Euxine was described by him as Sarmatia Europe (Figure 2.2). His narration of the tribes living in Sarmatia is fairly good but he fell into confusion about the rivers (excepting Ister) which fell into the Euxine Sea. Moreover, he exaggerated the shape and size of Palus-Maeotis (Sea of Azov) which was stretched up to the parallel of Baltic Sea, i.e. 55° N latitude. He considered Palus-Maeotis as the boundary between Europe and Asiatic Sarmatic (the land between Don and Volga). The Tanais (Don) river – the accepted boundary between Europe and Asia – was plotted farther northward in the higher latitudes. Ptolemy was the first to identify and to plot Rha (Volga) river on his map.[32]

He was more accurate about Caspian, showing it as an inland sea. Herodotus (the Greek historian and geographer) was the first who propounded this idea, but from the Alexandrian period this concept was discarded. Ptolemy was, however, in error in regard to its shape, supposing its greatest length to be from west to east. Its north-south extension has also been reduced, though he was aware that river Rha (Volga) discharges its water into the Caspian Sea.

Ptolemy's knowledge of the coastal region, mountains and islands of the Caspian Sea was adequately reliable. Nevertheless, he made mistakes in the plotting of Italy, Gulf of Genova and the Sea of Adrias (Adriatic). His map of Greece was, however, far superior in comparison to his predecessors.

Ptolemy described the geography of Central Asia on either side of the *Imaus* (Altai) mountains meticulously. He gave an account of tribes like Abii, Alani, Hippophagi and Issedones. He possessed some definite information concerning the country that adjoined Scythia on the east (known to the Greeks as Sericasi) or the land of Seres (China). Sera, the easternmost point of the habitable world (Figure 2.2) was considered to be the great trading centre – famous for silk and silk products. The caravans used to reach Sera by crossing the difficult terrain and desert of Central Asia.[33] He plotted Echardes and Bantisus as the two main rivers of Sericas, traversing the whole land from west towards east (Figure 2.2). But, it is strange that his information about these two rivers should not have led him to the conclusion that they must have their outflow into the sea or ocean.

It was Ptolemy who plotted the Gangetic Gulf (Bay of Bengal) for the first time. He showed the source of the Ganges and its main tributaries in the Himalayas. The general direction of the western coast of India from north to south was correctly conceived and Taprobane (Ceylon) was placed opposite to the western coast. Ptolemy had a fairly correct notion of the general shape and outline of Taprobane, but he exaggerated its size and extended it up to 15°S latitude. The total sprawl of Ceylon has been shown in 12° latitude and pushed it two degree south of equator also. He gave a good account of the tribes of Taprobane which reveals that he had good hearsay information about these people.

His information about the trans-Gangetic region (India-Trans-Gangem) was vague and obscure. To the east of the Bay of Bengal, he places a land called Chryse (the golden island) or Peninsula of modern Malaya. To the farther east, he plotted Kattigare (Hanoi), near the coast of Annam. He introduced Maganus – the Gulf of Siam (Thailand). Ptolemy believed that the unknown land beyond ultimately joined the unknown parts of the East African coast, making the Indian Ocean a vast island sea (Figure 2.2). Thus, he predicted the Indian Ocean to be surrounded on all sides by land. It was a mere hypothesis, which has also been postulated by Hipparchus in days when these eastern seas were almost entirely unknown.

Ptolemy's account of Arabicus Sinus (Red Sea), shores of Erythrean (Arabian Sea), and Persicus Sinus (Persian Gulf) is much superior to that of his predecessors. This is especially conspicuous in respect of the projection from the coast of Oman – a marked feature in the geography of Arabia Felix (Peninsula of Arabia) which had been ignored or misunderstood by all the previous authorities.[34] But, the case is quite different so far as the interior parts of the Peninsula of Arabia are concerned. In fact, traversing Arabia Felix was difficult and very little hearsay information about the interior land was available.

The knowledge of the eastern coast of Africa was restricted up to the Cape of Prasum which Ptolemy placed as 15° 30'. The eastern coast of Africa was reasonably known as far as Cape Guardafui. He mentions several who had sailed from Aromata (Somali coast) past Cape Guardafui as far as Rhapton (Tanganyika). With regard to the continent of Africa, Ptolemy had better information than any of his predecessors, except Marinus. The highly controversial point about the source of the Nile which drew the attention of scholars of Alexandria from the days of Eratosthenes to Ptolemy was also solved. Ptolemy cites Nero, the leader of a Roman expedition, who reached up to the marshes of the Nile (9° 5' of equator). On the basis of the evidence gathered from Nero's records, Ptolemy established the source of the Nile in the Coloe lake (lake Tanza) situated in the highlands of Abyssinia, from which the Behr-el-Azrek (the Blue Nile) really derives its waters. He gave the source of the White Nile in the high peaks of Lunar Mountains. The snow-covered Lunar mountains are none other than the Mountains of Kilimanjaro (5,895 metres) and Kenia (5,652 metres) which lie between the lakes and the Indian Ocean (Figure 2.2).

Ptolemy, while giving the tables of latitudes and longitudes of Africa, introduced the Gir and Niger rivers. In the native language, 'Gir' means stream. He held that these two rivers had their source in a chain of mountains. Moreover, their course has been described to the north and western parts of Sahara and to the south of the Atlas mountains which shows that these rivers were in the western parts of Libya (Africa) (Figure 2.5). But their present courses are not traceable. However, these rivers should not be confused with the Niger river of Nigeria.

Figure 2.5 Map of North-Western Africa According to Ptolemy

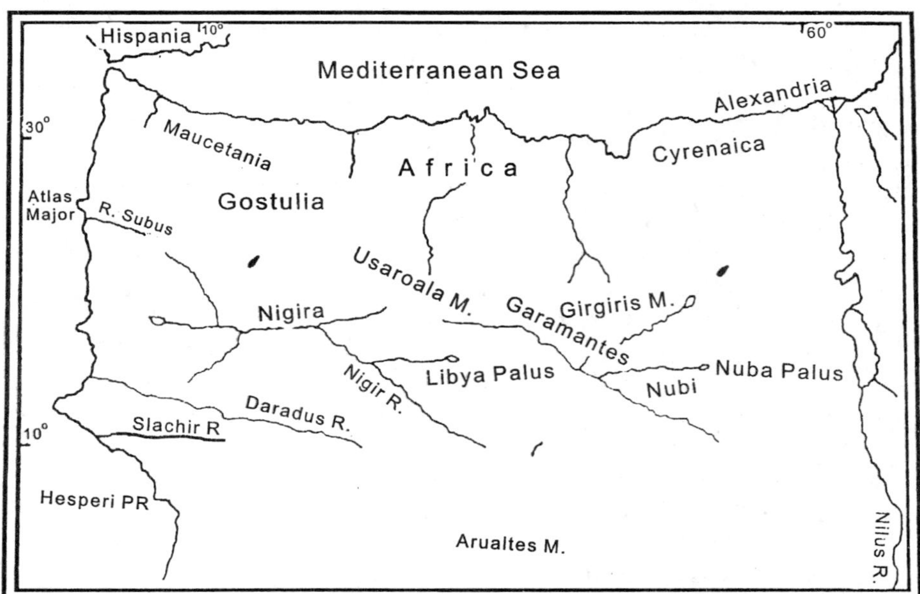

So far as the western coast of Africa is concerned, the Romans, especially their traders, were quite familiar with the coast of Mauretania (Morocco and Algeria). He described the Daradus and Stachir rivers and the Peninsula of

Hesperi. It is doubtful whether any point beyond Sierra Leone was ever visited by him. He also described the land between Sudan and equatorial Africa, giving it the name of Ethiopia.

The world map prepared by Ptolemy revealed exaggerated size of the land hemisphere. The Black Sea and the Sea of Azov are shown in distorted form. The Caspian Sea is shown as an inland lake. The map shows no connection between South-East Asia and Africa, making the Indian Ocean a land-locked sea (Figure 2.2).

Whatever the defects in the great works of Ptolemy are, we must bear in mind that the construction of such a scientific map on projection in the absence of reliable data and observed information was not an easy task. It was because of his efforts that the New World (North and South America) and the continents of Australia and Antarctica were discovered by the explorers of the 15th and the 18th centuries - after a lapse of more than 13 and 16 hundred years, respectively.

Ptolemy's Longitudes in the Mediterranean Sea

Some of the important longitudes of the prominent places of the Mediterranean Sea have been given in the following table. These longitudes were calculated by Ptolemy in relation to the prime meridian passing through the Ferro Island of the Fortunate Islands.[35]

	Place	*Longitude According to Ptolemy*	*Real Longitude East of Ferro Island (Prime Meridian)*
1.	Sacred Promontory	2° 30'	9° 20'
2.	Mouth of Baetis	5° 20'	12° 0'
3.	Carlis in Sardinia	32° 30'	27° 30'
4.	Lilybaeum in Sicily	37° 0'	30° 45'
5.	Rhodes	58° 20'	46° 45'
6.	Issus	69° 20'	54° 30'

Though the contribution of Greek and Roman geographers was enormous in the various branches of geography, their main concern may be summarised as under:

1. The detailed topographical description of places and their history, which Ptolemy called *chorography*.
2. The measurement of the earth and the production of maps.
3. A more philosophical interest in the relations between humanity and environment. They opine that the environment influences people and that people can only to some extent modify their environment.

DARK AGE (PERIOD OF TURMOIL) IN EUROPE

Among the Phoenicians, Greeks and Romans, the period of Claudius Ptolemy unquestionably marks the highest point to which geographical science ever reached. The decay and disintegration of the Roman Empire led to the decline in literature, science, and explorations in the European and South-West Asian parts of the world. This, however, does not mean that geographical knowledge at that time did not flourish in China, India and South-East Asia.

The period of about five hundred years, i.e. 200 to 700 AD, which followed the publication of Ptolemy's *The Guide to Geography* was an age of complication, turmoil and abridgement. During this period, not a single work of originality in any field of the sciences and humanities was written. There was continuous deterioration, both in the theory of geography and the practice of exploration from the glories of the Greek and Roman periods. For this retrogression, which was most rapid during the period 300 to 500 AD, the following conditions were responsible:

1. Parts of the Roman Empire, e.g. Dacia, Gaul and Spain, passed into barbarian hands. North Africa was seized by the Vandals. Consequently, travel even within the empire was hazardous.
2. The Middle and Far East, almost entirely passed into the hands of Persian, Arab ar Abyssinians. Moreover, the northern overland route through the Dariel Pass (Astrakhan) by way of the Caspian to Central Asia was too insecure for even the most adventurous.
3. Apart from the political instability, the decline of theoretical study of geography was largely due to the wrong thinking of the church. The attitude of most Christian writers at that time was not calculated to promote any form of scientific inquiry. Any scientific investigation about the shape and size of the earth was discouraged. The church started saying cosmography is less valuable because Moses – the servant of God – has said nothing concerning the shape and circumference of the earth. Unfortunately, Father Cassopdpris (early 6th century) and Father Isidore (early 7th century) described the earth as a flat surface. In other words, a literal interpretation of *Genesis* could not be brought into harmony with the Ptolemic system of the universe and the postulate of a spherical earth, and so progress in knowledge was ruled out.

This was the time of decadence of Roman power during which the Romans could not contribute anything new. During this period, religion had dominated the mind of people and they were not allowed by the Church to raise scientific questions. It is also known as the 'Dark Period' in the development of science in Europe. At best, scholars of this period made accurate but sterile copies of the works of the ancients, rejecting anything which did not conform to the dogmas of the Church. Such an intellectual environment stifled any development of critical scientific analysis. Concepts of the world which had been developed in Greek and Roman times were reshaped to conform to the teaching of the Church. The earth became a flat disc with Jerusalem at its centre.

Solinus (250 AD), who appears to have flourished in the 3rd century AD, gave a general geographical account of the world. The work of Solinus entitled *Collective Rerum Memorabilum* (the collection of wonderful matters) cannot be taken as a worthwhile geographical description of the world. In fact, the basic motive of Solinus was to collect 'all the wonderful things' and the geographical framework in which they were set. He, however, has been described as a plagiarist of Pliny and Pomponius.

Pomponius Mela (335–391 AD) was the last important geographer and historian of the time. He had military training and experience. He was also the last among the ancients to recognize the value of geography in relation to history and defence. He made free use of Ptolemy. But his attempt is considered as an abridgement of Ptolemy's work. The period from the 3rd century AD till the rise of Islam was influenced by Christianity. The Christian era was marked by the loss of ancient scientific concepts about the world as well as their replacement by unscientific, uncritical cosmogonies based largely on the scriptures. As stated above, during this period, travelling and explorations, owing to political instability, were hazardous. The missionary travels are the only source of knowledge for the regional account of different nations of that period.

Most of the correct classical concepts were forgotten and old errors reappeared about the map of the world and the habitable parts of the globe. Lactantius Firmanus (260–340 AD), one of the leading protagonists of Christianity, denied the concept of a spherical earth. The interpretation of the nature of the universe reached its fullest expression in the work of Cosmas of Alexandria (600 AD). His book *Christian Topography* written in about 550 AD, refuted all the pre-Christianity views on geography. He worked out on earth modelled in all respects upon 'Moses Tabernacle'. Cosmas, who was a merchant in early life, travelled fairly widely. He visited Ethiopia, Indian Ocean, Socotra, Persian Gulf and Ceylon. The work of Cosmas is, however, full of absurdities. About the shape of the earth, he deduced that it was flat, bounded on all sides by high walls. On these strong and high walls is 'supported the semi-spherical sky'. It was because of these erroneous concepts that Cosmas' *Christian Topography* could not influence later writers in the field of geography.

During the period of Christian Europe, there was a deterioration in the art of map-making. The fairly accurate delineations of the better-known coastlines of the Greeco-Roman period were lost, and instead maps became purely fancy. This was the period of the so-called *T-O maps*. The inhabited world was represented by a circular figure, surrounded by the ocean. The figure was oriented towards the east.[36] In the middle of the land area was a T-shaped arrangement of water bodies. The stem of the 'T' represented the Mediterranean. The top of the 'T' represented the Aegean and Black Seas on

Figure 2.6 The T-in-O Map (Orbis Terrarum)

the one hand, and the Nile river and Red Sea on the other. The three divisions – Europe, Asia and Africa – were accepted as standard. The centre of the inhabited world, just above the centre of the 'T', was Jerusalem. At the Far East, beyond the limit of the inhabited world, was Paradise.[37] Moreover, in all these maps were inserted the mythical places, beasts and dragons, such as the kingdom of the legendary Gog and Magog, who were non-believing menaces to the Christian world. This type of cartography continued to be in vogue for a pretty long time.

So far as the sprawl of the habitable world is concerned, many wrong ideas were put forward. There was vagueness in the east-west and north-south extension of the world. The sphericity and the nearly correct distances of places, latitudes and longitudes were ignored. New theories were postulated and raised on the weak foundations of little understood scriptural texts, in which there was nothing definite on the subject. The Christian monks tried to prove the concepts of Greeks and Romans as pseudo-scientific. Some of them like Lactantius Firmianus (260–340 AD) argued that the earth was not a sphere and that a spherical sky did not necessitate a spherical earth. Thus, the idea of the possibility of antipodes according to him was thoroughly absurd. Completely dominated by Christian supernaturalism, the map-makers of Dark Age made no serious attempt to show the world as it actually is. Instead, Firmainus followed an ideal pattern in his own mind, concentrating on artistic and symbolic expression. As early as the 4th century the world map of Romans had been christianized by the great church father, St. Jerome, who made a map in which he exaggerated the Holy Land beyond all real proportions. In the 8th century, a Spanish monk, named Beatus, prepared an

Figure 2.7 The World as Discovered by Ancient Explorers

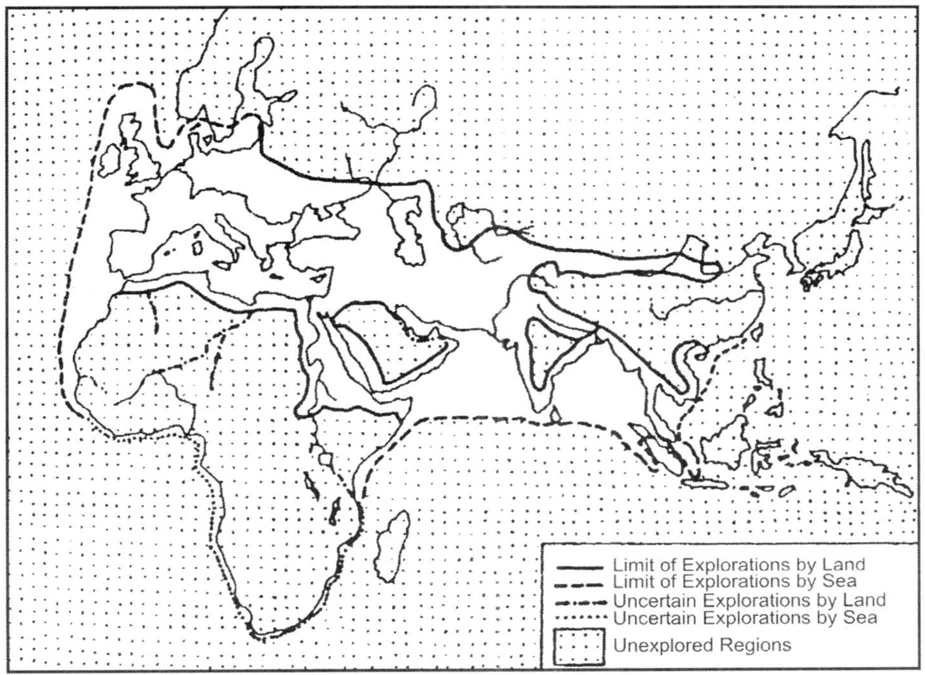

interesting version of the old Roman map. Later scribes who copied it paid slight respect to its geographical contents. Even the oval shape of the map itself was frequently distorted, sometimes to a rectangle, sometimes to a circle.

The typical world map of the Dark Age remained a disk, as it had been for the Romans. In its most extreme form, it is known as the 'T-in-O' (*Orbis Terrarum*), or the 'wheel' map. In this schematization, Asia was usually shown occupying the upper half of the 'O', with Europe and Africa more or less equally dividing the lower half. Jerusalem was generally placed in the centre following the Biblical text (Figures 2.6 and 2.7).

About the Dark Age, the German scholar Schmid summarizes: 'New countries were not discovered; the empire became smaller not greater; trade relations, thanks to the war in the east, the south and the north, became more and more restricted; besides, there was no longer any question of research in industry and of the spirit of discovery. Thus, the only books that were put together were compilations from older works'.[38]

Notes

1. Bunbury, E.H., 1959, *A History of Ancient Geography*, New York, p. 209.
2. Humboldt, A.V., *Cosmas*, Vol. II, p. 187 (Eng. Translation).
3. Strabo, *The Geography of Strabo*, Trans. H.L. Jones, 1917, New York, G.P. Putnam's & Sons.

4. Ibid., p. 670.
5. Bunbury, E.H., op. cit., p. 411.
6. Tozer, H.F., 1976, *A History of Ancient Geography*, p. 239.
7. Bunbury, E.H., op. cit., p. 218.
8. Lyell, S.C., *Principles of Geology*, Vol. I, pp. 24–25.
9. Tozer, H.F., op. cit., p. 250.
10. Strabo, vi,2, p. 269.
11. Bunbury, E.H., op. cit., p. 266.
12. Ibid., pp. 300–315.
13. Strabo, p. 72.
14. Bunbury, E.H., op. cit., p. 310.
15. Thomson, J.O., 1965, *History of Ancient Geography*, New York.
16. Heidel, W.A., 1937, *The Frame of Ancient Greek Map*, New York.
17. Tozer, H.F., op. cit., p. 267.
18. Forbiger, *Geographic*, Vol. I, p. 492.
19. Morgan, D. in *Dr. Smith's Dictionary of Ancient Biography*, Vol. III, p. 570.
20. Wildberg, *Ptolemaco Geographia est ars Delineandi Tabulas Geographicas.*
21. Tozor, H.F., op. cit., p. 344.
22. Ibid., p. 342.
23. Bunbury, E.H., op. cit., pp. 550–555.
24. Ibid., p. 577.
25. Supra, V., p. 175.
26. Tozer, H.F. op. cit., p. 344.
27. Plot, *Geoger*, I, p. 23.
28. Strabo, *The Geography of Strabo*, Translated H.L. Jones, New York, 1917.
29. Bradley, H., 1885, *Ptolemy's Geography of British Isles*, *Archaeologia*, Vol. 48, pp. 382–383.
30. Ibid., p. 586.
31. Ibid., p. 593.
32. Plot, VI, p. 16.
33. Ibid., p. 609.
34. Ibid., p. 636.
35. Smith, *Dictionary of Biography*, Vol. III.
36. Wright, J.K., 1925, The Geographical Lore at the Time of Crusades, New York, *American Geographical Society Research Series*, No.15, pp. 66–68.
37. James, P.E. and Martin, G.J., 1981, *All Possible Worlds,* pp. 43–44.
38. Schmid, W., Gesch,*d. Griechisch,* Liter, ii, 852.

3

Ancient Indian and Chinese Geographical Concepts

In order to further study the evolution of geographical thought, it would be worthwhile to have a look at the ancient Indian and Chinese geographical concepts. The present chapter gives an account of the major geographical concepts as developed and described by the classical Indian as well as Chinese scholars and geographers.

INDIAN GEOGRAPHICAL CONCEPTS

Indian geography has a long history. In fact, various geographical concepts have been developing in our country since the dawn of Indian civilization. Though a systematic account of the classical Indian geographical concepts is not available in a book form, yet some valuable geographical information is contained in Hindu mythology, philosophy, epics, history and sacred laws. Chronologically, the *Vaidikas*, the *Ramayana*, the *Mahabharata*, the works of Buddhists and Jains, and the *Puranas* are the main sources of ancient Indian geographical concepts.[1]

The ancient Indian scholars had accurate knowledge of topography, physiography, flora, fauna, natural resources, agriculture and other socio-economic activities of India and adjoining countries. They also had conjectured about the solar system and the universe. In the *Aitareya Brahmana* one may find materials regarding the regional geography of India. The *Satapatha Brahmana* furnishes a systematic description of the various branches of geography. The *(Vaidika Age)* inspired geographers and they produced valuable works in various branches of geography. In the *Ramayana*, inventory of mountains, rivers, plateaus, and important places has been made, while the epic of *Mahabharata* may serve as an encyclopedia of geographical knowledge, and the *Bhuvankosa* deals amongst other things, with climatology and meteorology in detail. The Buddhist *Jatakas* furnish quite a good knowledge of ancient geography. The term 'Bhogol' has been first used in the *Suryasiddhanta*. The author has succeeded in defining the concept of the earth's surface, the word known to the ancients, *Bharatavarsa* and its land and people and the concept of the ancients regarding village and town planning.

The ancient Indian geography hinges on religion and mythology. Every physical phenomenon, every major or spectacular landmark on the earth's surface has a religious background for Indians. Every mountain peak, every river, every crag, every huge and useful tree is sacred and is preserved in these traditions.

Apart from religious records, the travellers' accounts (religious, commercial, expedition) abound in the description of different regions of the world. The accounts of these travellers reveal that India had closer links with the neighbouring lands and Indian scholars were familiar with the geographical conditions of China, South-East Asia, South-West Asia, Central Asia, Mesopotamia and the Trans-Oxus Asia.

An in-depth study of the religious records, historical accounts and travellogues reveals that the ancient Indian scholars had fairly accurate concepts regarding cosmology and cosmography. They also had a good knowledge of the various *dwipas* (continents), mountain systems, rivers, fauna and flora of the *brahmatvarsa* (sub-continent of India) and the lands lying in its vicinity.

The work done by Varahamihira, Brahmagupta, Aryabhata, Bhaskaracharya, Bhattila, Utpala, Vijaynandi and others has substantially helped the development of astronomy, mathematical geography and cartography. *Suryasiddhanta* is one of the earliest doctrines or traditions in Astronomy. Its original version is by an unknown author. It describes the archeo-astronomy theories, principles and methods of the ancient Hindus. There is significant coverage of time, length of the year, and how planets move eastwards and sidereal revolution. It also gives the length of Earth's diameter and circumference of the Earth. Citation of *Suryasiddhanta* is also found in the works of Aryabhata. Thus, geography of the ancient time appears to have included astronomy in its sphere. In the *Padma Puranas* a difference has been made between *Bhogol* (geography), *Khogol* (the science of space) and *Jyotishakra* (astrology).[2]

Varahamihira (505–587 AD)

Varahamihira was an Indian astronomer, mathematician, and astrologer who lived in Ujjain. He was born in Avanti region (present Malwa region of Madhya Pradesh). He is considered to be one of the nine jewels (Navaratnas) of the court of legendary ruler Yashodharman Vikramaditya of Malwa. He contributed significantly in the field of astronomy and mathematical geography. He also described the causes of solar and lunar eclipses and their impact on human society.

Varahamihira

Brahmagupta (597–668 AD)

Brahmagupta was an Indian mathematician and astronomer. He wrote two important works on mathematics and astronomy, namely the *Brahmasphutasiddhanta* (Extensive Treatise of Brahma – 628), a theoretical treatise, and the *Khandakhadyaka*, a more practical text. Brahmagupta was the first to give rules to compute with zero.

Bhaskara (Bhaskaracharya) (1114–1185 AD)

Bhaskaracharya was an Indian mathematician and astronomer. He was born near Bijapur Karnataka. Bhaskara is said to have been the Head of an astronomical observatory at Ujjain, the leading mathematical centre of medieval India. He lived in Jalgaon district of Maharashtra.

Bhaskara (Bhaskaracharya)

His main work *Siddhanta Shiromani,* consists of four parts that deal with arithmetic, algebra, mathematics of the planets, and spheres respectively. He is particularly known for the discovery of the principles of differential calculus and its application to astronomical problems and computations. He was a pioneer in some of the principles of differential calculus. He was perhaps the first to conceive the differential coefficient and differential calculus.

Aryabhata (476–550 AD)

Aryabhata was a great mathematician, astronomer and geographer of the classical Indian period. His works include the *Aryabhatiya* (499 AD) when he was only 23 years old and the *Arya-Siddhanta*. These works deal mainly with mathematics and astronomy.

Aryabhata

Aryabhatiya provides no information about his place of birth. The only information comes from Bhaskara who describes that Aryabhata was an Asmakiya. Asmakiya was the land between the Narmada and Godavari rivers (present border region of Madhya Pradesh and Maharashtra).

His major work, Aryabhatiya is a compendium of mathematics and astronomy. This work is extensively referred to in the Indian mathematical literature and has survived even in modern times. The mathematical part of the Aryabhatiya covers arithmetic, algebra, plane, trigonometry, and spherical geometry. It also contains continued fractions, quadratic equations, sums of power series and table sines.

The *Arya-Siddhanta,* a lost work contains a description of several astronomical instruments. The gnomon (*Shanku-Yantra*), a shadow instrument (*Chhaya-Yantra*), possibly angle measuring devices, semi-circular and circular in shape, a cylindrical stick, an umbrella-shaped device called the *Chattatra-Yantra* and water clocks of at least two types, bow-shaped and cylindrical.

He used letters of alphabet to denote numbers, expressing quantities. Knowledge of zero was implicit in Aryabhata's place value system as a place holder for the powers of ten with null coefficient. Aryabhata gives the area of a triangle by stating 'for a triangle, the result of a perpendicular with the half-side is the area'. He also discussed the concept of sine in his work by the name of *ardhya-jya,* which literally means 'half-chord'. For simplicity, people started saying Jya.

Aryabhata's system of astronomy was called the Aud-Ayaka system in which days are reckoned from Uday, dawn at *Lanka* or 'equator'. In some texts, he seems to ascribe the apparent motions of the heavens (skies) to the Earth's rotation. He believed that the planets' orbits are elliptical rather than circular.

He correctly insisted that the Earth rotates about its axis daily, and the apparent movement of the stars is a relative motion caused by the rotation of the Earth, contrary to the prevailing view, that the sky rotates. In the same way that someone in a boat going forward sees an unmoving (object) going backward. The cause of rising and setting of stars is that the sphere of the stars together with the planets turns to west at the equator, constantly pushed by the cosmic wind.

Aryabhata described a geocentric model of the solar system, in which the Sun and Moon are each carried by epicycles. According to him the order of the planets in terms of distance from Earth is the Moon, Mercury, Venus, the Sun, Mars, Jupiter, Saturn, and the asterisms.

Solar and lunar eclipses were scientifically explained by Aryabhata. He states that the Moon and planets shine by reflected sunlight. Instead of the prevailing cosmogony in which eclipses were caused by pseudo-planetary demons *Rahu* and *Ketu,* he explained eclipses in terms of shadows cast by and falling on Earth. These will only occur when the Earth-Moon orbital plane intersects the Earth-Sun orbital plane, at points called lunar nodes. Thus, the lunar eclipse occurs when the Moon enters into the Earth's shadow.

The Universe and its Origin

The universe and its origin remained a point of speculation among all the ancient civilizations of Egypt, Babylonia, China, Greece and Rome. The ancient Indian scholars of the Vedic and Puranic periods gave considerable thought to this matter. The ancient Indian literature deals with many problems pertaining to cosmology and cosmography. For example, issues like whether matter existed prior to the creation of the world or whether the universe was fashioned out of a pre-existing substance or if it was made out of nothing are mentioned in the *Vedas* and the *Puranas*. The cosmology of the *Vedas* which has a strong bearing on the Puranic views may be summarized as (a) artistic origin of the universe, (b) mechanical origin, (c) instrumental origin, and (d) philosophical origin.[3]

The *Rigveda* mentions a number of gods who performed various functions during the process of creation. These gods were artists who contributed their skill to the construction and completion of the universe. They wove various materials into a pattern,[4] and shaped the universe by blasting and smelting.[5] The universe was compared to a house and *Rigveda* alludes to various stages in the construction of this universal house.

The views regarding the mechanical origin of the universe developed in the last phase of *Rigveda* period. It suggests the sacrifice (or disintegration) of the primeval body known as *adi-purusa* who is conceived as soul and the nucleus of the universe and an embodiment of the supreme spirit. The sky, the wind, the moon, the sun, and all the terrestrial elements were the result of dismemberment of *purusa* as a result of sacrifice ceremony.[6]

The philosophical theory of cosmogony has its origin in the song of creation which says that in the beginning there was neither being *(sat)*, or not-being *(asat)*. There was no atmosphere, no sky, no days, and no nights. The space was empty but for a unit which was born by its own nature, perhaps due to its inherent heat. This heat has been explained by Wilson as austerities, but it may conveniently be considered as a physical action in the process of formation of the universe.[7]

The instrumental origin of universe is based on the occurrence of parent bodies from which the universe was created. *Agni* (Fire), *Indra*, *Soma*, *Surya* (Sun), *Rudra* and the other gods are mentioned as having been instrumental in the creation of the earth and the heaven - the twin parents of the whole universe. The union of the earth and the heaven results in the birth of the sun which is the most important agent in the creation of the world. He is the soul of all that moveth or not moveth, the sun hath filled the air, and the earth and heaven. He was later identified with Rajapati, Viswakarma and sometimes with the golden egg and unborn being. The unborn being is also named as *atma* (soul) who produced the universe through an intermediary body.[8]

The universe has been described as *Brahmand* in the ancient Indian literature. It was conceived as very immense and wide which cannot be described. In the epics and *Puranas*, it was however, divided into seven upper and seven internal divisions.[9]

About the origin of the earth, it has been mentioned in the *Upanishads* that in the beginning death concealed all. Water was produced after worshiping death, from which the earth was originated. According to the *Puranas*, there was neither day nor night, neither light nor darkness and nothing else.[10]

Unlike our modern scientists, the ancient Indian astronomers believed in a geocentric universe. In the *Rigveda*, we come across the description of 34 heavenly bodies including the sun, the moon, five *grahas* (planets) and 37 constellations. The five planets have been described as the five gods. The astronomers of the Puranic period established nine planets, namely, the Sun, the Moon, the Mars, the Mercury, the Jupiter, the Venus, the Saturn, the *Rahu* and the *Ketu*. The astrophysical characters of some of the planets have been described in classical literature. *Budha* (Mercury) has been taken to be of green colour, *Shukra* (Venus) of white colour, *Mangala* (Mars) of red colour, *Brahaspati* (Jupiter) of yellow colour and *Sani* (Saturn) of black colour.[11]

Eclipses

The ancient Indian scholars were also conscious of the causes of *grahnas* (eclipses). It was because of this knowledge that they advocated performing of some rituals and ceremonies on the days when eclipses occurred. The Aryans considered an eclipse inauspicious and a herald of disaster. It was also believed that if a solar and a lunar eclipse occurred in the same month, it becomes more disastrous. Varahamihira had considered the effects of eclipse monthwise and emphasized the fact that eclipse in *Posa* (December) leads to famine and its occurrence in April and May results in good rainfall, while an eclipse in *Phaguna* (March) and *Asadh* (June) are inauspicious.[12]

Earth

The concept of *prithvi* (earth) is the most basic concept in the study of geography. The word *prithvi* (earth) has been used profusely in the *Vedas* and the *Puranas*. The word *Bhogol* (geography) in the ancient Indian literature signifies the spherical shape of the earth. The spherical shape of the earth was visualized by Aitareya Brahmana, who stated that sun neither sets, nor rises. We feel that it sets, but in reality, at the end of the day, it goes to the other side. Thus, it makes night on this side and day on the other. There is other evidence also like the shadow of the earth during lunar eclipse which is circular. From this it may be inferred that the earth is spherical in shape.

Size of Earth

Earth is an oblate spheroid slightly flattened at the poles; its equatorial diameter measures 12,757 km, and its polar diameter 12,713 km. In the Vedic and Puranic literature, no definite information regarding earth's dimensions is available, but later literature of the 5th and 6th centuries AD on astronomy gives somewhat convincing information[13] which is as follows:

Dimensions of the Earth

Source	*Yojanas*	*Miles*	*Kms*
Pancha Siddantika (verse-18)	1018.6	8148.8	13038
Aryabhata (verses)	1050.0	8400.0	13440

These dimensions were based on crude estimates. The real facts about the earth as known today are that its volume is 260,000,000,000 tonnes, the equatorial circumference 24,902 miles and meridinal circumference 24,860 miles and its estimated age according to the latest researches is at least 4,600 million years. The mass or weight of the earth has been calculated as 6,586,000,000,000,000,000,000 tonnes.[14] The estimates mentioned in *Suryasiddhanta* and ascertained by Aryabhata were very close to these established facts.

Latitudes and Longitudes (Akshansa and Deshantra)

The position of a point on the earth's surface in relation to equator, expressed as its angular distance from the equator, is known as latitude, while longitude is the angular distance of a given point measured in degrees east or west of the Greenwich Meridian (Prime Meridian).

The classical Indian astronomers were conscious of the importance of *akshansa* (latitudes) and *deshantra* (longitudes) in the determination of a point or place on the earth's surface. In the *Puranas*, there are references about latitudes *(akshansa)* and longitudes *(deshantra)*. On the basis of latitudes, they have divided the earth into various regions. For example, the Niraksadesa (hell) represents the equatorial belt while Meru (North Pole) is 90° latitude. Sri Lanka (Ceylon) is placed on the equator and on the North Pole is the mountain of Meru, with its antipode (Nadir) on the South Pole named as 'Badavanala'.[15] The longitude of Ujjain passing through Lanka and Mt. Meru was taken as the Prime Meridian by the Indian astronomers.

Cardinal Points

The *rishis* of *Rigveda* initially formulated the principle of four directions, i.e. *purva* (east), *paschima* (west), *uttra* (north) and *dakshina* (south). By adding Zenith (Meru) and Nadir (Badavanala), it was raised to six. Afterwards eight and ten directions are frequently mentioned in the Puranic literature. The designation of directions in the *Puranas* and subsequent literature *Saptapadarthi* is significant in the sense that it bears the original concept of the gods dominating in each of them. The ten directions and the ruling deity of each direction are given as under:[16]

Directions as per Puranic Literature

Direction		*Ruling Deity*
Hindi	*English*	
Purva	East	*Indra*
Agneyay	South-East	*Agni*
Dakshina	South	*Yama*
Nairitya	South-West	*Niriti*
Paschima	West	*Varuna (god of water)*
Vayavya	North-West	*Marut*
Uttara	North	*Kubera (god of wealth)*
Isana	North-East	*Isa*
Urdhva	Zenith	*Brahma*
Adhah	Nadir	*Sesanga*

The classical Indian astronomers were also conscious of the fact that local time of a place, depending upon the position of the sun or the moon in the sky, differs from that of other places situated along other meridians. They devised a method of calculating these differences. Some significant phenomena in the sky, like a lunar eclipse, was observed simultaneously from different places. The exact time, showing contact of eclipse or its totality, was recorded in terms of the local time of the individual places. A comparison of these records could provide the correct difference in local times and consequently the longitudinal difference between individual places.[17]

Origin

So far as the origin of the earth and the rock material of the earth crust is concerned, the ancient Indian scholars believed in solidification of earth from gaseous matter. The earth crust, according to them, was made of hard rocks *(sila)*, clayey *(bhumih)* and sandy *(asthma)*.[18]

The *Puranas* consider the earth to be floating on water like a sailing boat in a river. The Aryans considered the problem of the distribution of land and sea and held the view that more land surface was to be found in the Northern Hemisphere.

Earthquakes (Bhukampas)

For earthquakes, the word *'bhukampa'* has been used in the *Puranas*. It was assumed that earthquakes were caused by deities like *Vayu* (Air), *Agni* (Fire), *Indra* and *Varuna* (Water). This shows that the ancient *rishis* and scholars had a fairly good knowledge about the origin of the earthquakes. Similarly, they had some knowledge about the origin of volcanoes (*jawalamukhis*).

Atmosphere, Weather and Climate

The evidences in the Vedic and Puranic literatures clearly reveal that the Aryans were quite familiar with atmosphere, weather and climate. According to them, the earth was surrounded by *antriksa* (space/atmosphere) which exists between the earth and the heaven. The *Rigveda* mentions that the thickness of the atmosphere cannot be traversed by birds. Moreover, the *Ramayana* furnishes a lot of information regarding the atmosphere. Later on, Bhaskaracharya had conceived the thickness of the atmosphere to be 12 *yojanas* (154 km) round the earth in which winds, clouds, lightning, rain, fog and frost occur.

The *Rigveda* also mentions five seasons, i.e. *vasanta* (spring), *grisma* (summer), *Varsa* (rainy season), *sarad* (autumn), and *hemanta* (winter).[19] In the *Ramayana*, Valmiki has, however, referred to six seasons *(ritus)* in India which are as follows:

	Ritu (Season)	*Months*
1.	*Vasanta* (Spring)	*Chaitra-Vaisaka* (March–April)
2.	*Grisma* (Summer)	*Hyestha-Asadh* (May–June)
3.	*Varsa* (Rainy Season)	*Sarvana-Bhadrapada* (July–August)
4.	*Sarad* (Autumn)	*Asvina-Kartika* (September–October)
5.	*Hemanta* (Winter)	*Margosirsa-Pausa* (November–December)
6.	*Sisira* (Severe Winter)	*Megha-Phaguna* (January–February)

From the above *ritus* (seasons), it becomes clear that Indian in ancient times had a good knowledge of seasons, especially those of Northern India.

Continents (Dwipas)

In the early period of human civilization, owing to poor means of transportation and communication, knowledge of various parts of the world was very limited and it grew at snail's pace. To explore the unknown parts of the world is an inherent habit of man. In fact, the ancient Indian explorers and travellers made voyages, travels, pilgrimages and military expeditions to acquire knowledge about the unknown parts of the world.

As per references found in the *Puranas*, the landmass of the earth was divisible into several *dwipas* (continents). The word *dwipa* has been interpreted differently by different scholars.

Originally, a *dwipa* signified a land bounded by water (ocean, sea, river, lake or by a combination of these water bodies) on all sides. Thus, *dwipa* was equally applied to an island, a peninsula or a *doab* (land between two rivers).[20] The *Puranas* appear to have further extended the meaning of the term *dwipa* to include any land which was ordinarily inaccessible or detached by virtue of its being surrounded by water, sand, swamp, or even high mountains or thick forest. Thus, the Puranic *dwipa*, by accident or design, came to signify a natural region – either physiographic or climatic.[21]

The known world during the Puranic period was divided into seven *dwipas*. The areal stretch of these *dwipas* has been given in Figure 3.1. The seven *dwipas* have been described briefly in the following paras:

1. Jambu Dwipa

The name Jambu Dwipa has been derived from the Jambu *(Eugenia jambolana)* tree. In the opinion of some of the ancient Indian scholars, it embraces the whole of the Northern Hemisphere, lying to the north of Salt Sea (Figure 3.1). Jambu Dwipa is surrounded by the Salt Ocean and lies in the heart of the concentric sequence of the *dwipas*. This insular *dwipa* is further divided into sub-regions called *varsas* (realms) – the dwelling seats of *rishis* (observers). Ilavrila (Pamir region) is the central *varsa* (realm) and the Meru (Pamir), west of which is the Ketumala *varsa* and in the east lies the Bhadrasva *varsa*. Kimpurusa (Tibetan Plateau) lies to the south of the Ilavarila *varsa*.[22]

In general, the Jambu Dwipa is comparatively lower on the south and north flanks and highly elevated in the middle. The Meru lying in the heart of the Jambu Dwipa is considered to be heaven in the Pamir. The Sita river flows to the east of Meru (Pamir Knot). This river resembles the Yarkand river which has even today been referred to as *Sito* by the Chinese. The river Suvamksu (Amu-Darya) flows west of the Pamirs which is also called as *Bakshu* in Mongolia, *Potsu* in China and *Paksu* in Tibet. This river debouches into the Aral Sea.

The Bhadra, the river of the north is the Syr-Darya of today, flows northwards and debouches into the Aral Sea. South of the Pamir is the Kishan Ganga, flowing from the Gangabal Lake and Harmukh Glacier (about 70 km north of Srinagar in Kashmir).

Figure 3.1 The Puranic Dwipas

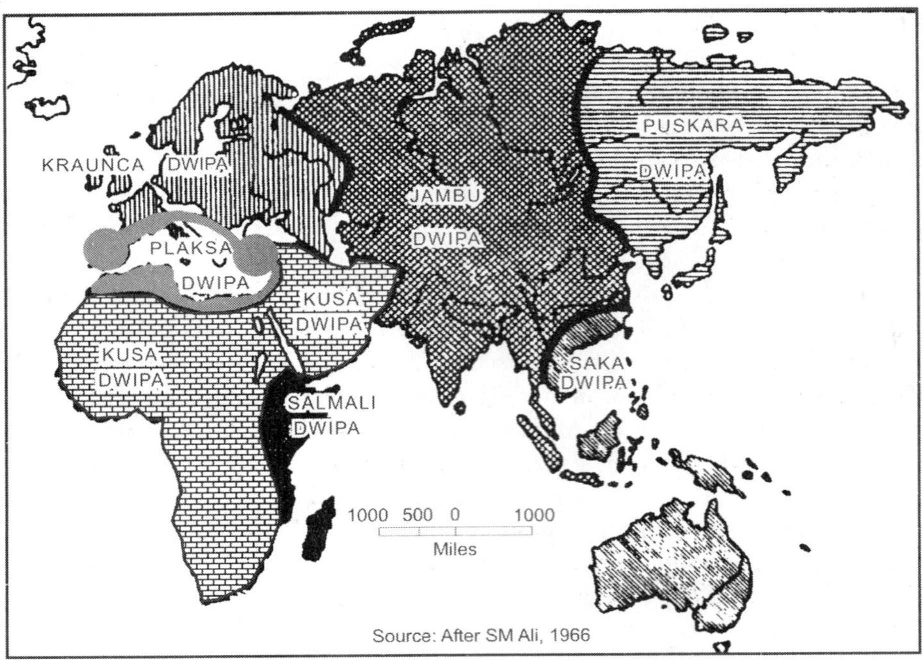

The other geographical features of the Jambu Dwipa include the Nishad (Hindukush) extending from the Pamir knot to *Kohe-Baba* (Baba mountain) in the west of Kabul. It is said to be a three-peaked mountain *(Trisringa)* which is visible from Peshawar (Pakistan). The Vaidurya (Badakhshan) Mountain lies to the west of Pamir Knot.

2. Kusa Dwipa

Kusa Dwipa has derived its name from *kusa* grass or *poa* grass. This *dwipa* stretches over Iran, Iraq and the fringing lands of the hot deserts, i.e. the south-west corner of the landmass round Meru which is left out in the regional pattern of Jambu Dwipa. This is a land of grasses and characterized by seasonal droughts. It contains seven major rivers and thousands of their branches that flow when god Indra pours down rain. In other words, these tributaries are seasonal in character. The mountains of Kusa Dwipa are covered with herbs, trees and creepers. Its mountains and rocks are full of minerals and precious stones. The presiding deity of this *dwipa* is Agni (Fire).

3. Plaska Dwipa

Plaska Dwipa has derived its name from Plaska tree. Wilford identified this tree with figure. One would, therefore, without hesitation, identify this *dwipa* with the basin and the surrounding lands of the Mediterranean Sea (Figure 3.1).

4. Puskara Dwipa

Puskara Dwipa is the land of horrors, devoid of purity, cruel and leading to the destruction of the soul. It is the land of demon, full of awful hollows which are twenty in number.[23]

The name of this *dwipa* has been derived from the fact that it is surrounded by Puskara (lakes of lotuses). This *dwipa* is bounded by a huge circular chain of lakes. The people living in Puskara Dwipa are nomads, hunters and in general primitive and savage. One side of the *dwipa* is a dry desert and the other side is suitable for human occupation. It promises a paradise for those who approach the *dwipa* from one direction, while it presents the appearance of a wasteland if one enters it from the opposite direction. Such phenomena of knife-edge boundaries between two regions of strong contrast are not uncommon. Puskara Dwipa is surrounded by an ocean of fresh water and surrounds the sea of milk. This region sprawls over the eastern and north-eastern Siberia (Russia). These countries contain numerous lakes, support nomadic people who live by hunting and are washed by Arctic waters and Bering Sea which have fresh water and low salinity.

5. Salmali Dwipa

Salmali Dwipa has derived its name from silk cotton tree. It consists of the tropical part of Africa bordering to the west of the Indian Ocean. It includes Madagascar, the *Zenj* of the Arab and Persian geographers. The main characteristic of this *dwipa* is *salmali* (silk cotton) tree. This tree is commonly found on the margins of equatorial regions of monsoon lands with moderate rainfall. It is a region of high cloudiness. Consequently, no star, planet or moon is visible. The people of this *dwipa* are essentially food gatherers and not food producers. The vegetation cover produces enough food to satisfy their needs.[24]

6. Kraunca Dwipa

The *Mahabharata* locates Kraunca Dwipa in the north and west of Meru (Pamir). This *dwipa* is watered by thousands of streams in addition to the seven important rivers which carry great volume of water. The *dwipa*, therefore, is definitely a humid region with abundant rainfall. The entire North-West Europe, including British Isles, thus constitutes the parts of this *dwipa*.

7. Saka Dwipa

Saka Dwipa has been identified as a wide stretch of land to the south-east of the Jambu Dwipa, covering the present Myanmar (Burma), Thailand, Vietnam, Malaysia, Indonesia and the islands of the South-East Asia (East Indies). This has hot and moist climate with thick cover of evergreen forests (Figure 3.1).

Apart from the *dwipas* (continents), the classical Indian scholars also tried to delineate the boundaries and frontiers of the Indian sub-continent. In the Vedic and Puranic literature, India has been given different names but *Bharatvarsa* is the one most commonly used in them.

Bharatvarsa

Bharatvarsa is commonly identified with the Indian sub-continent.[25] But, in fact, no comprehensive designation was given to the Indian sub-continent in ancient Indian or foreign literature. 'Sapta Saindhava' was the name given to the Punjab plains by Vedic Aryans. 'Aryavarta' was the designation of Aryan domain in the days of Baudhayana and Manu; the word 'Ind' or 'Indu' (Hindu) was applied by Darius and Herodotus to the Indus Valley of the upper Gangetic region with which they were acquainted. It is only in or about the 4th century BC that Katyayna and Megesthnese gave an account of approximately the whole country down to the Padya region in the extreme south. The epics also mention the Pandya realm in the south and the peninsula and islands beyond the Bay of Bengal.[26]

In the Puranic literature, the entire country from the Himalayas to Kanyakumari (Cape Comorin) is designated by a single name – *Bharatvarsa* (India). *Bharatvarsa* bears the testimony of geographical and historical significance, revealing the basic concept of Aryans.[27]

The etymological meaning of the word *Bharatvarsa* gives a clear conception of its various characteristics and its historical significance. It symbolizes a fundamental unity which was certainly perceived and understood by those who coined the term *Bharatvarsa* derived from *Bharata* – a sovereign king. The Altareya Brahmana refers to his coronation ceremony, subsequent conquests and *Asvamedha* sacrifice.[28]

According to one school of thought, Bharat is the name of Manu, who creates and sustains people in *Bharatvarsa*. Some of the *Puranas* mention the name 'Bharata' after king Bharat – the son of Rishabhadeva and the grandson of Nabhi. Thus, *Bharatvarsa* was divided into *nava khandas* (nine divisions). These divisions were separated by seas. Out of the nine divisions, eight have been shown as parts of Greater India while the ninth is surrounded by the sea.[29]

Culturally, caste has been the strongest element in ancient Hindu way of life. Caste is basically a system of functional stratification of society maintained by religious sanction. *Bharatvarsa* was inhabited by four castes, i.e. Brahmins, Kshatriyas, Vaisyas and Sudras. At the apex of the caste system were Brahmins – the priestly caste of Hindu religion. The second caste consisted of Kshatriyas who were warriors, cultivators and artisans. The third caste, Vaisyas, consisted of traders and businessmen, while the Sudras used to do inferior services. The Sudras were excluded from the main street and were obliged to live outside the main village, commonly in much inferior dwellings of thatch and matting.

Mountains and Rivers

The *Vedas*, epics and *Puranas* have mentioned a series of mountains in *Bharatvarsa*. Himavat, Uttra-Kuru, Utter-Madra, Trikakud (Hindukush), Vindhya, Paripatra, Durdura and Mahendra are the main mountains described in the ancient Indian literature.[30]

The Himalayas are mentioned as lying in the north, extending from west to east with its bend like a bow. Its regional divisions into Antargiri (inner Himalayas) and Bahirgiri (outer Himalayas) have been mentioned in the *Mahabharata*. The Kailash mountain has been said to be studded with diamonds, minerals and other precious stones. This was the abode of *apsaras* (nymphs) and *devas* (deities).

About the Vindhyas, it has been said that it is an extensive mountain with hundreds of peaks, variegated with trees and creepers. It stretches along the banks of the Narmada river up to Kaimur through Amarkantak. The Eastern Ghats were known as Mahendra-Mali and the Malabar coast includes the Nalla-Malai, Anna-Malai and Eta-Malai ranges. There are a number of other mountains also mentioned in the Puranic literature. Some of them are Sahya (Western Ghats in Maharashtra), Suktiman (Mountains of Khandera, Ajanta, Golkunda) and Rika (from Ken to Ton river – north of Vindhya).[31]

Apart from mountains, many drainage systems have been described in the ancient Vedic and Puranic literatures. The *Rigveda* has mentioned rivers like Ganga, Yamuna, Sarasvati, Sutudri (Sutlej), Parusni (Ravi), Asikni (Chenab), Vitasta (Jhelum), Arjikiya (upper part of Indus), Susoma (Savan), Sindhu (Indus), Kubha (Kabul), Gomati (Gomala), and Krumu (Kurrum). There are references to the Indus drainage system, to Narmada, Tapti (Tapi), Godavari, Krishna, Kaveri and Tungbhadra. It abounds in elaborate descriptions of the Ganga and the Brahmaputra rivers.

The Ganga

The river Ganga is said to flow from the Vindusarovar (Gangotri). In the initial phase, it was divisible into seven channels, out of which three channels, i.e. Haradini, Pavni and Nalni, flow eastward and the other three, i.e. Suchaksu, Sita and Sindhu, flow westward. The seventh channel, known as Ganga, follows a southern course in the great plains of India. It is joined by the Yamuna at Prayag (Allahabad). The Ganga, after passing through thousands of mountains and hills, irrigates hundreds of valleys and passes through thousands of forests and hundreds of caves. It then merges into the Southern Sea.[32]

Tsangpo

There is mention of river Lauhitya (Brahmaputra) which debouches at Dihang near Sadya (Assam). It has its source about 100 km (60 miles) east of Daya as per ancient Hindu literature.

Among the rivers of the south, the Narmada is said to take its course from the Amarkantak Hills. Its length is said to be 100 *yojna* (1,280 km) which tallies with the modern measurement. It is said to debouch into the Paschimodadhi (Arabian Sea). In the *Puranas,* it has also been mentioned that to the south of *Bharatvarsa* (India) there is an ocean *Mahasagra* (Indian Ocean) which is more than 10,000 *yojnas* in extent.[33] In the *Mahasagra,* there are numerous islands *(dwipas).*

From the foregoing paras, it is abundantly clear that the *rishis,* astronomers and scholars of the ancient Indian Vedic and Puranic period had well-developed concepts about cosmology, cosmogony, geography, and science of space. Their knowledge of the size, shape of the earth, continents, oceans, islands, eclipses, earthquakes, volcanoes, mountains, rivers, lakes, bays and peoples was appreciably correct and reliable. In fact, the Indian scholars contributed significantly to the growth and development of geography and its allied sciences.

CHINESE CONTRIBUTION TO GEOGRAPHY

Situated in the East Asia, quite far away from the Greeks and Romans, the Chinese made an important contribution to geography, between 200 and 1500 AD. In fact, they created great tradition of geographical scholarship by portraying information on trade routes and maps and by describing topography and life of the people living in different parts of the then known world. A brief account of their main contribution to geography and cartography has been given in the following paras.

With the collapse of the Western Roman Empire in 476 AD, European geography entered a period of dark stagnation. A few copies of Greek and Roman geographical texts survived, but the Germanic tribes that overran the vestiges of the Roman world had little use for such works. Indeed, many had already been lost, particularly at Alexandria, where the fire of 47 BC had destroyed some 400,000 manuscripts in the Great Library, and the disturbances of 391 AD had led to the loss of perhaps 300,000 more works in the Temple of Serapis. Parallel to, but totally separate from, the Greek and Roman world, a completely different culture of science had meanwhile evolved to the east in China. It was here that the subsequent focus of global intellectual and scientific activity was to be encountered, particularly under the Tang (618–970 AD) and Southern Sung (1127–1279 AD) dynasties, the latter of which was described so magnificently by the Venetian Marco Polo. Once again, the emergence of a tradition of geographical writing at this time can be seen in part to have been influenced by military conquests and the need for the emperors to have a sound knowledge of their lands in order to retain their positions of power. Moreover, the development of paper, magnetic needle, trough compass, marine compass and new surveying and cartographic skills enabled the Chinese to produce maps of a quality far surpassing anything being produced in Medieval Europe. The Chinese used co-ordinates and triangulation to produce beautiful maps of China and neighbouring countries. When Marco Polo (1254–1322 AD) wrote an account of his travels of China describing the high level of Chinese learning, his book was widely discounted as a fictitious adventure.

The earliest Chinese geographical document is reputed to be the Yu Kung 'Tribute of Yu' chapter within the *Shu Ching* (Historical Classic) dating from the 5th century BC. This provides an inventory of the Chou Empire, mainly in terms of its physical geography, and lists the traditional nine provinces, their kinds of soils, their characteristic products, and the waterways running through them. Other ancient travellers' guides, such as the *Shan Hai Ching*, much of which dates from the 4th century BC, can also be considered to be geographical, but most include mythological and magical elements together with details of semi-human races and peoples. The father of Chinese geography was Phei Hsiu, who was appointed Minister of Public Works by the Chinese emperor in 267 AD. Needham and Wang Ling (1970) suggest that there were five main types of Chinese geography:

1. Anthropological geographies, known as *Chih Kung Thu* (Illustrations of the Tribute-Bearing Peoples), dating from the mid-6th century AD.
2. Descriptions of the folk customs of the countries to the south of China *(Feng Thu Chi)* and descriptions of unfamiliar regions *(I Wu Chih)* both dating from the 2nd century AD.
3. Hydrographic books and coastal descriptions, such as the *Shui Ching* (Waterways Classic).
4. Local topographies or gazetteers, such as the *Hua Yang Kuo Chih* (Historical Geography of Szechuan) which was mostly written from the 4th century AD onwards.
5. Geographical encyclopaedias compiled from the Chin dynasty (3rd and 4th centuries AD) onwards, in a style similar to that of Strabo.

Chinese geography was also closely tied to astronomy and cartography. Astronomy played a central role in Chinese science because of its religious concern with cosmic unity, and its links with astrology. Moreover, knowledge of astronomy and the compilation of agrarian calendars were also a way in which the state could control the productive capacity of the population. Traditionally, there was an ancient belief in China that the heavens were round and the earth square, but by the 2nd century AD, three main schools of cosmology and astronomy had emerged:

1. The Kai Thien theory which envisaged the heavens as a hemisphere covering the earth which was shaped like an upturned bowl.
2. The Hun Thien school, which corresponded with the Greek view of heavenly spheres revolving round the globe.

3. The Hsuan Yeh teaching which envisaged an infinite space in which the heavenly bodies floated freely about.

Among the most important contributors to Chinese astronomy and cartography were Chang Heng (78–139 AD) and Phei Hsiu (224–271 AD). Although maps from as early as the 3rd century BC are recorded in China, it was these two scholars who first developed a scientific method of cartography based on a rectangular co-ordinate system. The corpus of cartographic work in the ensuing centuries was then brought together by Chu Ssu-Pen (1273–1337 AD), who used it to summarize the wealth of new information which had become available as a result of the Mongol unification of Asia. His map of China, prepared between 1311 and 1320, is a remarkable achievement, and remained a basic work of reference for over two centuries. Although Chu Ssu-Pen was cautious about depicting lands distant from China, it is evident that he had a level of knowledge surpassing anything current in Europe at that time. He thus recognized that Africa was a southward pointing triangle, whereas on contemporary European and Arabic maps it was always represented as pointing eastwards.

Notes

1. Ali, S.M., 1966, *The Geography of Puranas*, People's Publishing House, New Delhi, p. 15.
2. Dube, B., 1967, *Geographical Concepts in Ancient India*, The National Geography Society of India, BHU, Varanasi.
3. Majumdar, S.N., 1924, *Canningham's Ancient Geography of India*, Calcutta.
4. Ibid.
5. Dey, N.L., 1927, *Geographical Dictionary of Ancient and Medieval India*, London.
6. Sircar, D.C., 1960, *Studies in the Ancient and Medieval India.*
7. Wilson, H.H., *Visnu Purana*, Translation.
8. Das, A.C., 1921, *Rigvedic India*, Calcutta.
9. Bhargava, M.L., 1964, *The Geography of Righvedic India*, Lucknow.
10. *Suryasiddhanta*, *Bhugoladhyaya*, Verse 53.
11. Dube, B., 1967, op. cit., p. 11.
12. Ibid., pp. 20–25.
13. Ibid., p. 30.
14. *The Reader's Digest, Great World Atlas*, London, p. 144.
15. Sidhanta Siromani, Bhuvanakosa, Goladhyaya, p. 15.
16. Dube, B., 1967, op. cit., p.34.
17. Burges, E., 1936, *Suryasiddhanta* (English Translation), Calcutta, p. 47.
18. Law, B.C., 1934, *Historical Geography of Ancient India*, Paris.
19. Sircar, D.C., 1960, *Studies in the Geography of Ancient and Medieval India*, Varanasi.

20. Ali, S.M., 1965, op. cit., pp. 30–70.
21. Ibid.
22. Ibid.
23. Ibid.
24. Ibid.
25. Dube, B., op. cit., pp. 80–90.
26. Ali, S.M., op. cit., p. 85.
27. Sircar, D.C., 1960, op. cit.
28. Ray, H.C., 'The Study of Ancient Indian Geography', *Indian Historical Quarterly*.
29. Ibid.
30. Ali, S.M., op. cit., pp. 109–130.
31. Ibid.
32. Ibid.
33. Dubey, B., op. cit.

4

Arab Geographical Thought

The followers of Prophet Mohammad, from 8th to 13th century, made significant contribution to the field of geography. They embarked on a conquest of the world outside of Arabia. In 641 AD, they conquered Persia, and in 642 AD, took control of Egypt. They swept westward across the Sahara and by 732 AD, all the Great Desert was under their control. They crossed through the Iberian Peninsula in France. For some 900 years, the Muslims ruled most of Spain and Portugal. The Muslim rule was also extended to Central Asia, Northern China, India, eastern coast of Africa, Malaysia and some islands of the Arabian Sea and South-East Asia.

The preceding period of the Arabs has been rightly called as the 'Dark Age' in the European and Christian parts of the world. The Greek and Roman achievements which reached their zenith in the works of Ptolemy were forgotten. Many incredible and ridiculous stories about the shape and size of the earth were in currency. Against this background was the intellectual curiosity, integrity and catholicity of the Arab geographers which led to the great achievements of Islamic civilization.

Some primitive geographical notions were inherited by Arabs from Jews and Christians. The Arab geographical literature came to the world in 800 AD. In 762 AD, Muslims founded the new city of Baghdad and for more than a century it remained the centre of the intellectual world. With the patronage of Caliph Harun-al-Rashid, an academy called 'Baitul-Hikma' was established. In this academy, scholars from all over the world were invited to teach and assist the Arab academics and to help them in the translation of the Greek, Latin, Persian and Sanskrit works into Arabic.[1] The main factors responsible for the growth and development of geographical knowledge in Arab world may be summarized as under:

1. Open Mind and Inquisitive Nature

The Arabs took help of Jacobites, Nestorians, Christians, Jews, Greeks, Persians and Indians in the translation of forgotten literature produced by Greeks and Roman scholars. Caliph Al-Mamun, who ascended the throne in

813 AD, gave great support to scholars. He favoured the Nestorians, Christians and Jews of his kingdom, who till then were the custodians of Greek sciences. He collected at great expense Greek works, and constituted a library. The learned scholars of his court translated the works of Euclid, Archimedes and Aristotle. Ptolemy's *Almagast* and his treatise of geography were also translated into Arabic.[2] The Arabs invited Indian scholars to Baghdad to learn Indian mathematics and numerals. They also studied the works of Aryabhata and went through *Suryasiddhanta* – the Sanskrit treatise containing the principles of trigonometry. The Arabic translation of the *Sidhanta* was a landmark in the history of Arab astronomy. Al-Khwarizmi prepared two editions of *Sindhind* and summarized it. He also summarized the great astronomical works which were then available in Arabic in his *zij*. Thus, Arabs collected geographical ideas from Greeks, Romans, Iranians, Chinese and Indians. On the basis of their observations, explorations and studies, they developed their own concepts and theories with great vigour and speed. They produced innumerable books on various aspects of geography and allied sciences and carried their knowledge to far-off lands from the shores of the Atlantic to the Pacific and Indian Ocean.[3]

2. *Islamic Brotherhood*

A further stimulus to geographical research was given by the very immensity of Arab Empire so long as it was still undivided. There was a period during which the travellers could pass from the confines of China to the Pillars of Hercules (Spain), from the banks of Indus to the Cilician Gate (Turkey), from the Oxus to the shores of the Atlantic, without stepping outside the boundaries of the territory ruled over by the *Khalifa* (Caliph) in Damuscus or Baghdad. Even after this vast empire broke up into separate principalities, the journey of the Muslim traveller was facilitated by that brotherhood of Islam which gives to the Muslim world its cosmopolitan character, and enables community of faith of wipe out all differences of race, origin, nationality and language. However, many hundreds of miles the Muslim might journey from his native town, he could confidently hope for a welcome and generous hospitality at the hands of his co-religious brothers, especially if he had any reputation for piety or religious knowledge, and he might even chance to come across a fellow townsman, even though his wanderings had carried him into the land of the infidels far beyond the boundaries of the Muslim Empire. Thus Ibn-Battuta, an energetic traveller of the 14th century, to whom reference will be made later, tells us how on his arrival in a town in China, which he calls Kanjanfu, the Muslim merchants there came out to receive him with flags and a band of musicians with trumpets, drums and horns, bringing horses for him and his party, so that they rode into the city in a triumphal procession. This incidence was a characteristic of Muslim society during the Middle Age; it also reveals the enterprise that merchants and travellers showed in journeying such enormous distances and the facilities which their co-religionists provided for those who braved the perils of such arduous journeys.

3. The Pilgrimage

Among the duties incumbent upon every Muslim, provided only that he had health and sufficient wealth for the expenses of the journey, was that making the pilgrimage to Mecca at least once in his lifetime. Consequently, throughout the whole of Islamic era, except on the few occasions when political disturbance has prevented, there had been a stream of pilgrims setting their faces towards the holy city (Mecca) in which their religion first had its origin, from every part of the Islamic world – Egypt, Syria, Mesopotamia, Persia, Turkistan, India, Malaysia, China, Sudan, Morocco, Spain, Portugal and France. These pilgrims had to face great risks and undergo much toil and trouble in order to attain the fulfilment of their pious aim. It was this religious obligation that thousands of devout persons who have undertaken the journey despite all obstacles of age, poverty and ill-health.

4. Trade and Commerce

The next reason that stimulated travelling in Islamic world was trade and commerce. In Islamic society, the merchant enjoys respect and consideration that is closely connected with the origin of this faith; for Prophet Mohammad, the founder of Islam, who had himself been a merchant, and thus conferred upon the profession of the trader an elevation and a dignity which has gained for him an entrance into the highest society. Several sayings traditionally ascribed to the Prophet assigned an honourable position to the merchant in the Muslim hierarchy, for example, 'on the Day of Judgement an honest truthful Muslim merchant will take rank with the martyrs of the faith', and in another tradition the Prophet says that the truthful merchant will sit under the shadow of the throne of God on the Day of Judgement. The Prophet commends the merchants to his successors for 'they are the couriers of the world and the trusty servants of God upon earth'. The greatest of the early *Khalifas* (Caliphs), Umar Farooqi, said: 'There is no place where I would be more gladly overtaken by death than in the market place, buying and selling for my family'. The same great ruler is reported to have set honesty in the commercial life above the punctual fulfilment of religious duties, as a test of worth and excellence of a man. The Arab traders did commerce in precious stones – pearls, diamonds, turquoise, cornalian, onyx, coral, etc. – then scents, such as musk, amber, camphor, sandalwood and cloves. According to them (Arab traders), the best amber came from South-East Arabia, the next best from Spain and Morocco; the best aloes come from India. These geographical references show how wide the medieval merchants threw their nets. The first-hand experience of the Arab traders provided enormous reliable information about the geography of the distant places.

5. Marine Adventures

The Arabs carried most of the trade by land, but they were equally adventurists in sea traffic and trade. There are numerous manuals for mariners dealing specially with the Red Sea, the Persian Gulf, the Arabian Sea, the Indian Ocean, the Bay of Bengal, the Sea of Malacca and the Sea of China. Their adventures served as a great source of information about the salinity of seas and oceans, the climate, winds and lifestyles of the distant peoples. It was because of this source that Al-Masudi gave a reliable geographical account of the countries and nations he visited. The sea adventures served as the basis for the well-known story of 'Sindbad – the Sailor'. The Arabs voyages to Far East (China) added to the geographical knowledge of the Arabs substantially.

The Arab geographers carefully preserved the ancient geographical knowledge at the Arab Universities of Spain, North Africa and South-West Asia. Moreover, Arab traders travelled widely and gathered information which could be used by scholars to fill the gaps on Ptolemy's original map.

The Arabs made an outstanding contribution to the fields of mathematical, physical and regional geography. Their achievements in climatology, oceanography, geomorphology, linear measurements, determination of cardinal points, limits of habitable world, sprawl of continents and oceans are highly commendable.

The Arabs who were largely influenced by the Greek traditions adopted the Greek ideas about the shape and size of the earth. The early Arabs considered the earth as the centre of the universe, round which revolved the seven planets. The planets, in order of their distance from the earth were, Moon, Mercury, Venus, Sun, Mars, Jupiter and Saturn.[4] It was imagined that each of these planets occupied a particular 'sky' or using the expression of Ptolemy, a separate 'sphere' and that each 'sky' was under the authority of the 'planet' to which it had been assigned. In this hypothesis, all the stars formed the seventh sky or the 'highest heaven'. They believed that the first of the seven skies, closest to the earth, was of green emerald, the second of white silver, the third of red ruby, the fourth of white pearl, the fifth of red gold, the sixth of yellow topaz, and the seventh of a mass of fire *(nur)*. Similarly, there were seven earths, one inside the other, the lowest of them being hell.[5]

About the circumference of the earth, Ptolemy, in 157 AD, had established that the length of equator was 24,000 miles. During the time of Al-Mamum, the circumference of the earth was calculated as 20,160 miles, while Al-Battani fixed this figure at 27,000 miles. According to Arabs, the western limit of the habitable world was at the end of the Mediterranean Sea, the eastern at Sila (Japan), the northern in the land of Yajuj Majuj (Siberia) and the southern to the south of equator.

About climate, the Arab scholars made some valuable observations. In 921 AD, Al-Balakhi gathered climatic data and information from Arab travellers and prepared the first climatic atlas of the world entitled, *Kitabul-Ashkal*. Al-Masudi gave a detailed description of the Indian monsoons. In 985 AD, Al-Maqdisi (945–88 AD) offered a new division of the world into fourteen climatic regions. He recognized that climate varied not only by latitude but also by position east and west. He also presented the idea that the Southern Hemisphere was mostly on open ocean and that most of the world's land area was in the Northern Hemisphere.[6]

Figure 4.1 Arabic Cartograms used in Medieval Arabic School Maps

The Arab geographers offered important observations regarding the processes shaping the world's landforms. Al-Biruni wrote his great *Kitab-al-Hind* (Geography of India) in 1030 AD. In this book, he recognized the significance of the rounded stones he found in the alluvial deposits in the south of the Himalayas. The stones became rounded, he pointed out, as they were rolled along the torrential mountain streams. Furthermore, he recognized that the alluvial material dropped close to the mountain was relatively coarse in texture and that alluvial material became finer in texture farther way from the mountain.[7] Ibn-Sina (Avicenna) keenly observed the work of the agents of denudation and weathering in the mountains and held that the mountain streams eroded the slopes. He formulated the theory that mountains were being constantly worn down by rivers and that the highest peaks occurred where the rocks were especially resistant to erosion.[8] Mountains are raised up, he pointed out, and are immediately exposed to this process of wearing down, a process that goes on slowly and steadily.[9] Ibn-Sina also noted the presence of fossils in the rocks in high mountains, which he interpreted as examples of nature's effort to create living plants and animals that had ended in failure.

The Arab geographers borrowed from the Greeks the division of the globe of the earth into five zones; each zone representing particular temperature conditions. The five zones are: (i) the torrid zone, situated between the two tropics; (ii) the two frigid zones, placed near the poles; and (iii) the two temperate zones which occupied intermediate positions.[10] According to the Arabs, only a quarter of the world was inhabited, the rest was covered by water or rendered uninhabitable, either by excess of heat or by excess of cold. The Persians divided the world into seven empires or nations, called by the indigenous name of *kishwars* and at other times by the name of the 'climates'. These seven empires (*kishwars*) are: China, India, Arabia, Turan or Turkistan, the Roman Empire, Africa and Iran.[11]

The prime meridian, plotted by Ptolemy, was adopted by the Arab geographers also for the calculation of time and longitudes. This meridian used to pass through the Fortunate Islands. Abu-Mashar and some others had put the prime meridian at the eastern extremity. This idea of the eastern prime meridian was borrowed from the Indian scholars. The Indian scholars used to place India in the middle of the earth and therefore they made the prime meridian pass directly through the city of Ujjain which was the capital of Malwa and the principal intellectual centre of India.[12] The Indian scholars showed this meridian as passing through Lanka, Ujjain and the Mt. Meru (North Pole). In Arabic, the word 'Ujjain' has got corrupted and is written as Ozain, Ozin, Arin.[13]

For the determination of latitudes, the Arabs like Eratosthenes and other Greeks made use of sun's shadow when it happened to be on the meridian. The shadow was marked by a column (gnomon). The Arabs improved over the works of Greeks and Ibn-Yunus in his astronomical tables observed that the shadow, taken with perpendicular gnomon, did not correspond to the height of the centre of the sun, but to its upper limb.

The phenomenon of tides was also observed by the Arab navigators and scholars. They proved that the tides are caused by the gravitational pull of the sun and the moon. Al-Masudi, who navigated in the Caspian Sea, recorded the spring and neap tides, while Al-Biruni wrote in his book, *Kitab-al-Hind*, that the Indians believe that the tides are caused by the moon.

The Arabs were the first who put forward the idea of the periodic nature of the monsoons. The world 'monsoon', in fact, has been derived from the Arabic word *mausam* which means season. Many of the local winds which blow in the desert of Arabia, Egypt, Algeria and Libya have been described by Arab geographers. Al-Masudi observed variations in the colours of ocean water and attributed it to the variation in the salinity of water and the presence of marine vegetation.[14]

The Arab historical geographers like Ibn-Khaldun, Al-Biruni, and Al-Masudi described the influence of climate on vegetation and the style of life of the people. According to Ibn-Khaldun, the people of the warmer climates are known for their passionate nature. Those of the colder climate incline to stolidity and lack of vivacity. Those in the temperate climates excel in wisdom and are neither excessively passionate nor markedly stolid. He also explained that the Negroes are black because they live in the warm and humid climatic zones while the people of temperate and cold regions are whitish in colour. Similarly, people try to build their house and settlements on the southern slopes close to the springs and water sources.

There are numerous Arab writers and scholars who have contributed to the various branches of geography. But, here, the contribution of important amongst them is being presented.

Ibn-Hawqal-Abu-al-Mohammad Qasim (912–978 AD)

Ibn-Hawqal was a resident of Baghdad. He was probably born in Nasibin (Nisibis) in upper Mesopotamia (at Jezera). His real name was Mohammad Abdul-Qasim. From his childhood, he was interested in books of voyages, explorations, travelogues and itineraries, and the modes of life of distant tribes and nations. He started his travels in May 943 AD and visited on foot different countries of the world, within the orbit of the authority of Holy Quran.[15] The countries and regions visited by him include North Africa, southern edge of Sahara, Egypt, Syria, Armenia, Azarbaijan, al-Jazira, Iraq, Khuzistan, Fars (Iran), Khawarizm and Transoxania and finally Sicily after which we lose trace of him.

The treatise of Ibn-Hawqal is entitled *A Book of Routes and Realms*. In the preface to his book, Ibn-Hawqal remarks: 'I have described the earth in length and breadth and I have written about the Muslim countries. I have indicated the limits of each region, the cities and the districts included therein, the rivers which irrigate it, the bodies of water which modify the surface, the resources which are available, the various kinds of taxes which are imposed there, the routes which traverse it, the distances which separate it from the adjacent countries, the types of commerce which is successfully carried on there; and I have reassembled all the information which has made geography an interesting science for princes and persons of all classes'.

Besides the Arab world, Ibn-Hawqal has given an account of the European countries and established that the Caspian Sea did not have any connection with the Northern Ocean. The Black Sea communicated with the Northern Sea by a channel which may be an arm of the sea. According to him, Europe was an island. About Africa, he wrote that its eastern coast adjoining the Red Sea turns to the east. He also described countries and peoples bordering the Islamic world. His description about the Turks, the Khazars, the towns of Southern Italy, the Sudanese and the Nibians are highly reliable. He described the major crops and agricultural techniques of different regions and was the only Arab geographer of the period who really sketched a vivid picture of production.

Al-Masudi (896–956 AD)

Al-Masudi was born in Baghdad towards the end of the 9th century, most probably in 896 AD but the exact year of his birth is not known. He died in 956 AD at Fustat, near Cairo in Egypt. He was a descendent of Abdullah Ibn-Masud, a companion of Prophet Mohammad.

Al-Masudi

Al-Masudi was not only a geographer, but also a historian, a world traveller and a prolific writer. He travelled far and wide in the countries of Asia and Africa, especially Persia, Transoxiana, Sham (Syria), Armenia, Azarbaijan, Caspian Sea, Volga region, Central Asia, India, Lanka, Qanbalu (Madagascar), Oman, southern parts of

Arabia, Greek Empire, Spain and Egypt. Besides enriching geography and history, Al-Masudi contributed to cosmology, meteorology, oceanography, study of landforms, astronomy, Islamic law, and the Arabic folklore. As a writer, he had extremely diverse interests and an exceptional fecundity. His famous works are: (i) *Kitab-Muraj-al-Dhahab,* (ii) *Kitab-al-Tanbhwal-Ishraf,* (iii) *Kitab-Akhbar-al-Zaman* (in 30 volumes), and (iv) *Kitab-al-Ausat.* Most of these works, except *Kitab-Muraj-al-Dhahab* (Golden Meadows), are lost.

Al-Masudi made an in-depth study of the Greek and Roman sources and gathered information through travels. He sought to overcome local and regional prejudices and visited numerous places to investigate the geographical reality and facts for himself and for his writings. He tried to describe the geographical reality as he saw it. Some of the major contributions made by Al-Masudi to the various branches of geography have been described below.

The Christians, during the Dark Age, tried to prove that the earth is flat, triangular, and rectangular, twice as long west and east as to north and south, which is surrounded by water on all sides.[16] The medieval European mind, clouded with religious fanaticism, was not prepared to accept the idea of sphericity. Al-Masudi had a clear conception of the sphericity of the earth. He believed that the surface of the sea is curved, since when a ship approached the land, the coast and the objects thereon gradually became more and more visible.[17] He compared the merits and demerits of the earth being spherical instead of flat, saying that had it been flat, all lands would have remained eternally submerged under the sea.[18]

Al-Masudi also tried to determine limits of the oceans and continents and followed the Greek tradition, taking Japan and the Eternal Islands as the eastern and western limits of the world, respectively. To determine the southern limit, he navigated up to Sofala and agreed with Al-Battani that the shape of Africa was approximately the same as we know it today.[19]

Expressing his opinion about the Encircling Ocean, he stated that according to many authors the Encircling Ocean is the principal sea and that all other seas are derived from it; in the east it is connected with the China Sea. About the Arabian Sea, Al-Masudi considered it as the largest in the world. Moreover, he gives details regarding the location and size of the seven seas of the east. These seven seas were situated between the Arab states and China. The Arab traders who were having trade relations with China had to cross the seven seas. The first of these seven seas was the Persian Gulf. The names of the seven seas were as under: (i) Sea of Persia, (ii) Larevy Sea, (iii) Sea of Herkend, (iv) Sea of Shelahet or of Kalabar, (v) Sea of Kedrenj, (vi) Sea of Senf, and (vii) Sea of Senjy.[20]

The Sea of Persia comprised the Persian Gulf and the Sea of Makran. The Sea of Larevy stretched from the Delta of the Indus to the Cape Comorin (Arabian Sea). The third sea is that of Herkend (Hari Kund or Bay of Bengal). The Sea of Shelahet is the Sea of Malacca. The fifth sea Kedrenj or Kerdenj is on the eastern coast of the Peninsula of Malacca in the south of Gulf of Siam (Thailand). The sixth sea Senf corresponds to the coast of Vietnam (Cochin, China). The seventh and the last sea was that of Senjy, or the Sea of China which, according to Al-Masudi, stretched indefinitely towards the north and east.

During the period of Al-Masudi some important questions, like whether the Caspian Sea was connected with the northern ocean, or the Caspian Sea and Black Sea were interconnected, exercised the minds of many scholars. These controversies were coming down from the time of Hecataeus and Herodotus. Al-Masudi made independent observations and did not follow the Greeks and Romans blindly. After navigating he resolved that the Caspian Sea is not connected with any of the seas. He established that the Oxus flowed into the Aral Sea, which was mentioned for the first time. Volga river was described by him as an active commercial highway. Al-Masudi gives to the Atlantic Ocean the name of Dark Green Sea and was of the opinion that the Atlantic Ocean and the Indian Ocean are connected with each other.

One of the most important contributions of Al-Masudi lies in the field of physical geography. Modern ideas of geomorphology include both the comparative study of landforms and analytical study of processes involved in their formation.[21] Landforms are visualized to pass through a cycle of development from the youthful stage to maturity and finally to the old stage – the stage of peneplanation. Al-Masudi appreciated the role of cycle of erosion and adjustment of streams to structure in the evolution of landforms when he says, 'there is no place on the earth that is always covered with water, nor one that is always land, but a constant revolution takes place, effected by the rivers, which are always shifting, for places watered by rivers have a time of youth and decreptitude, like animals and plants with this reference that growth and decay in plants and animals manifest themselves in all parts at once so that they flourish and wither at the same time. But the earth grows and declines part by part'.[22] Al-Masudi's observations become all the more significant when it is considered that the role of rivers in the evolution of physical landscape has begun to receive adequate attention only during the last two hundred years or so.

Al-Masudi, who sailed in several seas, has described the weather conditions a voyager faced while sailing. About oceans, he showed acquaintance with the different problems and theories current in his time, relating to shapes and limits of different seas.[23] Regarding the Indian Ocean, he had a better idea than Ptolemy as Al-Masudi held that the Indian Ocean is connected with the Atlantic Ocean. He opined that salt in the seas and oceans comes from the land.

Al-Masudi tried to solve the problem concerning the exact source of the Nile river. He rejected the idea of Ptolemy that the Indus was connected with the Nile river. He described the source of the Nile in the mountains of Abyssinia.[24]

Al-Masudi was one of the well-known climatologists of his time. He gave a good account of the periodic winds (monsoons) of the Sea of Larevy (Arabian Sea) and Herkend (Bay of Bengal). He has made interesting remarks about the utility of winds as a source of energy. He has given the example of the windmills that he found in the desert of Sajistan on the western frontier of India.[25]

In the field of human geography, Al-Masudi tried to correlate man with environment. While describing the effect of environment on the mode of life and attitudes of people, Al-Masudi says: 'The powers of the earth vary in their influence on man on account of three causes, viz., water, natural vegetation and topography'. In the land where water is abundant, humidity predominates, people are frank and humourous, and where water is absent, dryness predominates; again in the land where vegetation is dense heat prevails, and if the region is devoid of natural vegetation, the reverse is the case.[26] Further, speaking of the selection of sites of human habitation, he emphasized the importance of the nature of the surrounding country, elevations and depressions; nearness to mountains and seas and lastly the nature of the soil.[27]

While dilating upon the character of the nomads of Arabia, Al-Masudi has observed that it is desirable for affluent people to go round the world. The nomads pondered over the problem of making their houses in cities and came to the conclusion that life in city changed man's character, dissuaded him from moving round, diminished his courage and suppressed his urge for progress. It was for these reasons that the Arabs preferred life in the open countryside where the air was pure and free from pollution. The Arabs are, therefore, marked by strength of resolution, wisdom and physical fitness. They are distinguished in acts of charity and possess intelligence of a high order, for these qualities are generated by a pure and clean environment.[28] The influence of environment on man can be appreciated from the following lines in which Al-Masudi stated: 'As for the northern quarter, which is farther away from the Sun, in the extreme north, and which is the abode of Saqaliba (the Slavs), the Afranja (Franks) and the neighbouring races, and where the influence of the sun is rather alleviated and the region abounds in cold, moisture and snow, the people are characterized by good physique, rude behaviour, slow wit, harsh tongue, white complexion, thick flesh, blue eyes, thin skin, curly and red hairs'. All these characteristics are found due to the predominance of moisture in their lands and their cold nature does not encourage firmness of religious belief. Those living farther north are characterized by dullness of mind, harsh behaviour and barbarism. These characteristics increase proportionately as we proceed farther north. He vividly examined the impact of environment on the physical and intellectual properties of man and cited the example of the Turks. In his opinion, the Turks who emigrated to India lost their national characteristics and acquired new characteristics suited to the new environment. Further evidence of his conception of adaptation to environment is found in his observations of animals and plants, who adopt the natural colour of the physical environment in which they dwell or grow.[29]

Al-Masudi contributed appreciably to the field of regional geography also. He gave a fairly reliable account of Al-Sham (Syria), Persia, Central Asia, Georgia, Mesopotamia and the countries he travelled.

About Sham (Syria), Al-Masudi writes that this country is hilly and the abode of clouds, winds, mists, fog, and heavy rains, where trees are numerous and rivers are perennial.[30]

On the basis of language, he divided the habitable world into seven nationalities: (i) Persians, (ii) Chaldaeans (Arabs), (iii) Greeks, (iv) Egyptians and Libyans, (v) Turks, (vi) Hindus, and (vii) Chinese.

In brief, Al-Masudi was such a geographer who examined the real document of geography, i.e. earth and compared the knowledge he acquired from books with the actual conditions on the ground.

Al-Biruni (973–1039 AD)

Al-Biruni

Al-Biruni's full name was Abu-Rayhan Mohammad Ibn Ahmad. He passed his youth on the banks of the river Oxus in the city of Khwarizm (Khiva) situated in the Republic of Uzbaikistan. The prince and ruler of Khwarizm professed ardent zeal for science and arts, and encouraged scholars like Al-Biruni to study the various branches of knowledge and inspired people to travel different parts of the world. Al-Biruni, through his dedication and vast knowledge, achieved great scholarship in philosophy, religion, linguistics, mathematics, physics, astronomy, chronology, medicine, and various languages and literatures. He was conversant in Khwarezmian, Persian, Arabic, Sanskrit and also knew Greek, Hebrew and Syriac. He was a man endowed with creative genius, sagacity, wisdom, sincerity, and commitment to inductive reasoning. His humour, courage, enterprise, objectivity, honesty, prodigious industry and intellectual skills were unprecedented. He is given the titles 'founder of Indology' and the 'first anthropologist'. He was also given the title Al-Ustadh (The Master) for his remarkable description of the 11th century India.

Al-Biruni was one of those prodigious minds at work in the medieval world whose creative, versatile, scientific, and international outlook, coupled with universality of thought, amazed the modern world. He was free from dogmatic conservatism as well as emotional bondages. This helped him to remain free from prejudices. He was more of a synthesizer than a dogmatic scholar, a keen observer of comparative studies par excellence. Al-Biruni's position as a scientist and scholar may be appreciated from the fact that the eleventh century has been regarded as the 'Age of Al-Biruni'.

According to some historians, Al-Biruni was born on 4 September 973 AD (3rd of Dhul-haj, 362 A.H.). His father and mother expired at an early age. Although a Tajik by race, he was Persian by culture. There is a strong controversy about the birth of Abu-Rayhan. It revolves round the identification, interpretation and meaning of the word 'Biruni', a part of

Abu-Rayhan's name. Is Biruni a city? Where was it situated? Or is Biruni a suburb of Khwarizm (Khiva)? Or does it denote one who was born in Khwarizm or one who lived outside the city of Khwarizm? This confusion has led some later authorities to find a logical explanation for this term. According to Samani's *Kitab-al-Ansab*, the people of Khwarizm called foreigners *Beruni* (Biruni) in Persian and for this reason Abu-Rayhan was called Al-Biruni. The well-known historian, Yaqut, has opined that perhaps Biruni meant one who lived outside the city or in the countryside. Except Abu-Rayhan, no other person was given this appellation which means that it was not in general use. Abu-Rayhan's stay in Khwarizm was also not short for his first 23 years were spent in that region under Al-i-Iraqi, and a further period of 8–10 years under the Ma'munids. In all he spent nearly 30 years in Khwarizm. Moreover, he was born in the suburb of Kath – a town in the region of Khwarizm. His other contemporaries like Ibn-Sina stayed for far shorter period in Khwarizm but none was ever styled Al-Biruni, even though Ibn-Sina was a Persian by origin. The use of Al-Biruni with the name of Abu-Rayhan, therefore, appears to refer to his birthplace, a town or a suburb settlement of Khwarizm.

Unfortunately, Abu-Rayhan has not left behind any autobiographical account. There are some scanty references in his writings but these do not throw light on his education and early life. One may presume that he received the traditional *Maktab* and *Madarasah* education. Al-Biruni was a great scholar and possessed encyclopaedic mind. He was always in search of new knowledge. He was a liberal and profoundly interested in science. In his age orthodox reaction had already set in and Abu-Rayhan was wary of being accused of heresy while writing *al-Hind*. Abu-Rayhan, born with a great analytical mind and keen comprehension, was irresistibly drawn to the study of mathematics. Travelling was always regarded as a part of education in Islam. But, by the 11th century, numerous rich libraries established by rulers had greatly reduced the need to travel to distant places for acquiring knowledge. However, when Abu-Rayhan wanted to study Indian sciences, he had to travel, visit, and use all the means at his disposal to have access to the scattered sources in Western India.

The most noteworthy trait in Abu-Rayhan's character was his insatiable thirst for knowledge. Like a thirsty man he returned again and again to the fountain of knowledge. Even when his life was ebbing out he would not let the scarce minutes go by uselessly. Very little is known about the family of Al-Biruni. Probably, he had no children and this partly explains his polarized devotion throughout his life. Free from the burden of parenthood he lavished almost parental love on his studies and books. He studied the Greek books through Arabic translation. He was versed in Persian, Turkish, Syriac and Sanskrit. He was familiar with the Roman language of the Byzantine Empire. He had friendly contacts with Syriac and Christian intellectuals. He was acquainted with the Arabic translation of Indian works. Subsequently, when political developments brought him into contact with the Hindus of the subcontinent, he made full use of this opportunity. He learned Sanskrit when he had already completed 45 years of age. He commanded a vocabulary of nearly 2,500 Sanskrit words. The facility with which he discusses and explains Indian doctrines shows his full command over the subject. His proficiency in Sanskrit literature is also corroborated by the fact that he was able, while delving into the nature of God, to clearly explain the foundations of the Advaita School. He distinguishes between the beliefs of the educated Hindus and the common people. It is clear from his works that he made astronomical observations in the cities of Ghazna, Kabul, Lamghan, Peshawar, and Multan. He was a witness to the Muslim conquest of the city of Nagarkot situated at the foot of the Himalayas. This city was famous for an ancient Hindu (idol) temple. Probably, he accompanied the soldiers of Mahmud, up to Mathura and Kanauj on the banks of the Jamuna (Yamuna) and the Ganga, respectively. He died at Ghazna in the year 430 A.H. (1039 AD).

The prolific writer, Al-Biruni, has written a number of books and has dwelt upon a great variety of subjects. Among the main works of Al-Biruni include *Kitab-al-Hind, Al-Qanun-al Masudi* (The Canon of King Masud), *Vestige of the Past Athar-al-Bagiya, Tarikhul-Hind, Kitab-al-Jamakir,* and *Kitab-al-Saydna*. He translated from Sanskrit into Arabic the original title of *Patanjali* which contains valuable information on India and China. He wrote 27 books on geography, four each on cartography, geodesy, and climatology, and seven books on comets, meteors and surveying. Al-Biruni's academic interests and activities encompassed a wide variety of subjects, ranging from abstract theories of philosophy to the practical sciences of mathematics, geography, geology, physics, astronomy and medicine. His main field of study, however, was astronomy. He devised his own method of determining the radius of the earth by means of the observation of the height of mountain and carried it out at Nandana in Pind-Dadan-Khan, Pakistan.

Al-Biruni's age was characterized by orthodox reaction. There were people who regarded astronomy as heresy. This prejudice was similar to people's opposition to logic on the plea that its terminology belonged to pagan Greek literature and language, although the adoption of Greek terms was mainly the fault of the translators. In much the same way, there were people who ignored geography as something without any utility, though the Holy Quran is full of episodes of travels and adventures, e.g. Prophet Abraham's journey from Ur, Moses' journey from Egypt and the *hijrah* of the Prophet of Islam (peace be on him). After analysing these unscientific tendencies, Al-Biruni produced convincing arguments for establishing the claims of physical sciences. He reminded the opponents of astronomy that God asks people to contemplate on the marvels of the earth and heavens, believing that all the phenomena of nature reveal truth of the highest import.

He provided illustrations of the daily use of mathematical and astronomical knowledge. This knowledge helped in ascertaining the influences of the sun and the moon in the form of what we know as the seasons and tides. Knowledge of stars and their positions is of considerable help in setting directions during travels and journeys. Similarly, it is very helpful in ascertaining the correct directions of *qiblah* and the timings of prayers and the latitudes and longitudes of cities. In this way, astronomy was shown by him to be a useful, functional and applied science and in conformity with the injunctions of Islam.

As astronomy is interrelated with a number of other sciences such as cosmogony, mathematics, and geography, Al-Biruni's magnum opus, the *Qanun-al-Masudi* is modelled on the pattern of the *Almagast* of Ptolemy. His astronomical theories are of significant bearing and, therefore, have been discussed hereunder.

Al-Biruni considered the universe to be situated on the outermost surface of a limited sphere. A detailed study of the origin of the universe was made by Al-Biruni in his book *al-Tahdid*. The geocentric and heliocentric controversy engaged the mind of Al-Biruni. Some modern scholars have criticized him for accepting the geocentric theory. However, in that age when telescope and modern precision instruments were lacking, it was difficult to arrive at any definite conclusion. He was not ready to accept the heliocentric theory without definite scientific evidence. Until an alternative theory was conclusively proved, it was but logical to believe and accept the older theory or explanation. Al-Biruni wrote a separate book, *Kitab-al-Tatbiq fi Tahqiq, Harkatah al Shams* on the movement of the sun.

After discussing the basic problems relating to the sphericity of the skies and the earth, the geocentric theory, the nature of the eastern and western notions of the heavens, Al-Biruni goes on to define the imaginary circles and signs so often referred to in astronomy and geography, i.e. the poles, the equator, longitudes and latitudes, obliquity and the signs of the zodiac.

He devoted special attention to the study of time and dates. He studied the calendars of different nations. He also inquired about the differences in the time of day and night in different regions and the continued long day at the poles. The problem of finding the correct timings of prayers led him to conduct research over a long period beginning with the writing of the *Chronology* to *Qanun-al-Masudi*. He wrote a book (*Risalah*) about day and night which also proved the duration of a six month day at the poles. He also compiled a small treatise on the Indian calendar determining division of time.

About the sun, he asserted that it is a fiery body for the solar eruption which is noticeable during the total eclipses. Al-Biruni believed in the geocentric theory and regarded the sun as moving round the earth.

Al-Biruni had his reservations about Ptolemy's view that the distance of the sun from the earth was 286 times the latter's circumference. He, however, found the sun immeasurable with the instruments of that age and its distance remained an object for conjecture.

In his monumental book, *Qanun-al-Masudi*, he presented a masterly exposition of both the solar and lunar eclipses. He described the obliquity of the eclipse as the angle formed by the intersection of the celestial equator and the ecliptic. Earlier, the Greek, Indian and Chinese astronomers found it to be

24° 51' 20". Al-Biruni himself took measurements at Khwarizm and Ghazna and found the figure to be 23° 35' which is very close to the actual obliquity. He also discussed the reasons and timings of dawn and twilight. He found that twilight (morning and evening) occurs when the sun is 18° below the horizon. Modern researches have confirmed Al-Biruni's findings.

About the moon, he asserted that it does not move in a perfect circle. Its maximum and minimum distances differ appreciably. It changes its path and is variable. Al-Biruni stated that the moon returned to its former position in relation to fixed stars but minute differences occur and accumulate. He discussed the lunar month on a synodic basis, i.e. by referring to its position, and return to it, in relation to the sun.

Al-Biruni measured the longest and the shortest distance of the moon and the earth. These were 63° 32' 40" and 31° 55' 55" of the earth's diameter. However, he was not sure of the diameter of the moon. In this matter, he followed Ptolemy and accepted his value of the moon's diameter as 31' 20" of the earth's diameter. Here again, his scientific insight led him to choose the correct figure, for Ptolemy's value is nearer to the modern value of 31' 17".

About the tides, he opined that the increase and decrease in the height of the ebbs and tides occurred on the basis of changes in the phases of the moon. He gave a very vivid description of the tides at Somnath and traced the latter's etymology to the moon.

About the stars, he was of the view that it was practically impossible to determine the number of heavenly bodies (stars) even in a small portion of the sky. He was also aware of the limitations of the instruments of his age. Among the ancient astronomers, Hipparchus was the first to catalogue 850 stars. Ptolemy also worked on this basis. Al-Biruni adopted the Greek nomenclature of 48 figures and 12 constellations arranged on a belt.

He rejected Aristotle's contention that the 'Milky Way' was under the sphere of planets and correctly estimated it to belong to the highest sphere of the stars. He also attacked Aristotle for believing that stars cause injury to eyesight and are responsible for sorrow and misfortune. This shows that he was basically rational in approach and did not attach any superstition to natural phenomena. He thought these stars moved to the east on a central axis and parallel to the zodiac.

He believed that as there was no way to find out the parallel of the fixed stars it was impossible to determine their distance and magnitude. The Greeks thought that the stellar sphere was next to the most distant planet. Ptolemy regarded the distance as 19,666 times the earth's radius. Mars was accepted as one and a half times the sun's diameter. Al-Biruni used Indian figures about the distance and magnitude of the stars.

Regarding the planets, Al-Biruni followed Ptolemy taking his works to be the most authentic and correct. From the earth towards the stars, the planets were arranged by him in the following ascending order: Moon, Mercury, Venus, Sun, Mars, Jupiter and Saturn.

Al-Biruni was of the view that the Greeks were more exact in their sciences and observations. The Indians, however, were better equipped in solar and lunar studies and the eclipses. What he basically aimed at was the exposition of the scientific method backed by firm belief in natural laws. He insisted upon continuous observation, collection of reliable data and successful application of all these principles.

Though, Al-Biruni dedicated himself only to astronomy, yet he excelled in mathematics also. In that age, mathematics consisted of arithmetic, geometry, physics and music. Algebra was added to this only after the age of Al-Khwarizm. While Al-Biruni excelled in geometry and arithmetic, he possessed considerable knowledge of algebra too. He was also interested in physics, though he had no interest in music. In his book, *Kitab-al-Hind*, he discussed Indian beliefs, Hindu literature, grammar, metre, chess, etc. but totally ignored Indian music.

Besides having expertise in spherical trigonometry, Al-Biruni was an adept in Indian arithmetic. He wrote *Rashikat-al-Hind* (The Zodiac in India). He was also familiar with the different methods of arithmetic propounded by Brahma-Siddhanta.

Al-Biruni had special interest in geomorphology and paleontology. He compared the different fossils discovered in the plains of Arabia, Jurjan and Khwarizm along the Caspian Sea. His studies pointed to the existence of sea at these places in some bygone age while history possessed no such record. According to him, the Indo-Gangetic plain was formed by the silt brought by the rivers.

He also discussed the occurrence of floods and springs. His study of the changes in the course of rivers of Jurjaniyah and Balkh and Oxus shows his deep insight in geomorphological processes. He found that the course of the Oxus had undergone a change since the days of Ptolemy – a period of 800 years and he also explained how the life of the people living in the region had been affected by these changes.

He correctly estimated the known habitable world as greater in length, i.e. from China in the east to Morocco and Spain in the west. The seas limited the inhabitable world. The known world was divided into age-old, seven-fold divisions of seven *aqalim*.

Al-Biruni also had an accurate idea of the different bays, gulfs and smaller seas. He referred to the Ice Sea in the north-east of Europe and in the west of Tangier and Spain. He also mentioned the Sea of Warang (Norsemen), i.e. probably the Baltic. In the south of Europe, he was aware of the presence of a sea in the form of gulfs up to Sicily and Bulgaria (Mediterranean Sea). The Indian Ocean, he mentioned as being pricked by islands and felt that it met the oceans in the east and possibly below Africa in the west. The Indian Ocean also had its links with the Klymsa Sea (Red Sea) and with the Persian Gulf. He referred to the seas of China and mentioned the fact that in the east the seas were named after the islands or the countries.

The great geographer was also aware of the huge mountain range known in India as Himavant (the Himalayas) which spread across the length of the known world like a spinal column.

He also mentioned the Warangs and their predatory habits. There was mineral industry in North Europe. He referred to the Sawaras, Bulgars, Russians, Slavs and Azovs in the west and to the country of Frank and Galicia, situated beyond the Roman Empire at the western arm of Europe.

Regarding Africa, he was convinced that it lay and extended far into the south. He referred to the 'Mountain of Moon' situated near the equator which was the source of the Nile river. He analysed the causes of floods in the Nile and attributed them to the heavy rains in the upper reaches of the Nile.

Al-Biruni's knowledge of Asia was quite extensive and fairly accurate. In his opinion, the Great Central Mountain (the Himalayas) was the source of most of the perennial rivers of Asia. He provided detailed information about the land of the Turks, identified in the Augarer river, and about the region of the Baikal Lake in Eastern Siberia.

He wrote extensively and accurately about the geography of India. His estimate of India's extent from the forts of lower Kashmir to the Deccan Peninsula is amazingly close to the real dimensions of the subcontinent. He had a definite idea of its peninsular form. The mountains of Himavant and Meru (Pamir) surrounded it in the north. He said that the Eastern and Western Ghats controlled the distribution of rainfall in peninsular India. He provided detailed information about the sources of rivers. However, excepting the Indus, his information about the other rivers is limited to the location of their sources, based on hearsay and the knowledge derived from ancient books, e.g. *Matsya Purana*. He was the first person to provide correct information about the Indus, its origin, course and floods. His knowledge of the geography of Punjab and Afghanistan was based on his personal observations. He also described the rivers of Gherwand, Nur, Kaira, Sharvat, Sawa Panchir, Bitur (Afghanistan), Biyatta (Jhelum), Chandrahara (Chenab), Irwa (Ravi) and Shaltladar (Sutlej). The five tributaries of the Indus, according to him, meet the river at Pancanade (Panchanda) in Punjab near Multan.

Al-Biruni provided valuable information about North-Western India, particularly Kashmir. For Gilgit, he said that it was two-days journey from Kashmir. About Kashmir, he said that it lay on a flat fertile plateau, surrounded by inaccessible mountains. The southern and eastern parts of the country belonged to the Hindus, the west to the various Muslim kings, the north and eastern parts to the Turks of Khotan (Khatan) and Tibet. The best access to Kashmir was through the Jhelum gorge.

He also described the city of Qannauj – the city traditionally associated with the Pandavas. Moreover, he acquired considerable knowledge of the terrain and people of the Indo-Gangetic plains.

He gave an accurate account of the seasons of India. He described the nature of the monsoon which brought rainfall to the greater parts of the subcontinent during the summer season. He explains how Kashmir and Punjab receive rainfall during the winter season.

Al-Biruni also discussed the origin of castes in Hindu society, idolatry and the Hindu scriptures. His study of *Samkhya*, the *Gita*, *Patanjali*, *Vishnu Dharma* and of some of the *Puranas*, coupled with his acquired knowledge of the *Vedas*, provided Al-Biruni with a unique opportunity to give the first objective description of Hindu beliefs. Al-Biruni found a dualism in Hindu beliefs, i.e. the beliefs held by the educated (scholars) and the beliefs of the ignorant masses. This cleavage became wider with a dualism in linguistics. The language of the masses was quite different from that of the learned. Thus, the educated disapproved of idolatry but the masses believed in it.

In brief, Al-Biruni excelled in philosophy, religion, cosmology, astronomy, geography, geodesy, stratography, geomorphology, mathematics, science, medicine and several languages. He also contributed appreciably in the field of chronology, computation of years and dates. At the same time, he had a clear concept of the ideal historian. His correct view and reasoning led him to think that the institution of *varna* (caste), based on inequality, was the main obstacle in a rapport between the Hindus and Muslims. The condition of Indian learning, language, script, centres of learning was also brought out. The gigantic labour, scientific reasoning, and untiring efforts made Al-Biruni one of the most outstanding geographers of the medieval period.

Al-Idrisi (1099–1165 AD)

Al-Idrisi was one of the leading geographers of the 12th century. His name was Abu-Abd-Allah Muhammad. Al-Idrisi was, in fact, his surname, since he belonged to the family of Caliph Idris – the ruler of Cardova (Spain). He was a great Muslim geographer, cartographer, Egyptologist and traveller who lived in Sicily and died in Sicily. Very little is known about the early life of Al-Idrisi except that he was born at Ceuta. From his writings, it appears that he visited Lisbon, Andalusia (Spain), France, England, Sicily, Morocco, Constantine, Asia Minor (Turkey) and the interior parts of Africa.

Al-Idrisi

Al-Idrisi got education at the University of Cardova in Spain. At the invitation of Roger II, the ruler of Sicily, he reached Palermo, where he lived for a long time and wrote a new geography. In 1154 AD, he completed a book with the title *Amusements for Him Who Desires to Travel Around the World*. He corrected the wrong notion of an enclosed Indian Ocean as also the erroneous concept regarding the Caspian Sea which was earlier supposed to be an arm of the World Ocean. He defined with precision the courses of many rivers including those of Danube and Niger. The land of Yajuj and Majuj (Gog and Magog of the Bible) has been described by him as situated in the Siberia region. Regarding the Southern Hemisphere, he holds the view that owing to intense heat it is not habitable. Moreover, Al-Idrisi showed that the Greek division of the world into five climatic zones (two cold, two temperate and one torrid) did not correspond to reality, and had suggested a more sophisticated and refined world climatic system.

The most important contribution of Al-Idrisi to geography was his world map. On the world map he plotted the various geographical features and for this purpose he utilized the Islamic and the Christian sources. His map was based on a rough, rectangular projection. The Asiatic part of the map is very rich in information. It represents correctly the Caspian and Aral seas, which were misrepresented in the antiquity. In the shape of Africa, we find the influence of Ptolemy, although Africa and China are not connected. The map is oriented with the south on top – a common feature of Islamic maps. He gave an account of Europe in the first half of the 12th century that is unrivalled. He corrected the idea of an enclosed Indian Ocean and the idea of the Caspian Sea as a gulf of the World Ocean.

Ibn-Battuta (1304–1368 AD)

Ibn-Battuta

Abdallah Muhammad, having the surname Ibn-Battuta, was one of the great Arab travellers. He was born at the beginning of the 14th century in Tangier at the entrance of the Mediterranean Sea. He was of Berber origin and was not an Arab. However, he had received instruction in Islam and its tenets. He was from a family that produced a number of Muslim judges (*Qazis*). He received the traditional education in his native town Tangier. In search of education, he stayed at Egypt, Syria and Hejaz, and met there with leading scholars, *Sufis* and saints. A number of diplomas and degrees were conferred on him, mainly in Damuscus. Goaded by a desire to see new countries, he left his homeland in 1325 AD, at the age of 21 to make the usual pilgrimage to Mecca. After performing *Haj*, he visited successively Egypt, Syria, Iraq, Persia, Arabia, Somalia, Mogadishu, Zanzibar, Asia Minor (Turkey), Qipchaq land (steppes beyond the Caspian), Constantinople, Khwarizm, Bukhara, India, the Maldives, Ceylon, Sumatra and China.[31] His travels also took him to many parts of Arabia, Yemen, Aden,

Sanaa, Taizz, Oman, Zalya, Mogadishu, Somalia, and never before visited, for example, Ethiopia. He also navigated along the eastern coast of Africa as far as Kilwa – 10 degree south of the equator. At Kilwa, he learned about the Arab trading post at Sofala in Mozambique, south of the modern port of Deira situated more than 20 degree south of the equator. Ibn-Battuta confirmed what Ibn-Hawqal had implied – that the torrid zone in East Africa was not torrid, and that it was inhabited by the numerous native tribes.[32] Ibn-Battuta refuted Aristotle's thesis that hot regions of the world would be too warm for human habitation. Thus, he showed that Aristotle had been wrong in believing that it was too hot here for human habitation, in what the Greeks had called the 'torrid zone'.

From Mozambique, Ibn-Battuta returned to Mecca, and set on again to visit Baghdad, Esfahan, Zagros Mountains in Persia and the land around the Black Sea. He travelled in the Asia Minor (Turkey), Russian steppes and then eventually to Constantinople, Astrakhan, Aral sea, Bukhara and Sumarkand. Then, he crossed the Hindu Kush mountains through Afghanistan into India. In India, he served at the court of Muhammad Bin Tughlaq (1325–1351 AD) and travelled the country widely. Upon his arrival in Sindh, Ibn-Battuta mentions the Indian rhinoceros that lived on the banks of the Indus. The Sultan of Delhi appointed him his ambassador to Yuan Dynasty of China, but delay kept him from reaching China for several more years, during which time he visited Khambat (Gujarat), Sharavathi River, Uttara Kannada, Kozhikode (Calicut), Madurai. He became the Chief *Qazi* (Judge) of Maldives Islands, and married into the royal family of King Omar.

En route the coast at the start of journey to China, Ibn-Battuta and his party were attacked by a group of bandits. Separated from his companions, he was robbed and nearly lost his life. He also visited the Maldives Islands, Ceylon, Chittagong, Sylhet, Bengal, Kamrup (Assam), Dacca, and Sumatra. From Sumatra he travelled to Malacca and Vietnam and finally reached Quanzhou in Fujian Province of China (1345 AD). In Quanzhou, Ibn-Battuta was welcomed by the local Muslim *Qazi* 'Fanzhang' (Judge) and the leaders of local Muslim merchants.

After the completion of his assignment in China, he returned to India and finally left for Faiz (the capital of Morocco), through Egypt, Alexandria and Tunis in 1350 AD, but his travels did not end. He made a trip to Sardinia, Garanada, Spain and then crossed the Sahara to Timbuktu on the Niger river where he gathered important information about the Muslim Negro tribes living in that part of the world. He arrived back to Morocco early in 1354 AD. In all he travelled for about 28 years and covered more than 75,000 miles during this period (Figure 4.2).

About 600 years ago, it was probably Ibn-Battuta who pointed out that the climate along the equator was less extreme than the climate in the so-called temperate zone in North Africa. He confirmed that the torrid zone in East Africa was not torrid and that it was occupied by numerous native tribes which justified the establishment of Arab trading posts. He, however, was not so much interested in physical surroundings as in human beings. He was quick

Figure 4.2 Travels of Ibn-Battuta

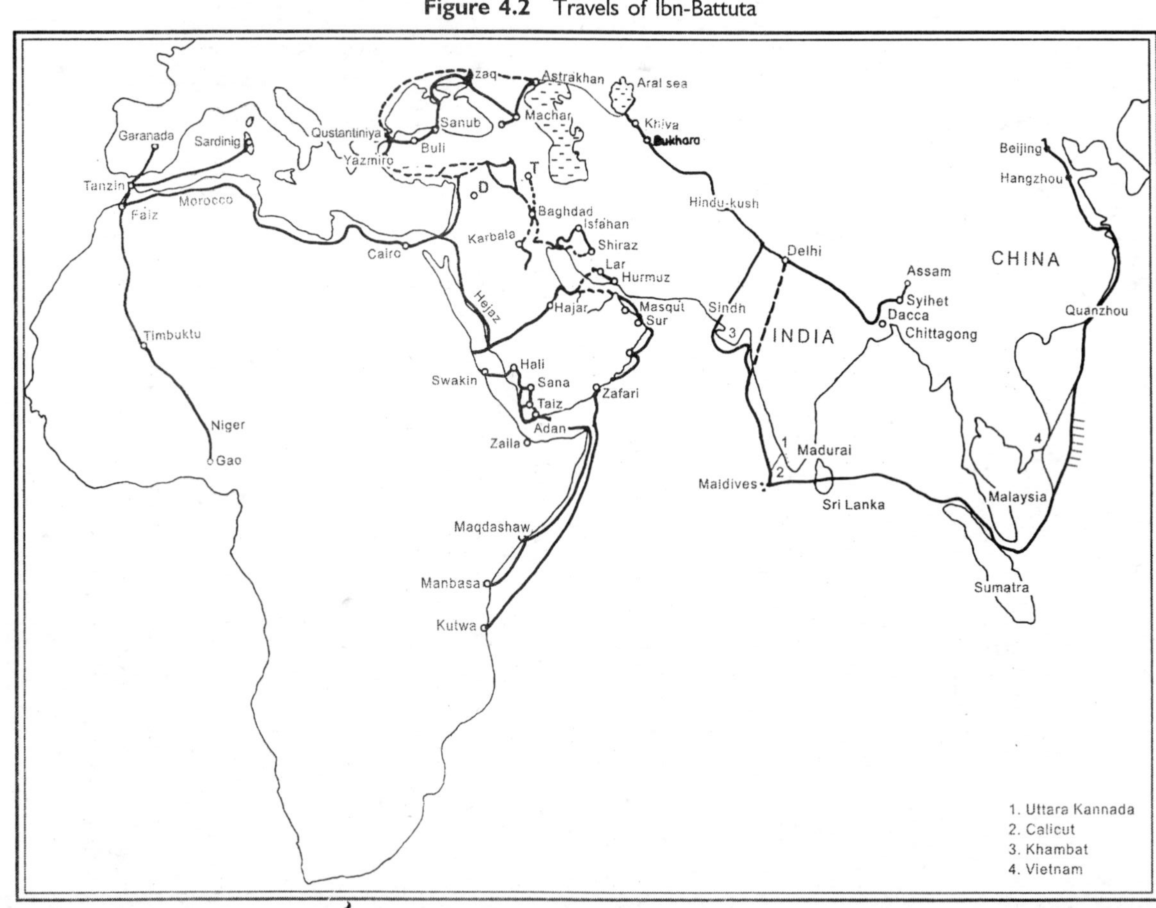

at noticing manners, customs, traits and traditions, means of communications, resources and industries. There are many facts of anthropological interest in his writings. His book *Rihlah* throws light on the soils, agriculture, economy and political history of the then Muslim world.

He was deeply rooted in orthodox Islam but, like many of his contemporaries, he oscillated between the pursuit of its legislative formalism and an adherence to the mystic path and succeeded in combining both. He did not offer any profound philosophy but accepted life as it came to him, leaving to posterity a true picture of himself and his time. In fact, he was a man of restless energy and curiosity, clear-sightedness and determination to enjoy life; at the same time he was a devout observer of the practices of his religion, with a particular devotion for the saints.

Ibn-Khaldun (27 May1332–19 March 1406 AD)

Ibn-Khaldun

Ibn-Khaldun may be considered as the last great historian Arab who made a significant contribution to geography. He is regarded to be the founding father of modern historiography, sociology and economics. He was born on the Mediterranean Sea coast of North-West Africa. Ibn Khaldun's life is relatively well-documented, as he wrote an autobiography in which numerous documents regarding his life are quoted word for word. However, the autobiography has little to say about his private life. In his autobiography, Ibn-Khaldun traces his descent back to the time of Prophet Mohammad through an Arab tribe from Yemen, specifically the Hadhramaut, which came to Spain in the 8th century at the beginning of Islamic conquest. Most of the time he lived in the cities of Algeria, Tunisia and Spain. The later years of his life he passed in Egypt. At the age of 45, he completed his monumental work known as *Muqaddimah*.[33] This masterpiece deals with description and discussion of human society in its various aspects. The work has been divided into six sections: (i) civilization, geography and anthropology; (ii) discussion of nomadic culture and its comparison with sedentary culture, sociological and historical causes and consequences of the conflicts arising continually from the fundamental opposition between the two cultures; (iii) dynasties, kingdoms, etc.; (iv) life in villages and cities, how should cities be organized; (v) professions, means of livelihood; and (vi) classification of sciences. The European scholars also acknowledged the significance of the book and considered Ibn-Khaldun as one of the greatest philosophers to come out of the Muslim world.

The *Muqaddimah* begins with a discussion of man's physical environment and its influence on the living styles of people. He has discussed the various stages of social organization, identifying the desert nomads as the most primitive and the purest. He has suggested that the sedentary city dweller is dependent on luxuries and becomes morally soft. He has also discussed forms of government, describing a sequence of stages that mark the rise of a dynasty to power, followed by its decline through corruption to its fall.[34] Ibn-Khaldun is best regarded as a historian, a philosopher of history, and a proto-sociologist.[35] Despite his great cyclic view of world history, the *Muqaddimah* provides a good overview of the state of Arabic thinking concerning geography in the 14th century. Although the main focus of Ibn-Khaldun's writing concerned the process of state formation and decline, he developed his ideas through the consideration of the physical environment which he saw as forcing people to live together in social and political groups. Central to his argument was the view that states develop through a natural sequence of growth, maturity, decline and fall, because group solidarity is inevitably eroded by the process of civilization. The same concept was later adopted by Ratzel in the 19th century.

He has maintained that the Northern Hemisphere is more densely populated than the Southern Hemisphere. Further, the population along the equator is thin, but away from the equator, there is greater concentration of population up to 64 parallels. Farther away, there is once again very little or no population at all. Intense heat of the equatorial belt was considered by him as the discouraging factor for the concentration of population. A harmonious blending of heat and cold in the temperate regions, according to him, is conducive to human growth and settlements. Away from the temperate regions, the excessive cold of the Polar areas is again a deterrent to human growth.

About the origin of settlements, he has stated that 'those who settle are attracted by the fertility of land and by the seas with which they can defend themselves against invaders'. In course of time, the population increases, and so too the pressure on land. Various handicrafts begin to develop and it becomes a permanent settlement. The settlement grows in course of time and becomes a city – which shows a definite social and socio-economic structure, in ways of division of labour, dearth and abundance, supply and demand. The origin of all cities has always been these small settlements. Moreover, he has tried to explain various human activities with the help of environment. He explains why the Arabs are nomads and why the Negroes are of an easy disposition.

In *Muqaddimah*, he has also discussed various ways of conducting commerce, evolving and encouraging crafts, and studying science. In physical geography, he accepted the traditional zoning of climate, running parallel to the equator. He stressed the fact that people who live close to the equator turn black. The physical environment influences the character and attitudes of people, which get modified under different cultural traditions. He suggested that warlike nomads often founded large states, but after a while the nomads were absorbed by their permanently settled subjects. As peasants and town people, the rulers lost their warlike spirit and eventually their kingdoms fell apart. Ibn-Khaldun both predicted and lived to see the collapse of the Islamic state he lived in. At the fall of Damascus in 1400 AD he actually met Tamerlane, the conqueror and devastator. It was unfortunate that the works of the Arab scholars were not translated into Latin or any other language until the 19th century. As a result, scholars in other parts of the world were unable to make use of their observations. Thus, Ibn-Khaldun can be considered the first environmental determinist who tried to correlate man with environment in a scientific way.

Notes

1. Ahmad, N., 1947, *Muslim Contribution to Geography*, Lahore, p. 135.
2. Wright, J.K., 1925, 'The Geographical Lore at the Time of Crusades', New York, American Geographical Society, *Research Series,* No.15.
3. James, P.E. and Martin, G.J., 1981, *All Possible Worlds*, p. 50.
4. Al-Masudi, *Muruj*, Vol. I, p. 49.
5. Ali, S.N., 1960, 'Some Geographical Ideas of Al-Masudi', *Al-Masudi-Millenary Commemoration Volume*, edited by Ahmad, S.M. and Rahman, A., Aligarh.
6. James, P.E. and Martin, G.J., op. cit.
7. Kimble, G.H.T., 1938, *Geography of Middle Ages*, London, Methuen.
8. James, P.E. and Martin G.J., op. cit.
9. Ibid.
10. Ali, S.M., 1976, *Arab Geographical Thought*, Aligarh, p. 17.
11. Ibid., 45.
12. Ibid.
13. S.M. Ali, *Arab Geographers*, Aligarh, p. 17.
14. Ahmad, N., 1947, op. cit., p. 65.
15. Sykes, P., 1961, *A History of Exploration from the Earliest Times to the Present Day*, New York.
16. Dickinson, R.E. and Howarth, O.T.R., 1933, *The Makers of Geography*, Oxford, p. 44.
17. Ahmad, S.M., 1953, 'Al-Masudi's Contribution to Medieval Geography', *Islamic Culture*, Vol. XXVII, No. 2, Hyderabad, p. 68.
18. Ali, S.M., op. cit., p. 87.
19. Ibid.

20. Ibid.
21. Shafi, M., 1960, 'Al-Masudi as a Geographer', *Al-Masudi-Millenary Commemoration Volume*, Aligarh, p. 76.
22. Ahmad, S.M., op. cit., p. 68.
23. Beazley, op. cit., p. 464.
24. *Tabih*, 28.
25. Silvestre, M.de Sacy, *Charastomathie*, p. 17.
26. Tanbin, p. 28.
27. Ibid., p. 29.
28. *Muruj*, ed. Barbier de Meynard I, pp. 243–245.
29. Ahmad, S.M., 1954, op. cit., p. 285.
30. Alvi, S.M.Z., 1960, 'Al-Masudi's Conception of the Relationship Between Man and Environment', *Al-Masudi-Millenary Commemoration Volume*, Aligarh, p. 94.
31. Ibn-Battuta, *The Travels of Ibn-Battuta, A.D. 1325–1354,* Trans. Defremery C. and Sanguinetti B.R., 1958, Cambridge University Press.
32. Ibid.
33. Ibn-Khaldun, *The Muqaddimah*, Trans. Franz Rosenthal, 1958, New York.
34. Ali, S.M., op. cit., p. 160.
35. Morgan, 1988, p. 202.

5

The Impact of Explorations and Discoveries

Man, throughout the history, has always been a wanderer, searcher and explorer. As early as 700 BC, Phoenician and Carthaginian traders were seeking new lands in the Mediterranean and beyond for their merchandise. About 470 BC ago, Hanno, the Carthaginian, sailed with large fleet as far as Sierra Leone, bringing back tales of gorillas and of a 'land of fire' (This was probably an account of the grass fires lighted before the rains in many parts of Africa).

In 330 BC, the Greek, Pytheas, sailed around Britain and into the North Sea. The most important exploration of this point was made by Alexander the Great in the years 330–323 BC, when accompanied with experts to record details of countries through Persia to India, and like a true explorer, returned by a different route.

In the 1st and 2nd centuries BC, the Romans in order to expand their empire, penetrated up to the Nile, as far north as Baltic, and westward across Europe. Norsemen discovered Iceland in 867 AD, then Greenland in 982 AD, and finally by their reaching the mainland of North America about five centuries later.

At about the same time, the Arabs were trading far afield in the Indian Ocean, ranging from Spain to India, Malaysia, Indonesia and China and as far as Madagascar. Their greatest traveller was Ibn-Battuta, who visited every Muslim country in a remarkable series of journeys that lasted for almost 30 years. Buddhist missionaries, passing to and fro from India to China across the desert of Takla-Makan, had come across what came to be known as the Jade Route, along which for centuries traders carried jade from the Himalayas to China in exchange of silk.

The explorations, discoveries, scientific inventions and scholarly works of the Arabs provided a new and more realistic picture of the Arab world as also of other nations. Marco Polo's adventures in Central Asia and China opened up new vistas in geographical literature. The Renaissance in Europe provided fresh impetus for widening the existing horizons. A copy of Ptolemy's *Geography*, which had been preserved in a library in Byzantium (Istanbul) was discovered and brought to Italy where it was translated into Latin in 1406 AD. After the invention of printing in the 1450s, copies of this book, including reconstructed maps, were printed in Bologna, Rome and Ulm, and made a great impact on contemporary scholars. Moreover, there were definite improvements in the art of navigation including wide adoption of magnetic compass. Consequently, there was a concomitant renewal of interest in world exploration, mapping and geographical description. In 1507 AD, German cartographer Martin Waldseemuler (1470–1521 AD) produced map of the world that clearly indicated both North and South America. With these devices, the navigators became more confident. Thus, the stage was set for the next step in the search for the discovery of the unknown parts of the world. Incidentally, this was the period of Muslim decadence. The Arab Empire started disintegrating. Prince Henry, also known as 'Henry the Navigator', defeated the Arabs at Ceuta and captured the Strait of Gibraltar. He established the first institute of Geography at Sargres near Cape St. Vincent. At this institute, he invited geographers, cartographers, mathematicians, and astronomers belonging to all faiths.[1]

As a result of the impetus given by Henry, many of the navigators offered themselves to explore the western coast of Africa, and Cape Bojador and the Gulf of Guinea. In 1492, Christopher Columbus reached the New World (America), and Vasco da Gama, through the Cape of Good Hope, entered into the Indian Ocean from the Atlantic, and sailed with the Arabs to reach India in 1498 AD. In 1520 AD, Ferdinand Magellan was the first to reach Asia by sailing west. He was the first to circumnavigate the world in 1520 AD. From 1768 to 1778 AD, Captain Cook sailed extensively in the Indian and Pacific Oceans to locate the *Terra-Incognita* (unknown land) of Ptolemy. In his venture, he discovered many new islands and the continent of Australia (Figure 5.1). Columbus, Vasco da Gama, Magellan, Cook and almost all the expeditionary leaders relied on Ptolemy's calculations about the latitudes and circumference of the earth, but some of their more famous discoveries disproved Ptolemy's calculations and changed the picture of the world he had established. There were new developments in cartography – new projections, especially that of Mercator (1569 AD), were invented; the first globe of the world was made and new and accurate maps of the world and countries were prepared.

Most of the cartographers of the 15th century lived either in Venice or Genoa because it was from these two places that Europeans departed on voyages to the Eastern Mediterranean to pick up cargoes of valuable items from the east. It was during this period that many of the erroneous geographical concepts were eliminated. The coasts of the Pacific Ocean were determined and the shape of the earth was finalized. Again, it was during this time that the fantastic stories of strange creatures gradually gave way to sober and reliable accounts.

Up to the 15th century, most of the scholars took earth as the centre of the universe and tried to study the universe with a geocentric approach. But,

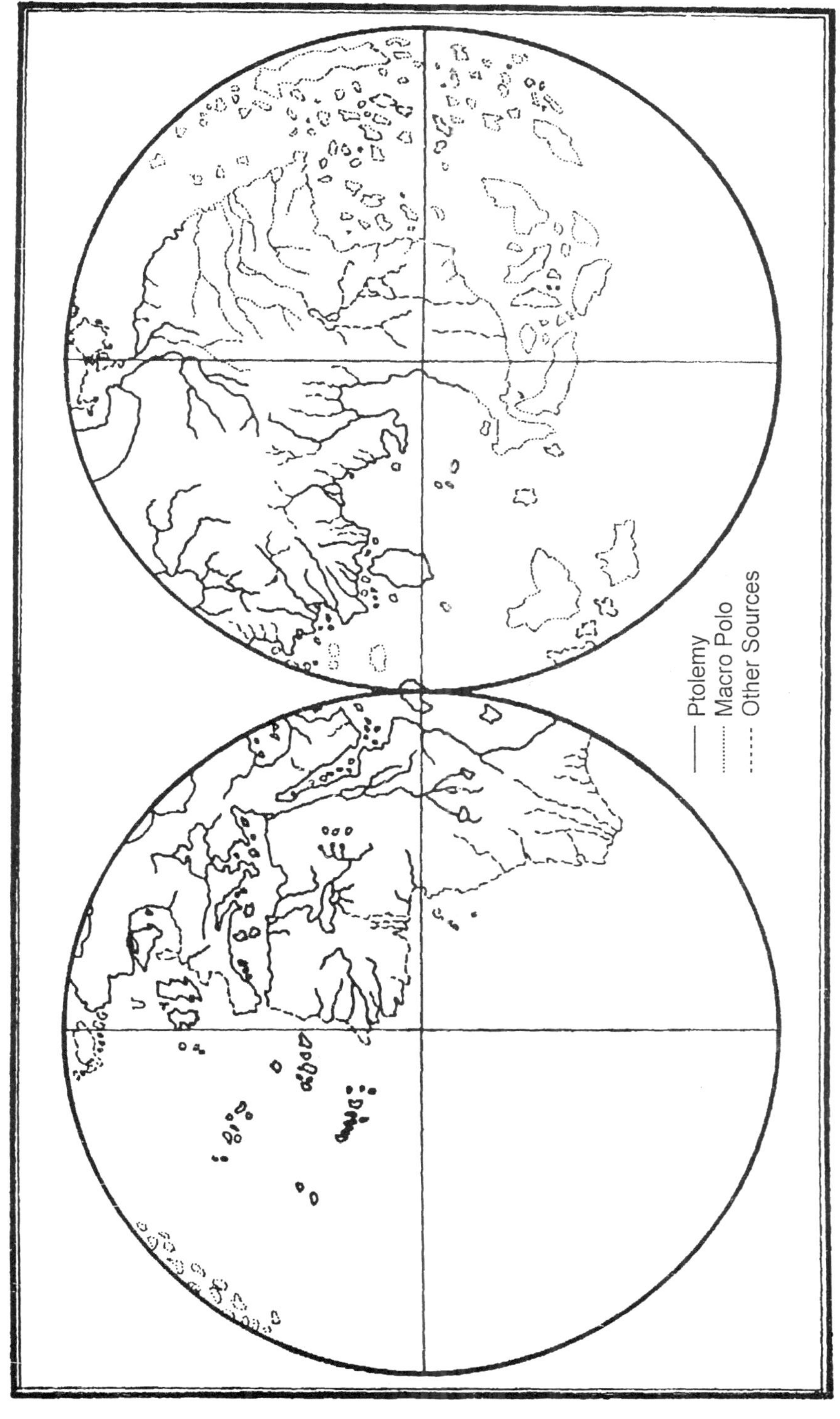

Figure 5.1 Behaim's Globe and its Sources

after the Great Age of Discovery, the heliocentric idea got strengthened. The Polish scholar, Nicholus Copernicus, who, between 1497 and 1529 AD, carried out numerous observations of the planets, moon and stars, established that all the planets rotate round the sun. This is known as the heliocentric concept of the universe. In 1618 AD, Kepler proved that planetary motions were elliptical rather than circular. In 1623 AD, Galileo presented proof that Copernicus was right about the heliocentric universe. In 1686 AD, Issac Newton presented his laws of gravitation. These developments led to a scientific revolution and a period of specialization started which gave birth to taxonomy of the branching of physical, biological and social sciences. The printing machine was invented, consequently many books describing explorations, voyages, discoveries and geography of the newly discovered lands were published in Europe. But in all the writings of this period, the earth was taken as the home of man in which the mode of life of the people is closely influenced by the physical environment. The works of Sebastian, Munster, Cluverius and Carpenter were highly appreciated as their geographical writings gave a new dimension to the discipline.[2]

The impact of explorations and discoveries was maximum on the art of map-making. For example, in 1500 AD, Juan de la Cosa drew a map using observations from the first three voyages of Columbus and also from John Cobot's voyage to North America.[3] But the first world map that shows America as a separate continent and not as the eastern part of Asia was drawn by Martin Waldseemuller in 1507 AD. He also made use of the name, America, for the first time, either because he thought that Amerigo Vespucci had reached the new continent before Columbus or because Amerigo was the first explorer definitely to identify the newly discovered lands as a separate continent. As a result of this decision by Waldseemuller, the new continent was not named after the European who first reported seeing it.[4] Although this new map of the world prepared by Waldseemuller was called 'Carta-Marina', it was in no way more useful for navigation than the other maps of this period that were made on the Portolano principle of design. Explorers had already found that when they followed straight lines on these maps for long voyages, they did not arrive at expected destination.[5]

In order to overcome the shortcomings of the earlier maps and to construct more useful maps for navigators and explorers, Peter Aspian produced in 1530 AD a heart-shaped map. On this map also, distance and directions were distorted. This problem was finally overcome by Gerardus Mercator who designed the Mercator Projection. The improved form of Mercator Projection which was designed in 1569 AD became a perfect solution of the problems of navigators and enabled them to reach the desired destination by sailing on the great circle (shortest) route. In Mercator Projection, the shape of the continents was, however, distorted and in the higher latitudes it was highly exaggerated. For example, on this projection, Greenland is shown larger than the size of the subcontinent of India.

Figure 5.2 The Great Explorations

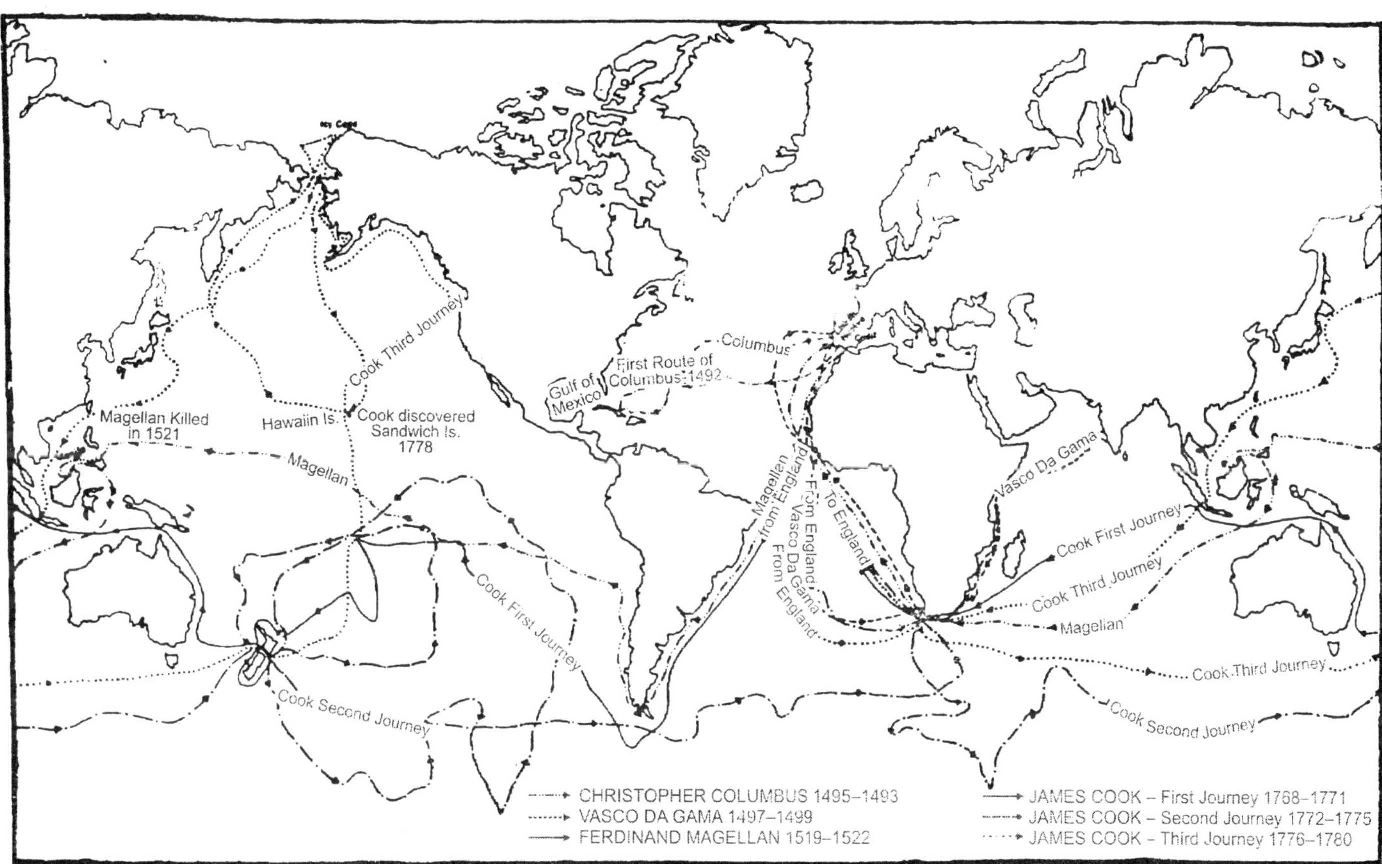

Moreover, Mercator did not explain the method used in making this projection. It was in 1599 AD when Edward Wright produced trigonometric tables that made it possible for other people to reproduce Mercator's projection. In the later parts of the 16th century, Amsterdam became a major centre for the publication of atlases and wall maps, both of which became very popular, especially in the 17th century.[6] Later on, the printing of atlases and wall maps at different scales became quite popular in France and England. In France, the first producer of atlases was Nicholas Sanson d'Abbeville, who, in the 17th century, founded a 'dynasty of cartographers' which produced maps and atlases for over a century.[7] The Great Age of Discovery, thus, brought about many changes in the world outline and removed many erroneous concepts about the shape and size of the earth, about the dimensions of the continents and countries and thus the subject of geography started growing on scientific lines. Consequently, many new theories about the origin of the earth and continents and man-nature relationship were advocated.

The discovery of classical cartography played a significant role in the European expansion of political and economic power in the 15th and 16th centuries.

Few European explorers or scholars regarded themselves as geographers. They preferred the term *Cosmography* (the descriptive science of the globe and its relation to the universe) which was used in a number of treatises, often modelled on the works of Strabo and Ptolemy.

The Renaissance saw the revival of another branch of ancient geography, that of *topographic description*. Account of voyages provided raw material for encyclopaedic works on the world, or on parts of it, often called *Chorographies*. Thus geography was defined either as *Cosmography*, the science of the globe, or it could be *Chorography*, the descriptive science of single regions. *Cosmography* involved the description of the universe (cosmos) and included the modern sciences of environment, biology, cartography, geography, geophysics, and anthropology. *Chorography* (choros = place) described places in general and larger areas whereas *topography* was concerned with particular places at a smaller scale.

Before an attempt is made to discuss the development of modern geographical concepts, it would be of immense utility to describe briefly the explorations and discoveries made by Marco Polo, Columbus, Vasco da Gama and Captain Cook.

Marco Polo (1254–1324 AD)

Marco Polo was the famous and outstanding traveller of the 13th century. His family had trade with the Middle East for a long time. His father Nicolo and uncle Maffeo had been to China at the time of Kublai Khan. Marco Polo was born most probably in 1254 in a family of noble and aristocrat traders. In 1260, the two brothers (Nicolo and Maffeo Polo) took leave of their families, including the six-years old Marco Polo, and embarked on the first great oriental trading expedition of the Polos. From Venice, they went to Constantinople (Istanbul) and from there they sailed across the Black Sea to the Crimea. Then, their quest for trading profit took them by land far north up to the Volga and eastward into Bokhara (Uzbekistan) where they entered the domain of the great emperor of the Tatars – Kublai Khan, the Mongol leader whose grandfather was Genghis Khan. Kublai Khan cordially welcomed the newcomers. Having passed nine years in the Orient, they arrived at Acre (Palestine) in 1269. On arriving home, Nicolo found that his wife had died, and Marco was now a strapping youth of fifteen.

Marco Polo accompanied his father and uncle on their journey to the Great Khan. The Polos started off their journey in 1271. They reached Palestine, where the papal legate, Teobaldo of Piacenza, gave them letters for the Mongol emperor. From Palestine they went to Acre on Accho (N. Israel).

From Acre, the travellers proceeded to Ayas (Gulf of Iskenderum in Turkey). They passed through Erzurum (Eastern Turkey), and Tabriz (Northern Iran). Subsequently, they crossed inhospitable deserts infested with brigands, before reaching Hormuz on the Persian Gulf. There, the Polos decided not to risk a sea passage to India and beyond but to proceed overland to the Mongol capital.

Soon they were on the road again traveling through deserts of 'surpassing aridity' toward the Khorasan region in what is now Eastern Iran. Turning gradually to the north-east they reached more hospitable lands; Badakhshan (Balascian) in Afghanistan, in particular, pleased the travellers. Marco Polo seems to suggest that they remained here for a year; long illness (possibly malaria) was cured by the benign climate of the district. It has also been supposed that Marco Polo, during this period, visited territories to the south (other parts of Afghanistan, Kafiristan in the Hindu-Kush), Chitral (in what is now Pakistan, and perhaps Kashmir). It is however, difficult to establish which districts he traversed and which he may be describing through information gathered in route.

Leaving Badakhshan the Polos proceeded up the Vakhan (Vocan) valleys toward the Pamirs, which were eventually crossed by an itinerary that has long been a subject of discussion and conjecture. Descending on the north-eastern side of the chain, they reached Kashgar (Cascar) in what is now the Sinkiang – Uighur Autonomous Region of China. By now, the Polos were on the main Silk Road, and their advance can be followed along the oases to the south and east of Takla Makan Desert – Yarkand, Khotan, Charchan and Lop Nor (Lop Lake) – stepping stones on the way to Sha-chou on the border of China, a place now called Tun-huang (Kansu province).

Before reaching Sha-chou, the Polos had travelled primarily among Muslim peoples and they also encountered sprinklings of Nestorian Christians, Buddhists and Zoroastrians; now, however, they entered the vast province of Kansu (Tangut) where an entirely different civilization, mainly Buddhist in religion but partly Chinese in culture, prevailed. Finally, they reached the Mongol capital, and presented the sacred oil from Jerusalem and the papal letters to their patron – Kublai Khan.

For the next 16 or 17 years, the Polos lived in the emperor's dominions, which included among other countries, Cathy (now North China) and Mangi or Manzi (now South China). Possibly, they moved with the court from the summer residence, Shang-tu, to the winter one, Ta-tu or 'Taidu' – modern Peking.

Marco Polo himself reached Cathy very young (at the age of about 20 years). Although he knew little or no Chinese, he spoke some of the many languages then used in East Asia – most probably Turkish and Arabized Persian. The emperor repeatedly sent him on fact finding missions to distant parts of the empire. One such journey took Marco to South-Western China, to Yunnan, and perhaps as far as Burma (Myanmar). On another occasion, he visited South-Eastern China as he gave a good description of Quinsay (now Hang-chou). There is ample evidence that Marco Polo considered himself as an adoptive son of his new country (Figure 5.3).

Some time around 1292 (1290 according to Otagi) a Mongol princess was to be sent by sea to Persia, then to become the consort (queen) of Arghun Khan. The Polos offered to accompany her. It seems that Kublai Khan was unwilling to let them go but finally granted permission. Kublai Khan was then nearing his eightieth year.

The princess, with some 600 courtiers and sailors, and the Polos boarded a fleet of ships which left the part Chuan-chu (Zaiton) and sailed southward. They touched Champa (Vietnam) as well as a number of islands and the Malaya Peninsula. On the Island of Sumatra, they stayed for five months to avoid monsoon storms. There, Marco Polo was much impressed by the fact that the North Star appeared to have dipped below the horizon. The fleet then passed near the Nicobar Island (Necuveran) and reached Ceylon (Seilan). The Chinese ship subsequently followed the west coast of India and the southern reaches of Persia, finally anchoring at Hormuz (Figure 5.3). The expedition then proceeded to Khorasan, handing over the princess not to Arghun Khan who had died, but to his son Mahmud Ghazan.

The Polos eventually departed for Europe and probably stayed for a few months at Tabriz. After some delays, they reached Constantinople (Istanbul) and finally Venice (1295).

The time when Marco Polo reached his home, Venice was at war with Genoa, and Marco Polo became commander of a fleet. In 1296, he was kept in prison, and while confined in Genoa he dictated in French his famous book entitled *Book of Various Enterprises*. It later became extremely popular and had a great effect on the geographic outlook of the time. Although regarded as partly fiction, it encouraged the exploration of Portuguese as well as those of Columbus. His *Il-Million* (The Million), known in English as the *Travels of Marco Polo*, became a geographical classic.

The Marco Polo's book *Il-Million* was an instant success. In a few months it spread throughout Italy. The book, however, raised many controversies and many of the scholars started saying that the book was a fiction, full of fables. As a result of such controversies, which have continued for nearly seven centuries, Marco's reputation has suffered dramatic ups and downs. For some he was a genius, a man of prodigious memory, a more conscientious observer, and a successful official at the Khan's court – a man at home in the cosmopolitan Asia of the Great Mongol rulers. For others, he was a braggart, who made too much of himself, a drifter ready to believe the gossip of ports and *bazaars* (markets); a man with little culture, scant imagination, and a total lack of humour; a man who failed among other things, to mention the Great Wall of China, the consumption of tea, and the ideographic script of the Far East (Figure 5.3).

Figure 5.3 Travels of Marco Polo

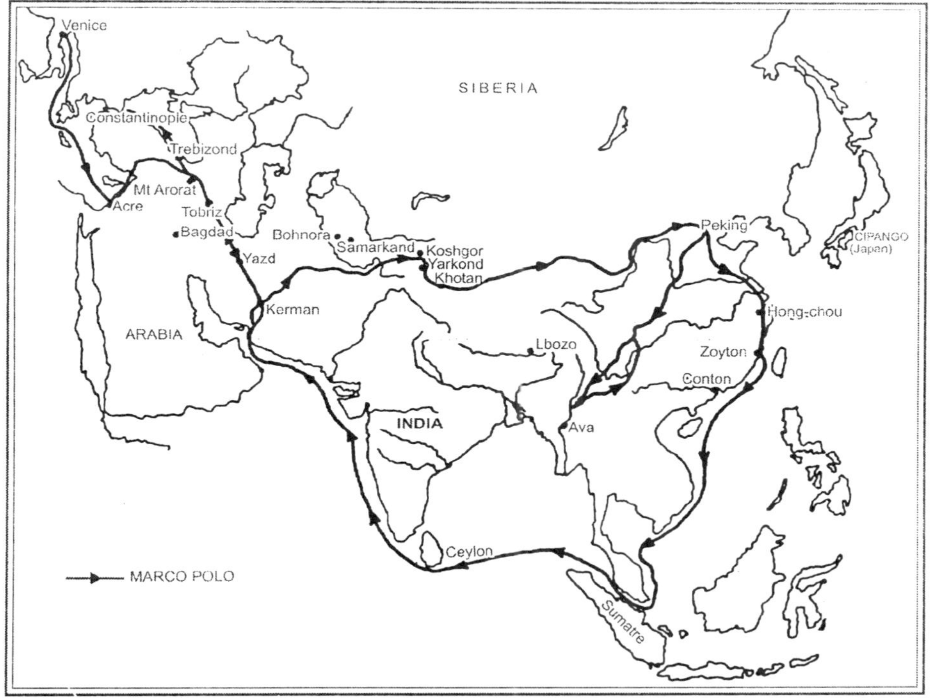

Sir Henry Yule, the great biographer of Marco Polo, fittingly comments about this great traveller and explorer: 'He was the first traveller to trace a route across the whole longitude of Asia, naming and describing kingdom after kingdom which he had seen with his own eyes; the Deserts of Persia, the flowering plateaux and wild gorges of Badakshan, the jade-bearing rivers of Khotan (Sinkiang), the Mongolian steppes, cradle of the power that had so lately threatened to swallow up the worldwide body or society of Christians, the new and brilliant Court that had been established at Cambaluc; the first traveller to reveal China in all its wealth and vastness, its mighty rivers, its huge cities, its rich manufactures, its swarming population, the inconceivably vast fleets that quickened its seas and its inland waters; to tell us of the nations on its borders with all their eccentricities of manners and worship; of Tibet with its sordid devotees; of Burma (Myanmar) with its golden pagodas and their tinking crowns; of Laos, of Siam (Thailand), of Cochin, China, of Japan, the Eastern Thule, with its rosy pearls and golden-roofed palaces; the first to speak of that Museum of Beauty and Wonder, still so imperfectly ransacked, the Indian Archipelago, source of those aromatics then so highly prized and whose origin was so dark; of Java, the Pearl of Islands; of Sumatra with its many kings, its strange costly products and its cannibal races; of the naked savages of Nicobar and Andaman; of Ceylon, the Isle of Gems, with its Sacred Mountain and its Tomb of Adam; of India the Great, not as a dreamland of Alexandrian fables, but as a country seen and partially explored, with its virtuous Brahmans, its diamonds and the strange tales of their acquisition, its sea-beds of pearl, and its powerful sun; the first in medieval times to give any district account of the secluded Christian Empire of Abyssinia (Somalia), and the Semi-Christian Island of Socotra (presently Yemen); to speak though indeed dimly, of Zanzibar with its Negroes and its ivory, and of the vast and distant Madagascar, bordering on the Dark Ocean of the South, with its Ruc and other monstrosities; and in a remotely opposite region, of Siberia and the Arctic Ocean, of dog-sledges, white bears, and reindeer-riding Tunguses'. Marco's story has inspired countless other explorers to set off and discover the world. Two centuries after Marco's passing, Christopher Columbus set off across the Atlantic in hopes of finding a new route to the Orient. With him was a copy of Morco Polo's book.

Christopher Columbus (1451–1506)

Christopher Columbus

Christopher Columbus, the Italian explorer and known as the discoverer of America, was probably born at Genoa (Italy) in about 1451. There is no definite record of his birth and several conflicting theories have arisen concerning his origin, but as none of these have been adequately substantiated, the above datum is still generally accepted. Columbus had little schooling, and his early years were spent in his father's trade as a weaver in Genoa. The young Columbus seems to have acquired a taste for navigation in Genoa and undoubtedly went to sea at an early age, probably on coastwise trading voyages. He started navigation at the age of fourteen. By 1476, he had settled in Lisbon and stayed with his brother, Bartholomew, who was a chart-maker and cartographer in the city. In 1478 he was married and lived for a few years on the island of Madeira (Portugal). After the death of his wife in

1482 he went to sea again for the Portuguese, sailing several times to the Guinea coast. In Lisbon, he further developed his knowledge of navigation and map-making and acquired valuable experience in voyages to Iceland (1477), Madeira (1478) and the west coast of Africa (1483).

He read several books on navigation and reached the conclusion that China was about 3,000 miles (5,000 km) west of the Canary Islands. The study of navigational theory gradually strengthened in Columbus the conviction that the earth was round. Soon he conceived the idea of sailing due west to Asia and thus reaching the land of gold and spices of Cipangu (Japan) and the East Indies by a shorter route than the prevalent route of sailing round Africa. Though he was not alone in believing the earth to be spherical, his concept simultaneously exaggerated the proximity of Eastern Asia and underestimated the size of the world as a whole.

For over a decade, Columbus tried to get financial support for his 'enterprise'. It was only in 1492 that he succeeded in persuading Ferdinand of Isabela of Spain to sponsor the expedition. He set out in 'Santa Maria' with two other small ships (Pinta and Nina) expecting to reach Japan.

Columbus sailed from Palos (Spain) on 3 August 1492 with a total crew of approximately ninety men. After refitting his ships in the Canary Islands, he set out on 6 September 1492 and sailed west across the Atlantic Ocean. He had serious difficulties in disciplining his crew as the voyage lengthened. Finally, on 12 October 1492, they sighted land and Columbus gratefully named the island San Salvador (Walting Island in the Bahamas). Thinking that he had reached the long-sought Indies, he named them 'West Indies'. Continuing his voyage, he discovered several islands and sailed along the north coast of Cuba to Hispaniola (Haiti), where he built the fort of La Navidad. Leaving 44 men to colonize Hispaniola, he sailed for home, reaching Palos on 15 March 1493. He was triumphantly acclaimed in Spain and many honours were showered upon him.

On 24 September 1493, Columbus sailed on his second voyage with seventeen ships, about 1,500 men and a large quantity of supplies. He landed at Dominica and discovered the islands of Marie Galante, Guadeloupe, Montserrat, Antigua, Santa Cruz, Puerto Rice and the Virgin Islands. Reaching Hispaniola in November 1493, Columbus found the fort of La Navidad in ruins. He established a new settlement a little further up the coast at Isabela, which became the first European town of the new world. Then, he sailed westward to Cuba which he took to be the Asiatic mainland. In March 1496, he returned to Spain, where he was again well-received (Figure 5.4).

After a stay of two years, Columbus started off on his third voyage. He sailed with six ships, and this time he took a more southerly direction. On this voyage, he discovered Trinidad and the mouth of Orinoc river – and gave it the name of 'Dragon's Mouth'. From here he returned to Haiti where his brother was ruling the little colony in his absence. But treachery and mutiny had been at work, and even Columbus could not improve the situation by his presence. He was a brilliant navigator but a poor statesman. To improve the

Figure 5.4 The Four Voyages of Christopher Columbus

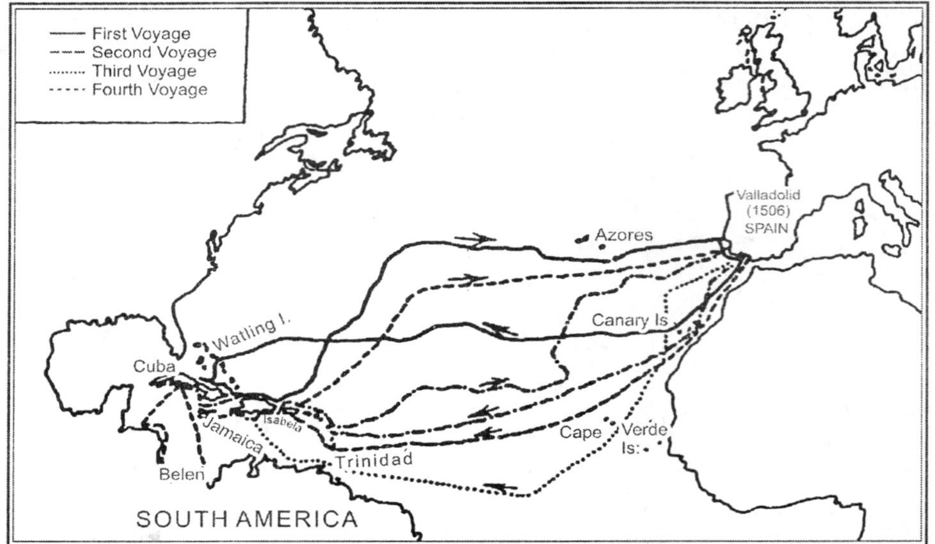

situation, a Spanish official was sent out to replace Columbus. This high-handed official put the poor navigator in chains and placed him on board of a ship for Spain. The queen pardoned him as the lined face of Columbus divulged his sufferings.

Columbus sailed on his fourth and last voyage on 9 May 1502, in an effort to find a westward passage to Asia. He discovered Honduras and followed south along the coast as far as Panama. Failing to find a passage, he attempted to establish a colony at Belen, but trouble with the natives forced him to abandon the project, and he returned to Jamaica. Had Columbus only sailed west he might have discovered Mexico with all its wealth. The bad weather discouraged him. The food was nearly finished, the biscuits were so full of maggots that the people could eat them only in the dark when they were not visible. Columbus himself seemed to be at the point of death. He reached Spain in 1504 to learn that the queen of Spain was dead. He was friendless, penniless and seriously sick. He believed that the lands he had discovered were a part of eastern Asia.

After 20 years of toil and peril he says pitifully, 'I don't own a roof in Spain'. The brilliant explorer passed away in Valladolid (N. Spain) in 1506.

Vasco da Gama (1460–1524)

When Columbus discovered the New World, the Portuguese were straining and exploring the sea route to India. To discover the sea route to India by east, Vasco da Gama was chosen by the ruler of Portugal.

Vasco da Gama was born in Sines in about 1460. He entered the service of the Portuguese court and army. As a young man, he gained experience in navigation and the management of a ship.

Vasco da Gama

Under the command of Vasco da Gama, a fleet of four small ships sailed for India in 1497. After crossing Cape Verde Islands, Vasco da Gama steered south-west into an unknown part of the South Atlantic. He did not know that at one time he was within 600 miles (1,000 km) of the coast of South America. Day after day, week after week passed in dreary monotony as they sailed the wide ocean that surrounds St. Helena. Having spent 96 days out of sight of land and sailed some 4,500 miles (8,222 km) they drifted on the south-west coast of Africa. It was a record voyage, for even Columbus had only sailed 2,600 miles (4,333 km) without seeing land. In November 1497, they entered a bread-bay (St. Helena), the name which is still in use. After a skirmish with coloured natives (Hottentots), the explorers sailed on, putting their trust in the Lord.

Because of stormy weather and high waves, the crew grew sick with fear and hardship and clamoured to go back to Portugal. All cried out to God for mercy upon their souls, for now they no longer took heed of their lives. At last the storm ceased and the sea grew calm. Vasco da Gama rounded the Cape of Good Hope and sailed north to the Zambezi river. Here they spent a month. It was at this place that for the first time in the history of discovery the disease of scurvy broke out. The hands and feet of the crew swelled, their gums grew over their teeth, which fell out so that they could not eat. This proved to be one of the scourges of early navigation. After arriving at Mozambique, they found four ships of Arab traders loaded with gold, silver, cloves, pepper, ginger, rubies and pearls which they had brought from the east (Figure 5.5).

With the help of the Arab traders, the expedition reached Mombasa (Kenya) and then visited the king of Melindi. From here they sailed across the Arabian Sea to reach the west coast of India. Seven months after the start from Portugal they reached Calicut (India). Vasco da Gama met the king of Calicut on 28 May 1498.

Vasco da Gama left India for Portugal with spices. He lost over 100 of his men including his brother – Paul, but he had discovered the sea route to India. When he returned to Lisbon he had sailed 24,000 miles (40,000 km) in two

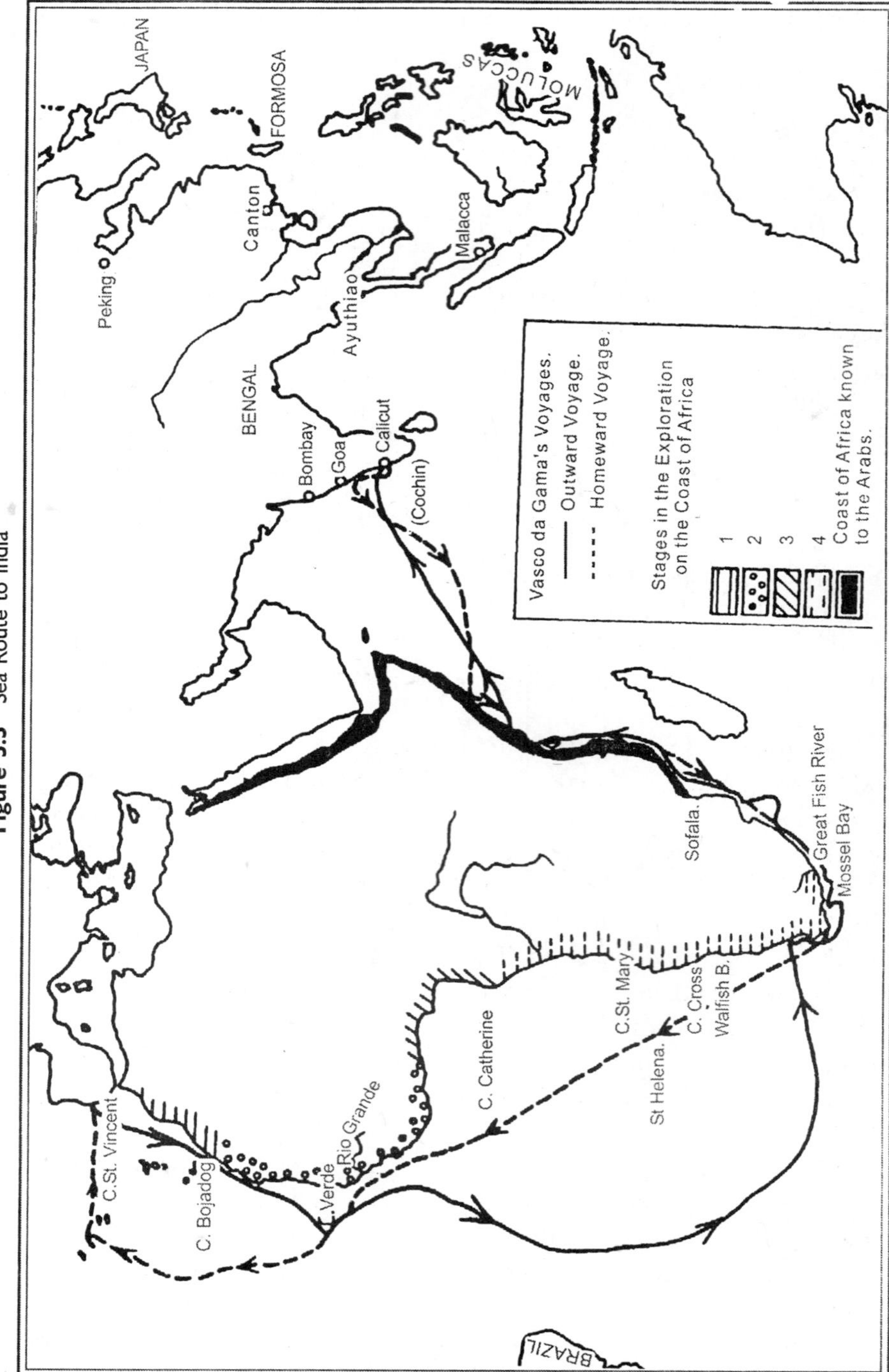

Figure 5.5 Sea Route to India

years; and of the 170 men who started with him, only 44 returned. The king of Portugal was overjoyed at his coming, and Vasco da Gama was received with extraordinary honour and warmth.

On a second voyage, from 1502 to 1503, Vasco da Gama established Portuguese colonies for King John III of Mozambique and Sofala (Mozambique) on the South-East African coast. He was appointed Viceroy of Portuguese Asia in 1524 and died at Cochin (India) on 24 December 1524.

Ferdinand Magellan (1480–1521)

Portuguese navigator Ferdinand Magellan was born at Sabrosa (Portugal) in about 1480. His parents were members of the Portuguese nobility, so after their deaths, when he was only 10 years old Magellan came under the guardianship of the queen. He served in the East Indies under Alfonse de Albuquerque, and took part in the capture of Malacca in 1511. In 1514, he renounced his citizenship and offered his services to the Emperor Charles V of Spain.

Ferdinand Magellan

Magellan and crew consisting of 230 men left Seville (Spain) on 20 September 1519 to seek a western route to the East Indies. The inclement stormy weather, incessant rains and diminished ration added to their miseries. The spirit of mutiny became apparent among the crew. Already the Spanish captain had murdered the Portuguese commander.

In November 1519, they reached the coast of Brazil, but the disloyal captains were not satisfied and one day the captain of one of the ships boarded the flagship and openly insulted Magellan. On the coast of Brazil, a large number of natives were converted to Christianity. He stayed in winter in the Bay of St. Julian in Patagonia, where on 1 April 1520 he crushed mutiny among his men. On 21 October 1520, he discovered the eastern entrance to the long strait now bearing his name (Magellan Strait) and sailed into the Pacific Ocean (Figure 5.6).

With the three little ships battered and worn, manned by crews gaunt and thin and shivering, Magellan took a northerly course to avoid the intense cold. No one had any idea of the vastness of the Pacific Ocean. In the Pacific Ocean, for a period of three months and twenty days, they met with no storm. They could not reach Asia even after voyaging for such a long period. They had little

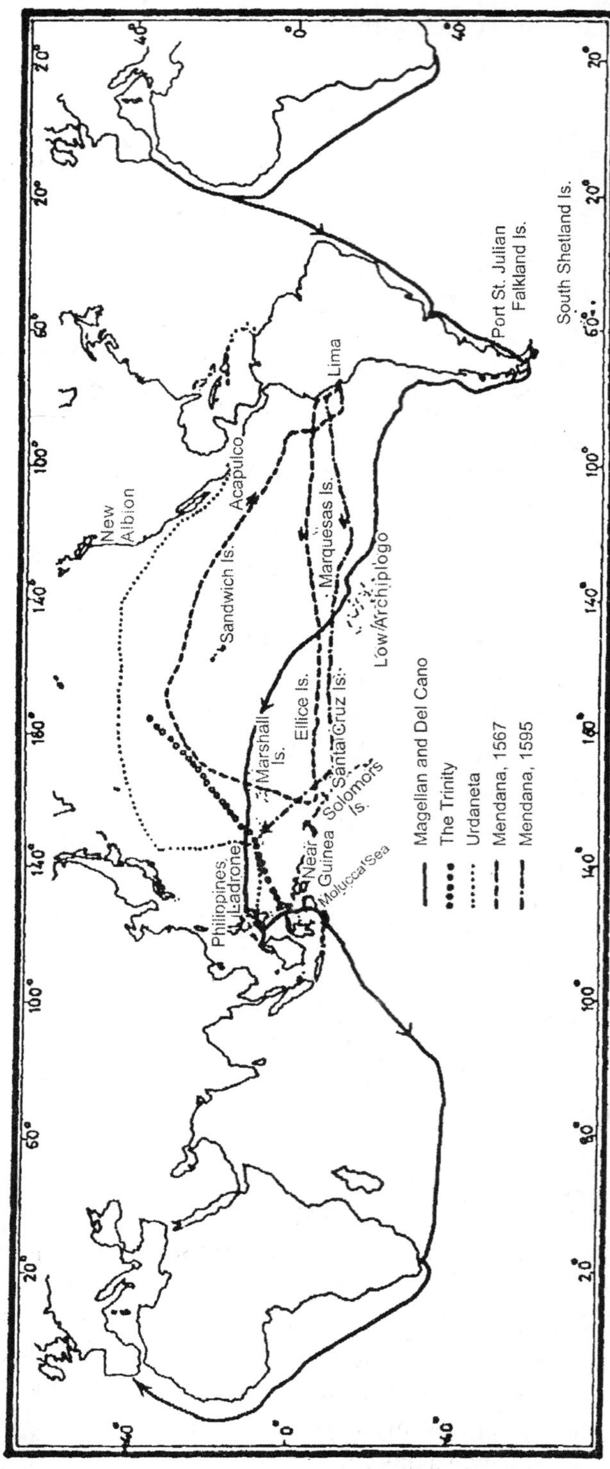

Figure 5.6 Exploration in the Pacific Ocean up to 1600

food left. 'We ate biscuit, but in truth it was biscuit no longer, but a powder full of worms. So great was the want of food that we were forced to eat that hides with which the main yard was covered to prevent the chafing against the rigging. These hides we exposed to the sun first to soften them by putting them overboard for four or five days, after which we put them on the embers and thus ate them. We had also to make use of sawdust for food, and rats became a great delicacy'. No wonder scurvy broke out in its worst form – 19 died and 13 lay too ill to work.

For 98 days they sailed across the unknown sea, till at last they came on a little group of islands peopled with savages – such expert thieves that Magellan called the new islands the 'Islands of Robbers'. The crew gathered fresh food here. The sailors were greatly refreshed, before they sailed. Then, they found the group of islands known as Philippines (after Philop II of Spain). Here, they met the merchants from China, who assured Magellan that the famous Spice Islands were not far off.

With a good supply of fresh food, the sailors grew better. Magellan developed friendship with the native king, and converted them to Christianity. Unfortunately, a quarrel occurred with one of the native kings. Magellan landed with armed men, only to be met by thousands of defiant natives. A desperate fight ensued. The explorer was wounded repeatedly till he was killed on 27 April 1521.

Such was the tragic tale of Ferdinand Magellan – the greatest of the ancient and modern explorers and navigators. Tragic because, after dauntless resolution and unwearied courage, he died in a miserable skirmish on the very eve of victory.

With grief and despair in their hearts, the remaining members of the crew (now only 115) crowded on the Trinidad and the Victoria for the homeward voyage. It was September 1522 when they reached the Spice Islands – the goal of all their hopes. Here, they took on board some precious cloves and birds, spent some pleasant months, and laden with spices, resumed their journey. The Victoria, with 18 men under the command of Juan Sebastian del Cano, completed the circumnavigation of the globe, and returned to Seville in 8 September 1922. In the meantime, Magellan's son had died and his wife Beatrix 'grievously sorrowing' had passed away on hearing the news of her husband's tragic end. Magellan's expedition, however, proved that the globe is round.

Captain James Cook (1728–1779)

In the 16th and 17th centuries, the English, French and Dutch sailors remained busy in finding a route to the East Indies (South-East Asia) either by north-east or by north-west. It is rightly believed that the Spanish and Portuguese sailors had touched on the west coast of Australia in the 16th century, but they gave no information about it except showing certain patches on map calling it Terra-Australia in their early maps. The rivals for sea power in the 17th century were England and Holland (present Netherlands). Both had recently started East India Companies.

Captain James Cook

In the mattes of exploring Australia, Antarctic and Arctic in the 18th and 19th centuries, England and Russia made incredible contributions. Dampier, Torres, Carpenter and Tasman touched different parts of Australia. Dampier returned to England in 1701 from his voyage to New Holland (Australia), but nearly 70 years passed before the English were prepared to send another expedition to investigate further the mysterious land in the south.

The English naval officer and explorer, James Cook was born in a small village near Middlesbrough in Yorkshire on 27 October 1728. He joined the navy in 1755, and from 1759 to 1767, was engaged in surveying the St. Lawrance river and the coasts of Labrador and Newfoundland. The accuracy of charts and observations Cook made of this region brought him to the attention of the Royal Society of London. In 1768, he was appointed to command Endeavour with 70 men to observe a transit of the planet Venus near Tahiti, and to conduct geographical research in the South Pacific Ocean. Having circumnavigated the whole of New Zealand, he opined that Tasman had wrongly assumed New Zealand to be a part of the great southern land (Terra-Incognita). Cook and his companions spent about six months around New Zealand and sailed away to the west and reached New Holland (Australia). He passed along the eastern coast of Australia and gave names to various capes and bays. On 10 May 1768 his ship hit against the Great Coral Barrier Reef. The ship developed a leak and they were threatened with total destruction. The sailors were also showing the signs of scurvy. The prospects of further exploration were not encouraging. Owing to the presence of the Great Barrier Reef, navigation became difficult and Captain Cook himself almost gave up hope. They, however, reached the northern promontory of the land which he named as York Cape. Captain Cook and his friends finally reached Batavia (Djakarta) in safety and were kindly received by the Dutch there (Figure 5.7).

Since leaving Plymouth (England) two years before, Cook had only lost seven men altogether, three by drowning, two frozen, one from consumption, one from poisoning but none from scurvy (a record in the history of navigation). But the climate of Batavia (Djakarta) now wrought havoc among the men. One after another died, and so many were weakened with fever that

Figure 5.7 The Age of Cook-II

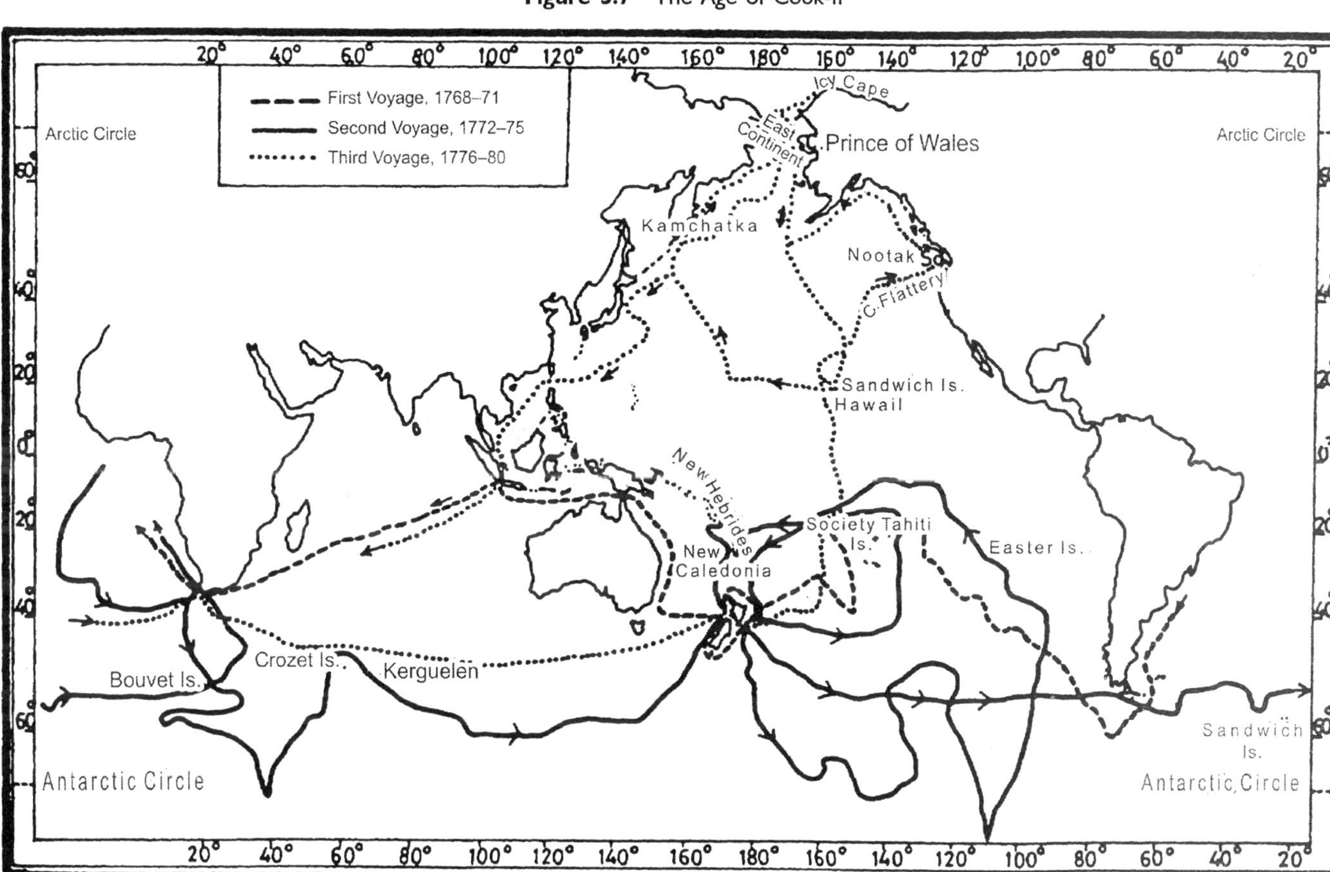

only 20 officers and men were left on duty at one time. Captain Cook and his team returned to London after a lapse of three years.

Captain Cook's next expedition, from 1772 to 1775, was undertaken to determine the extent of a possible Antarctic continent. Cook with the Resolution and Adventures circled the Antarctic region from New Zealand to Cape Horn, and also discovered New Georgia, Solomon Islands, New Caledonia, and Loyalty Islands. The Royal Society awarded him the Copley Medal for his successful measures for preventing scurvy and fever among members of the expedition.

In 1776, as a captain in the navy, he sailed in search of a passage from the North Pacific Ocean, across the American continent to the Atlantic Ocean. He made extensive explorations of the north-west coast of the North America with the Resolution and Discovery, and rediscovered the Sandwich or Hawaiian Islands en route. Upon his return to Hawaii he was stabbed and killed by natives on 14 February 1779 while trying to recover a stolen boat from a local leader.

Varenius (1622–1650)

Bernhard Varen, known as Varenius, was born in 1622 in a village near Hamburg in Germany. He studied philosophy, mathematics and physics at the University of Hamburg.[8] In 1647, he got an assignment as a private tutor at Amsterdam. At that time, Amsterdam was a busy trading centre and the Dutch traders were quite active in distant areas like South-East Asia, Pacific Ocean Islands and Japan. They were keen to learn more and more about the geographical conditions, products, articles of trades, cities, ports, and the socio-cultural conditions of the people with whom they were having trade relations. Varenius published a book in 1649 entitled *Description regni Laponiae et Siam.*[9] This book consists of five parts: (i) a description of Japan; (ii) a translation into Latin of a description of Siam (present Thailand) by Schouten; (iii) an essay on religions of Japan; (iv) some excerpts from the writings by Leo Africanus on religions in Africa; and (v) a short essay on government-dealing with places and people.

Verenius published his book *Geographia Generalis* in 1650. He was the first geographer to divide geography into 'general or universal geography' and 'special geography or particular'. His untimely death, at the age of 28 in 1650, prevented him from completing the second part (special geography) though his work on the geography and history of Japan indicates that he had enough interest in chorography (special geography). The German geographers changed the 'general' and 'special' geography with 'systematic' (*systematisch*) and 'regional' (*landerkunde*) geography respectively.

Varenius made two significant contributions to the development of geography. Firstly, he brought together contemporary knowledge of astronomy and cartography and subjected the different theories of his day to sound critical analysis. Secondly, he divided geography into 'general' and 'special' sections which led to the development of 'systematic' and 'regional' geography. Thus, Varenius was the first scholar who laid the foundation of the dichotomy of systematic vs. regional geography.[10]

The special geography which deals with the special features of different countries and regions may be termed as 'regional geography' while 'general geography' deals with the general laws of the subject. According to Varenius, general geography meant systematic geography. More stress on regional accounts resulted in enormous information and data about the meso and micro areal units. He emphasized the point that general geography depends on regional geography and the regional geography on general geography. Thus, they are interdependent. Moreover, the geographic generalis dealt with the whole world as a unit, but was restricted to the physical conditions, which could be understood through natural laws. On the contrary, special (regional) geography was primarily intended as a description of individual countries and world regions. It was difficult to establish laws in regional geography where people are involved as their behaviour is always unpredictable. The regional geography and its laws, nevertheless, help in the formulation of structured ideas, theories and hypotheses. The regional geography, according to him, also has great utility for government, administration, trade and commerce; while general geography provides the fundamentals which need to be applied in the special geography. In brief, the branching of geography into 'general' and 'special' does not mean that these are opposed to each other, rather they are mutually interdependent parts of a whole.[11]

Verenius further divided general geography into the following parts:

1. Absolute – the terrestrial part, which describes the shape, size of the earth and the physical geography of continents, seas and atmosphere.
2. Relative or planetary part – concerned with earth's relation to other stars, especially the sun and its influence on world climate.
3. The comparative section, which discusses the location of different places in relation to each other and the principles of navigation.

In the field of astronomy, he was in agreement with Copernicus, Kepler and Galileo, who established the heliocentric universe. The geographers prior to Verenius were mostly the followers of Ptolemy who believed in the geocentric concept of the universe, i.e. the earth is the centre of the universe. He was also the first scholar who advocated that the highest temperatures are not recorded in the equatorial belt but along the tropics in the hot deserts of the world.[12]

In the opinion of Verenius geography examines surface features, climate, water bodies, forests, deserts, minerals, animals and the human inhabitants. The cultural landscape, according to him, includes the description of the inhabitants, their appearance, arts, commerce, culture, language, religion, cities and government.

Immanuel Kant (1724–1804)

Immanuel Kant

Immanuel Kant was not only a great scholar of philosophy, he also made immense contribution to the development of natural sciences, especially astronomy, geology and geography. He precipitated the end of Christian theological way of processing geographical and cosmological facts and data and closed the era of Aristotle which had been dominant in Germany since Melanchthon. Further, he freed geography from its tight bonds with theology. Since Kant's geographical work has been handed down to posterity in a fragmentary state, his importance was hardly recognized for a considerable time. The latest research has revealed that the history of geography should be divided into two periods – before Kant and after Kant.

Kant was born on 22 April 1724 at Konigsberg, Prussia (Kaliningrad) in East Prussia (Russia), as the fourth child of a saddler and harness-maker of modest circumstances. His parents brought him up in strict adherence to the pietistic creed which was to prevail in Konigsberg until the Russian occupation in 1758. In his entire life, he never travelled more than ten miles (16 km) from Konigsberg.

He received his early education in the suburban hospital school. During his school days, Kant was so enthusiastic about Latin that he made up his mind to study classical philology. In later years, he was still renowned as a master of Latin. Kant was a solid, albeit unspectacular, student. Despite his upbringing in a religious household and still maintaining a belief in God, he was skeptical of religion in later life. Various commentators have labelled him as an agnostic. It is often held that Kant lived a very strict and predictable life, leading to oft-repeated story that neighbours would set their clocks by his daily walks. He never married, but was having strong emotional attachment to students and men.

In 1740, Kant matriculated in the university of his native city (Konigsberg). He was mainly interested in philosophy, mathematics and natural sciences. After finishing his studies at the University of Konigsberg, he spent several years (1746 to 1755) with some well-to-do families near Konigsberg as a private tutor. He was not firmly convinced of his qualifications for tutoring, but he used that period for further studies and eagerly abstracted books – an occupation which laid the foundation of his later fame as a man of wide reading.

On 12 June 1755, he was awarded Ph.D. on his thesis, *De Igne* (On Fire), which was highly praised by his teacher Teske. He became a lecturer at the age of 31. In order to qualify for as many professorial chairs of the faculty of Konigsberg philosophical faculty as possible and so to supplement the scanty fees from his lectures, he lectured with formidable frequency not only in the classical fields of philosophy, physics (natural science), mechanical science (mechanics, hydrostatics, hydrokinetics and aerometrics), but also in military engineering, such as fortification and pyrotechnics and finally mathematics. When, in later days, he concentrated on philosophy proper, he still lectured on physical geography. He also lectured on anthropology. Finally, he became a professor in 1770 at the age of 46 years. Earlier, he refused the professorship in the Universities of Erlangen and Jena as he was not willing to leave the city of Konigsberg. In 1780, he was appointed to the board of governors at the university and in 1786 he was elected rector (Vice Chancellor) of the university.

A scholar following an unassuming way of life, Kant always looked after the interest and well-being of his students beyond his delegation as teacher. He did everything to develop their personalities, and at any time he would assist them readily with advice, very often also with financial and practical helps.

Despite his moderate physical health, Kant was hardly ever really ill, and he was cruelly hit by the rapid physical and mental deterioration that beset him during the final six years of his life. He delivered his last lecture in 1796, and soon afterwards resigned his seat on the university's board of governors. He died on 12 February 1804. His unfinished final work, the fragmentary *Opus Postumum*, was, as its title suggests, published posthumously. Kant's mortal remains were entombed in 1880 in a plain neo-gothic Chapel near the Cathedral. On the occasion of his 200th birth anniversary, his tomb was replaced by a classical portico.

Kant gathered the data for his geographical lectures from many sources. In comparison to his abundant philosophical publications and also his tireless activities as a lecturer, the geographical written material he left to posterity appears to be scanty and meagre. His early work *General Natural History and Theory of the Heavens*, (1755) which often has been adopted for geographical work is a contribution to the speculative astronomy. Even his treatises on earthquakes (which deal mostly with the earth's interior) do not belong to geography proper. There are, however, geographical references in his works *Critique of Pure Reason* (1781), and *Critique of Judgement* (1790). He also wrote on the different races of man. His essay on 'Anthropology from Pragmatic Point of View' (1798) contains many geographical accounts of the races and ethnic groups of the world.

Kant was mainly interested in the physical geography. The physical geography of Kant included other geographies too including political, commercial, theological and regional geography. Kant and Humboldt, in their physical geography, included races of man and their physical works on the earth. Thus, Kant not only included man as one of the features encompassed in the earth surface but also considered man as one of the five principal agents affecting changes on the earth. Indeed, Kant's physical geography, both in purpose and in content, might be considered as 'anthropocentric' – a point of view which Ritter inherited from Kant. He used the term 'nature' to mean the objective world outside of the observer's mind. The physical geography of Kant was not as we now understand it, because it also discussed human racial groups, their physical activities on the earth and natural conditions in the widest sense of the term. Kant regarded geography as only an approach to the empirical knowledge that was necessary for his philosophical research. He was one of those scholars who believed that geography has to play a vital role in the progress and development of human society. His interest in physical geography arose through his philosophical investigation of the whole field of empirical knowledge. In fact, being a philosopher and thinker, he was a theoretical geographer who tried to classify empirical phenomena.

He pointed out that there are two different ways of grouping or classifying empirical phenomena for the purpose of studying them systematically. The phenomena may be classified either in accordance with nature, or in accordance with their position in time and space. The former is a logical classification and the latter a physical one. Logical classification lays the foundation for systematic science – the study of animals in zoology, the study of plants in botany, the study of rocks in geology, and the study of socio-cultural and ethnic groups in sociology. The physical classification gives the scientific basis for history and geography. History studies the phenomena which follow one after the other in time (chronological science), while geography studies phenomena which lie side by side in space (chorological science). History and geography are both essential sciences, standing alongside the systematic sciences. Without them man cannot achieve a full understanding of the world.

About the concept of space, he asserted that 'space' is a relative view which consists of system of relations among substance, and 'spatial magnitude' is therefore only a measure of intensity of acting forces exerted by the substance. In 1770, Kant said 'space' is not a thing or event. It is a kind of framework of things and events: something like a system of pigeon holes, or a filing system, for observation. Both space and time may thus be described 'as a force of reference' which is not based upon experience but intuitively used in experience and properly applicable to experience.

In the *General History of Nature of the Heavens* (1755), Kant laid out the Nebular Hypothesis, in which he deduced that the Solar System had formed from a large cloud of gas, a nebula. Kant also correctly deduced that the Milky Way was a large disc of stars, which he theorized had also formed from a (much larger) spinning cloud of gas. He further suggested the possibility that other nebulae might also be similarly large and distant discs of stars. These postulations opened new horizons of astronomy, for the first time extending astronomy beyond the solar system to galactic and extragalactic realms.

About knowledge, Kant's views were that it may be obtained either by the exercise of pure reason or through senses. Sene perception are of two types:

those perceived by the inner and those by the outer senses, and together they furnish the whole of man's empirical knowledge of the world. The world as perceived by the inner senses is *seels* (soul) or *mensch* (man), i.e. the self as perceived by the outer senses in nature. According to him, anthropology studies the soul of man; physical geography is thus the first part of knowledge of the world.

Since every person's experience is limited, both in time and space, each one must supplement his personal experience with that of others, always taking care to examine thoroughly the reliability of what is observed. Such borrowed indirect experiences are of two kinds – either narrative or descriptive. The first (narrative) is history, and the second (descriptive) is geography. Thus, empirical knowledge can also be classified in time and space. Study of phenomena in terms of space is geography and in time history. In this connection, Kant asserted:

> History differs from geography only in the consideration of time and space. The former is a report of phenomena that follow one another (*machemiander*) and has reference to time. The latter is report of phenomena besides each other (*nebeneinander*) in space. History is narrative, geography a description.
>
> Geography and history fill up the entire circumference of our perceptions: geography that of space, history that of time. (Hartshorne, 1939)

Kant also posed the question whether geography or history was first. He resolved that geography has existed at all periods and is the substructure of history; thus if there is an ancient history, so there must also be an ancient geography which helps to make clear the events of history. Physical geography is then, a general outline of nature, and constitutes not only the basis of history but also the foundation of all possible geographies.

Although Kant's views on geography were broadly similar to those of Von Humboldt and Hettner, they appear to have had 'no direct influence' other than 'as a form of confirmation' (Hartshorne, 1958; Buttner 1980). Indeed, they were not explicitly endorsed in any major programmatic statement of the scope of geography (in English) until Hartshorne's account of the *Nature of Geography* (1939), which accepted that the geography's basic task was essentially Kantian:

> Geography and history are alike in that they are integrating sciences concerned with studying the world. There is, therefore, a universal and mutual relation between them even though their bases of integration are in a sense opposite – geography in terms of earth spaces, history in terms of periods of time. (Hartshorne, 1939)

For Kant, geography was a *descriptive* or *taxonomic* discipline, rather than a science; it only had the status of a propaedeutic discipline. Kant used the word 'chorographic' meaning descriptive to describe geography. Hettner (1927) transformed this into 'chorologic' which refers to explanation rather than description.

According to his *Ankundingung* (1757), the earth can be studied and interpreted in the following five different ways.

1. *The Mathematical Study of the Earth*: This is an important aspect of geography inasmuch as it sees earth as a nearly spherical celestial body. It considers its size and shape and all imaginary circles that should be applied on its surface.
2. *Moral Geography*: It is also an important branch of geography which deals with the customs, traditions, rituals and character of man. For example, it examines the contrast of oriental civilization where patricide is most fearful one, with customs in Lappland where a father, if wounded while hunting, expects his son to kill him. Kant noted that 'in the mountains, men are actively and continuously bold, lovers of freedom and their homeland'.
3. *Political Geography*: In political geography, the consequences of interrelationship between nature and man, the condition of the nations and the peoples on the earth is evaluated in two ways, which interpenetrate. The first way is to see how the human condition developed through incidental causes, such as changes in government, annexation of territory and political intrigue. All this may be ephemeral but the second way deals with more permanent features, such as the position and situation of countries, their products, trade, customs and population. Both ways in their inevitable relationship contribute to the whole picture.
4. *Commercial Geography*: This branch of geography, according to Kant, examines the reasons why certain countries have a superfluity of one commodity while others have a deficiency – a condition that gives rise to international trade.
5. *Theological Geography*: It studies the changes theological problems undergo in different environments. Kant based the social branches of geography on their relations to physical geography: 'theological principles in many cases undergo important changes as a result of differences in the land' (Boden). For example, one would make a comparison of Christianity in the Far East with that in Europe and the variation of Christian beliefs in different parts of Europe.

About physical geography, he opined that it merely evaluates the natural composition of the terrestrial globe, and everything on it, i.e. the oceans, the firm ground, the mountains, the rivers, the atmosphere, man and animal, plants and minerals. Portrayal by cartographic means is unimportant; according to Kant, even though such maps exist they should be used merely for pedagogical purposes and for demonstration.

Thus, Kant was keenly interested in geography and described it as a scientific system of great educational value. He asserted that geography provides knowledge of men and the world, which is of great use for private and public conversation, to be of great interest to readers of newspapers and also politicians. Moreover, he freed geography from its tight bonds with theology. He was the first who provided an early statement of geography as *chorology*.

Thus, during the time of Kant, geographers started writing on mathematical, moral, political, commercial and theological geography. Branching of geography, specially into the physical and human geography, was started during this period. Kant saw man and his works in intimate association with the physical surroundings, and he also recognized human action as one of the main agencies of change on the face of the earth. But he made no distinction between human and natural processes. The period of Varenius and Kant is also known as the 'Classical Period of Modern Geographical Thought'.

Shifting Viewpoints in the Second Half of the 19th Century

There had been changing thrust in the nature of geography throughout its history. Since the term 'geography' means, and has meant, different things to different people in different times and places, there is no agreed upon consensus on what constitutes the nature and scope of geography. Prior to the period of Varenius and Kant, geography was *geo+graphy*, largely descriptive in character. Varenius divided geography into general (systematic/universal) and particular (special or regional) geography. The period of these scholars is often been called as the classical period of modern geography. Humboldt and Ritter are credited as the founders of modern geography.

Humboldt, who had an exceptional quality of sharp observation and travelled extensively in Europe, Asia and Americas, considered geography as a systematic discipline and also tried to develop universal laws and theories in the field of physical geography. On his return from the expedition of South America, he established the science of physical geography. Though Humboldt, who held no university post, had no immediate followers in academic ranks, his influence outside of Germany was vastly greater than that of Ritter. During the brief post-Ritterian period 'the real representatives of the true geographical science were the scientific travellers who took Humboldt for their model'.

Ritter, a teleologist, produced a galaxy of geographers who emphasized the 'historical' aspect and drifted from the systematic to the regional geography and the primary concern with man. Ernst Kapp, a prominent student of Ritter, developed special interest in political problems. Another follower of Ritter, Arnold Guyot, who held the first chair of geography in the United States at Princeton, was also a teleologist who described and interpreted the regional patterns on the lines of Ritter with a focus on regional geography.

The most successful student of Ritter was the French geographer Elisée Reclus, who derived his main principles and ideas about geography from Erdkunde. In the beginning he focused on systematic geography and then attempted a complete regional survey of the world. Reclus travelled extensively in North and South Americas to observe the landscapes. Reclus, who was a social anarchist, wrote systematic physical geography called *La Terre* (1866–67). He is well recognized for his 19 volumes on regional geography – *Nouvelle Geographie Universelle* (1875–94). His works give a historical account of humanity's life on earth and its resources. In other words,

Reclus was mainly interested in the human aspect of geography. It was because of his contribution to regional geography that Schmidt declared Reclus as the 'Ritter of France'.

Some of Ritter's students like Moltkeserved, who served at the military college, discussed the importance of geography in military science.

The foundation, which Humboldt and Ritter had established for geography, did not provide a clearly unified field. The followers of these stalwarts split the subject of geography in several directions and its position as a branch of knowledge was thereby brought into serious question. Following the death of Ritter there was no professor of geography in any German university and the return to university status and particularly the rapid subsequent growth was largely the work not of 'historical geographers' who followed Ritter, but of students who had been trained as geologists and tended to specialize in the study of non-human features of the earth, i.e. physical geography. With the rise in academic status of geography and the productive work of this period, the major problem of geographers was to overcome the apparent disunity in the methodology of the field and also establish its position as a single field of science to incorporate the physical and cultural features.

With the death of Humboldt and Ritter in 1859, and the publication of Darwin's classic work on the *Origin of Species*, the general scientific work was against their approaches to geography. It was at this time when Bucher attacked on the demarcation of natural boundaries and natural regions. The teleological approach of Ritter was rejected. The German geographers of that period were influenced by the approach of environmental determinism. Ratzel, in his book, *Anthropogeographie* highlighted the point that the history, culture and gamut of life are largely influenced by the physical environmental forces. He went to the extent of saying that 'similar locations lead to similar mode of life'. Darwin, Ratzel's concept of 'lebensraum' (living space) brought a revolutionary change in the political thinking of Europeans, especially that of Germans.

It was the impact of Darwin that the study of landforms (geomorphology) became the main domain of geographic investigation. It may be said as the 'geologification' period of geography. William Morris Davis developed the concept 'geographic cycle' (cycle of erosion). He developed an analogy between organic life and evolution of landforms. He advocated that the 'landforms evolve like the evolution of organic life'.

Miss Semple, one of the leading students of Ratzel, who was also influenced by environmental determinism declared in her book *Influences of Geographic Environment* (1911) that 'man is the product of the earth's surface'.

As a reaction to extreme generalization of environmental determinists, there developed the school of possibilism. The possibilists took man as an active agent in environment. They opined that physical environment provides options, the number of which increases as the knowledge and technology of a cultural group develops. They also asserted that nature is never more than an advisor. Febvre declared that 'there are no necessities but everywhere possibilities'. It was at this time when Vidal de la Blache developed the concept of *genre de vie* (lifestyles).

Subsequently, the shift in geographic work is generally regarded by German geographers primarily due to the work of Paschel and Richthofen. Paschel led geographers to study primarily the morphology of landforms. In his study, he also attempted to study the influence of landforms on human history. Richthofen, in his studies of China, published in 1877, also focused on the study of landforms. Penck, a trained geologist, emphasized on the importance of the study of landforms and declared it as the major concern of geographers.

The dichotomy of physical geography versus human geography was also a development of the second half of the 19th century. While Wegener declared geography as a pure natural science, the followers of French school of possibilism announced human geography as the thrust area of geography.

It may thus be summarized that there had been several shifting viewpoints in the growth and development of geography in the second half of the 19th century.

Notes

1. Wright, J.K., 1925, 'The Geographical Lore at the Time of Crusades', New York, American Geographical Society, *Research Series*, No.15.
2. James, P.E. and Martin, G.J., 1981, *All Possible Worlds*, pp. 60–65.
3. Ibid., p. 81.
4. Oliveira Martins, J.P., *The Golden Age of Prince Henry the Navigator*, Trans. J.J. and W.E. Reynolds, 1914, London, Chapman and Hall (Portuguese Title, Os Filhos de D Joao I, Lisbon, 1901).
5. Tooley, R.V., 1949, *Maps and Map Makers*, New York.
6. Bargrow, L. and Skelton, R.A., 1964, 'History of Cartography', *Research Series*, No. 8.
7. Tooley, R.V., op. cit., p. 61.
8. Dickinson, R.E. and Howarth, O.J.R., 1933, *The Making of Geography*, Oxford, Clarendon Press.
9. James, P.E. and Martin, G.J., op. cit., p. 96.
10. Ibid., p. 97.
11. Jensen, A.H., 1981, *Geography: Its History and Concepts*, p. 13.
12. In *Ankundingung* (1957) briefly enumerated the most important ones: Varenius, Lulofs, and Buffon, as well as the travel reports and the academic reference material from Paris and Stockholm.

Part II

Modern Geographical Thought

6

Founders of Modern Geographical Thought

After the Great Age of Discovery, two leading German scholars, viz., Alexander von Humboldt and Carl Ritter made valuable contributions to the fields of basic sciences, humanities and arts. The foundation of geography as a modern science was primarily laid by German scholars during the period from 1750 to 1850. The second half of this period, the time of Humboldt and Ritter, is known as the 'classical period of geography'. They lived at the same time in the same country – for over 30 years in the same city. Humboldt laid the foundation of plant geography and declared geography as a descriptive science, while Ritter is credited with introducing mankind into geographical studies – particularly humanity in relation to environment. He also opined that 'geography is not concerned with the individual plants but rather with the plant and animal cover'. He was the first to develop the concept of harmony in the interrelation of regional phenomena. He repeatedly emphasized the concept of the 'naturaganzen', i.e. unity of phenomena of many different categories. Both were contemporaries, and lived and worked at Berlin for more than three decades. These two scholars are regarded as the founders of modern geography although neither was trained as a geographer. In this chapter, a brief account of the contribution made by these scholars is given.

Alexander von Humboldt (1769–1859)

Alexander von Humboldt led the way in the expansion of geography in and outside of Germany. He was a scholar of great versatility, who contributed appreciably to the fields of geology, botany, zoology, physics, chemistry, anatomy, physiology, history, climatology, geomorphology and to all other branches of geography. He travelled about 4,000 miles (6,667 km) and in all his travels, however short, he made multitudinous observations. He performed all journeys with telescopes, sextants, cynometers (for measuring blueness of the sky), and barometers. With the help of these instruments, he measured accurately the temperature of air and ground, pressure, winds, latitudes, longitudes, elevations above the sea level, magnetic vibrations, nature of rocks, types of plants and their relations to climate, latitude, altitude and human attitude.

Alexander von Humboldt

Humboldt was born in an aristocratic family in Prussia. His father expired in 1779 when he was only ten years of age. After getting education in classical languages, economics, finance, history, technology and mathematics, Humboldt started his career in the army, but his mother prevailed upon him to study economics and to compete in the civil service examination. Later on he studied at Frankfurt in the University of Gottingen, where he studied botany, geology and mineralogy. He was taught by A.G. Werner – the famous geologist – who put forward the hypothesis that all the sedimentary rocks of the earth had been formed by precipitation under water and had been deposited in layers during the great flood. His interest in geography started with his acquaintance with George Forster – who was on Cook's second voyage around the world. Humboldt also attended lectures in physics, chemistry and mining. In the summer of 1790 he paid a short visit to England in the company of Forster. In 1772 and 1797 he was in Vienna. In 1795 he made a geological and botanical tour through Switzerland and Italy. The death of his mother, on November 1796 (when he was 27 years of age) set him free to follow the bent of his genius. In 1792, he was appointed Director of Mines in Prussia. He studied the effect of different rocks on magnetic declination and published his first paper in 1793. He became keenly interested in the rock structure of the Alps and visited Bavaria, Austria, Switzerland and Italy. In 1797, he resigned from the government job and planned his journeys to the new and unexplored lands. In Paris, he learned the art of handling various instruments of measurement like sextant, barometers and aneroid barometers.[1] He is frequently presented as a pioneer without predecessor, a lone figure who explored Latin America and returned to establish the science of physical geography.

Adventures and Explorations (Latin American Expedition)

Humboldt had a gift for exceptionally sharp observation and in fieldwork he was unsurpassed. In 1798, Humboldt along with a French botanist (A.

Bonpland), reached Madrid (Spain). On his way to Madrid, he made daily observations of temperatures and altitudes. He was the first to make an accurate measurement of the elevation of the Spanish Maseta.

In their expedition to Latin America (South America) they got the support from the King of Spain. They sailed on 5 June 1799. From Madrid, Humboldt and Bonpland reached Cumana Port in Venezuela (Figure 6.1). Along the sea coast they went to Caracas and explored the Valencia Lake. He noticed that the Valencia lake had shrunk and fields for cultivation of crops had been developed on its flat banks. He attributed the shrinkage of the lake to deforestation of the neighbouring lands. He established a positive relationship between the forests and rainfall. The idea that more forests mean more rainfall still significantly persists.[2]

In 1800, Humboldt's expedition explored the Orinoco river and its tributaries and established the truth of its connection with Amazon. The banks and basins of this river were uninhabited. During the venture, Alexander and his colleagues suffered many hardships. They had only banana, wild fruits and fish to eat, and were exposed to the bites of clouds of mosquitoes, ants, equatorial insects, man-eating fish and crocodiles. Even under these adverse conditions, Humboldt did measurements and established exact latitudes and longitudes of places. From this virgin region, he collected thousands of plant and rock specimens, which were transported to Cuba. In November 1800, they returned to Cumana (Cuba) and studied the economy and society of the people of Cuba.

In 1801, Humboldt and Bonpland arrived at Cartgona (Colombia) and from this port they went to Andes, Ecuador and Peru (Figure 6.1). Humboldt gave a scientific explanation of crops and the influence of altitude, temperature and vegetation on crops. His description of the vertical zones of the Northern Andes is a classic. He also examined the numerous volcanoes of Ecuador and descended in the craters of active volcanoes to collect gases emanating from within the earth. Moreover, Humboldt climbed Mountain Chimborazo Peak (6,267 m.) a world record at the time. He also visited Bogota, Quito, Lima and Callo in the Andes Mountains (Figure 6.1) and observed the influence of altitude on human body. It was Humboldt who explained the feeling of dizziness as resulting from low air pressure.[3] In fact, this disease is due to scarcity of oxygen at high altitude.

Traversing the Andes southward, the explorers reached Lima. On the coast of Peru, guana bird droppings were observed which have great manural value. Moreover, the cold water current of Peru was also observed and recorded for the first time. The temperature and velocity of this current were measured. This cold water current was subsequently named as Humboldt current. In March 1803, the expedition sailed from Guayaquil to the Mexico Port (Acapulco). He travelled in the different parts of Mexico and observed the impact of landforms on the cultural landscape. Staying for a short period at

Philadelphia and the White House Washington DC, Humboldt returned to Europe on 3 August 1904.

The adventurous nature of Humboldt did not permit him to stay at one place and therefore in 1806 he made a visit to Vesuvius volcano (Italy). After its completion he wrote his experiences and observations in 30 volumes in French, which were subsequently translated into a number of foreign languages. This encouraged many young scientists to investigate the geography of unexplored areas of the world. In his writings, Humboldt attributed the cause of prosperity of the inhabitants of Mexico to their better utilization of land resources. The idea of digging a canal across the Isthmus was also put forward by him.[4] In 1827, Humboldt finally shifted from Paris to Berlin.

In 1829, Humboldt was invited by the Russian Czar to visit the city of St. Petersburg (Leningrad) where he was entrusted with the task of exploring the virgin lands of Siberia across the Ural Mountains. From Petersburg, Humboldt, on a horse back, travelled through Voronezh, Kazan, Bogosloski, Tobolsk, Tara, Bersk, Omsk and reached up to the border of Mongolia (Figure 6.1). While returning he passed through Omsk, Orenburg and Astrakhan and made a survey of the coastal lowlands of the Caspian Sea. Throughout the Siberian expedition, a regular record of temperature and pressure was kept. On the basis of these observations, it was inferred that temperature on the same latitudes varies moving inward from the coast. It was on his advice that a number of meteorological stations were established by Czar in different parts of Russia. It was after this expedition that a world map showing isotherms was prepared for the first time. The concept of *continentality* was also established by Humboldt. Moreover, the term 'permafrost' was coined to explain the frozen characteristics of the Siberian soils. It was after this venture that the word 'climatology' appeared in geographical literature which deals with variations of atmosphere, temperature, humidity, barometric pressure, winds, atmospheric purity and the degree of visibility.[5] Thus, he contributed richly to a variety of fields. His major concern was, however, to correlate the physical environment with the human and organic phenomena. While dealing with man-nature interaction, he included man and his works but did not give adequate weight to man as the major determinant. Nevertheless, he portrayed towns, villages, fields, crops and transport linkages as elements of landscape. He also developed the concept of mountain sickness or soroches.

Looking closely at the rocks of the Andes, he decided that A.G. Werner was quite wrong about the origin of rocks and that granites and gneisses and other crystalline rocks were of volcanic origin. Humboldt made comparative study of Spain and Cuba. He attempted to develop a general picture on the distribution of average temperatures in the world in relation to the distribution of continents and oceans.

Alexander von Humboldt thought an approach to science was needed that could account for the harmony of nature among the diversity of physical world. For Humboldt 'the unity of nature' meant that it was the interrelation

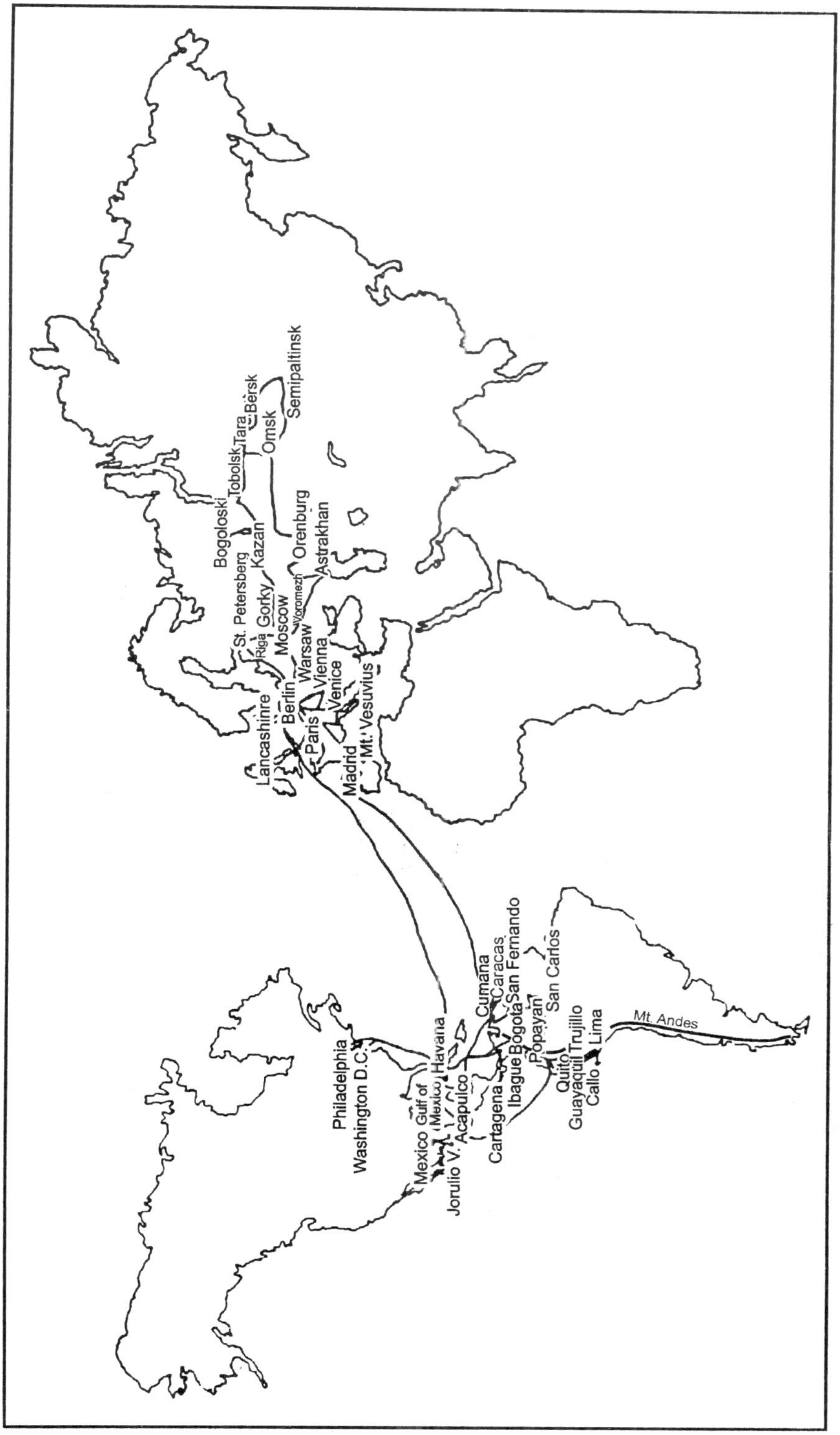

Figure 6.1 Humboldt's Travels in Europe, Russia and Americas

of all physical sciences – such as the conjoining between biology, meteorology, and geology. He viewed nature holistically, and tried to explain natural phenomena without the appeal to religious dogma.

In 1845, Humboldt's monumental work *Kosmos* was published, and was well received all over the world. It was later on translated in a number of foreign languages. *Kosmos*, a comprehensive account of the travels and expeditions of Humboldt, was written with the following four objectives: (i) the first is the definition and limitation of physical description of the world as a special and separate branch of knowledge; (ii) the second is the objective content, which is the actual and empirical aspect of nature's entity in the scientific form of a portrait of nature; (iii) the action of nature on the imaginative faculty and emotion becomes an incentive to nature studies through the means of travel, description, poetry, landscape, painting and display of contrasting groups of exotic plants; and (iv) the history of natural philosophy and the gradual concept pertaining to cosmos as an organic unit are dealt with. In brief, Humboldt's objective in writing *Kosmos* was to develop a universal science. Religions, he insisted, offer three different things to mankind: a lofty moral idealism, which is common to all religions; a geological dream regarding the origin of the earth; and a legend concerning the origins of the religion. He never mentioned the word 'God' in his writings.

He planned to write a series of books about the world with the following objective: Give a scientifically accurate picture of the structure of the universe. He saw nature as a whole and man as a part of nature. Humboldt believed that all the races of man had a common origin and that no race is necessarily inferior to the others. For the formulation, he thought that observation had to come first as he was believer of inductive method.

So far as the subject matter of *Kosmos* is concerned, in the first volume there is a general presentation of the whole picture of the universe. The second volume starts with a discussion of the portrayal of nature through the ages by landscape painters and then continues with a history of man's effort to discover and describe the earth since the time of ancient Egyptians. The third volume deals with the laws of celestial space which we would call astronomy. The fourth volume concerns with the earth, in which Humboldt considered man as a part of the earth.

While dealing with the subject matter of geography, Humboldt coined the term 'cosmography' and divided it into uranography and geography. In his opinion, uranography is the descriptive astronomy which deals with celestial bodies. Geography, on the other hand, was confined to physical geography which deals with the terrestrial part. Geography, according to him, is the description of the earth which deals with the interrelationship of phenomena that exist together in an area.[6] He was a pioneer of physical geography.

He considered nature as an organic whole born out of harmonious interrelationship between all living and non-living objects existing together in particular territories. He believed in 'unified universal science' encompassing of all physical, biological and social sciences. The foundation of universal science was the main objective of his *Kosmos*. He focused on geography as the discipline concerned with both inorganic as well as organic phenomena on the earth's surface as an interrelated entity. He advocated the concept of *zusammenhang* (hanging together). He was the pioneer of the concept that 'man everywhere becomes most essentially associated with terrestrial life'.

Humboldt believed in the inductive method and emphasized the importance of empirical method of research. He also made comparative study of the different geographical regions, especially that of the steppes and the deserts. He gave importance to geographical representation of data on maps and the utility of maps for geographical studies. He believed in the unity of nature and accepted the idea of inherent causality (causal connection).

Humboldt believed that all the races of man had a common origin and that no race is superior or inferior to others. Moreover, he stressed the need of casual observations of nature in the field and of careful measurement of observations. It was an approach towards theory-building and model-making.

In brief, Humboldt sought answers to a great variety of specific questions. For example, he attempted to develop a general picture of the distribution of average temperatures in the world in relation to the distribution of the continents and oceans. The influence of altitude in the tropical areas on plants, animals and human life was also studied by him. Humboldt concentrated largely, though not exclusively, on physical features, climate and vegetation. About the use of 'natural', he was the follower of Kant. The word 'natural' in its broadest sense included all the phenomena observed outside the observer's mind or the objective reality. It is because of these contributions that he is considered not only the founder of plant geography and climatology but also modern geography. Humboldt thus was the last master of universal science.

Humboldt made substantive contribution to systematic geography (Cosmos) and laid stress on the unity of nature. Ritter appreciated the classic work (*Kosmos*) of Humboldt, but 'being an agnostic, he (Humboldt) did not write a word of praise for the Creator (God)'.

Much is not known about Humboldt's private life which remains a mystery because he destroyed his private letters. He never married, but and enjoyed all the earthly pleasures according to one of his travelling companion (Francisco Jose Caldas), Humboldt formed emotional attachment to men. He accused him of frequenting where 'impure love reigned'. On 24 February 1857, Humboldt suffered a minor stroke. On 6 May 1859 he died quietly in Berlin at the age of 89.

Carl Ritter (1779–1859)

One of the contemporaries of Alexander von Humboldt and a scholar of diversified interests was Carl Ritter. He is also known as one of the founders of modern geographical thought. He was a dedicated fieldworker and believed in empirical research. Moreover, Ritter was a teleologist and had a strong belief in God and was not an agnostic like Humboldt.

Carl Ritter

Ritter had a vision of an ordered and harmonious universe. Thus, his approach was teleological.[7] As a teacher, he made it clear to his pupils how God's plan was revealed in the harmony of man and nature.

Ritter was born on 7 August 1779 in Berlin. His father was a physician who died when Ritter was only five years old. After getting his early education in a school at Schnepfenthal near Gotha through non-formal methods (advocated by Rousseau), he was taught by G. Salzamann and Guts Muths. At the university level, he opted for Greek and Latin and read history and geography widely. With his pupils Ritter made frequent trips around the city of Frankfurt and created in them keen interest for field studies. Later on, he reached Switzerland and Italy to make on the spot study of their physical and cultural landscapes. It was in 1807 that Ritter met Humboldt for the first time. Ritter was highly impressed by his versatility in observing natural and human phenomena in different parts of the world. In 1814, Ritter joined the University of Gottingen and studied geography, history, pedagogy, physics, chemistry, mineralogy and botany. In 1818, Ritter was appointed as Professor of History at the University of Frankfurt. After one year as Professor of History and Geography at the gymnasium in Frankfurt, Ritter was given the double position in the military college and the University of Frankfurt. Subsequently, when the first chair of geography was created in Germany, he was made as the first Professor of Geography in 1820 at the University of Berlin. He founded the Berlin Geographical Society. He served the department of geography of Berlin University for 39 years. In 1859, he breathed his last – the year in which Humboldt also expired and Darwin published his *Origin of Species*.

Ritter, in his class lectures, emphasized the point that geography is not a dry gazetteer of names of places, rivers, mountains, and trade routes. It is a subject of great importance which deals with man-nature interrelationship. He also developed the concept of 'unity in diversity'.[8] Ritter was chiefly concerned with studies of human geography. He believed, as did Vidal de la Blache much later, that earth and its inhabitants stand in the closest reciprocal relations, and one cannot be truly presented in all its relationships without the other. Hence history and geography must always remain inseparable. Land affects the inhabitants and the inhabitants affect the land.

Ritter declared geography to be *'Erdkunde'* or an earth science, which deals with local conditions and embraces the attributes of place with respect to topical, formal and material characteristics. The first attribute was topographical, i.e. it deals with natural divisions of the earth surface. The second included the distribution and movement of water, sea atmosphere, and the base of human life. The material conditions were described as the geographical aspect of natural history; this covered the distribution of minerals, plants and animals.[9]

As stated above, the scientific stance of Ritter was teleological (Greek *teleos* = purpose). Teleology seeks to understand events in relation to their underlying purposes. Teleological explanations are therefore often regarded as the opposite of mechanical (scientific) explanations, where the phenomena and observations are understood as outcomes of prime causes such as the 'laws of nature'. In the first volume of *Kosmos* (1845), Humboldt speaks of 'Ritter's great and inspired work'. He wrote of his *Erdkunde*, 'is to proceed from observation to observation, not from hypothesis to observation'. About the relationship of various facts of the earth, Ritter opined: 'We must ask the earth itself for its laws'. He was influenced by W.F. Hegel (1770–1831). He studied the working of nature in order to understand the purpose behind its order. His view of science sprang from his firm belief in God as the planner of the universe. He did not regard the shape of continents as accidental but rather as determined by God, so that their form and location enabled them to play the role designed by God for the development of human kind. Ritter regarded 'the earth as a whole as an 'organism' and the continents as 'individuals' or as 'organs' '. Ritter, a teleologist, was the first who made a major effort in modern times to divide the earth surface on universal consideration. Though his teleological approach was rejected as it was not scientific.

Ritter was also the founder of comparative method in regional geography.

Thus, according to Ritter, geography is that branch of science which deals with the globe in all its features, phenomena and relations as an independent unit and shows the connection of this unified 'whole' with man and with man's creator.[10] He claimed that the central principle of geography is 'the relation of all phenomena and forms of nature to the human race'. He makes the rhetorical claim that geography as the science of earth (in the sense of natural philosophy of cosmology) reaches far beyond the real objectives, namely, the description of the earth as the home of man.[11]

Ritter was the first great opponent of what may properly be called 'armchair geography'.

Principle of Unity in Diversity

The fundamental principle evolved by Ritter was 'unity in diversity'. According to him, there is a fundamental unity in the biotic and abiotic components of habitat in which man sculptures his cultural environment. In such an approach, all the physical and cultural components of environment are taken into consideration and their interrelationship is established in understanding the geography of an areal unit. This is a regional approach. Unity in diversity means that every naturally bounded area is a unity in respect of climate, production, culture, population and history. Ritter makes few deterministic observations; he seldom does more than repeat what Humboldt had already written and gives the same synthetical accounts of continents. The merit of Ritter's work comes not from his description of the continents but from his ability to deduce these from a system of laws governing 'the concept of regional association of terrestrial phenomena at various levels over the earth's surface'.

Ritter's method is said to be deductive because it deduces new conclusions from fundamental assumptions or from truths established by other methods. So far there is little to distinguish Ritter's ideas from Humboldt's and in the spatial arrangement of terrestrial phenomena, there is marked similarity between the two colleagues.

Ritter introduced many stimulating ideas. He stressed the idea of land and water hemispheres, the distinction between the rates of heating and cooling of land and water, the difference between the northern and the southern hemispheres in their proportion of land and water. He averred that there were differences between the continents. Africa had relatively short and the most regular of all coastlines and its interior had least contact with the sea, whereas Asia was better provided with sea inlets, but the interior had little marine contact and Europe was the most varied of all, with an ease of approach along its shoreline of comparatively great length. He identified each continent with a different race, having a different colour.[12] For example, according to him, Africa is a continent of black people, Europe of white people, Asia of yellow people and America of red people. This overgeneralization created much obscurity in the world of geography. About the universal and regional laws in geography his opinion was that 'the earth itself must be asked for its laws'. In brief, Ritter's theme was that the physical environment was capable of determining the course of human development. His ideas were strengthened by the publication of Darwin's *Origin of Species* in 1859 with its emphasis on the close relationships of organism and their habitat (environment).

Ritter emphasized repeatedly that he was teaching a 'new' scientific geography, in contrast to the traditional 'lifeless summary of facts about countries and cities, mingled with all sorts of scientific incongruities'. Ritter saw all of his studies of 'the earth and man as revealing more and more of God's plan'.

Ritter considered the earth as the home of man. He stressed that 'the outer shell of the earth is the area of geographical study and not the entire earth'.

Ritter is known for his *Erdkunde*, which conceptualized regional geography as contrasted with systematic geography of Humboldt.

Erdkunde

Ritter's monumental work is entitled as *Erdkunde*. *Erdkunde* is a comprehensive German word which stands for science of the earth in relation to nature and history. Ritter remarks that 'the earth and its inhabitants stand in the closest reciprocal relations and one cannot be truly presented in all its relationships without the other. Hence, history and geography must always remain inseparable'. Land influences the inhabitants and in turn the inhabitants transform the landscape. In Europe, for example, only in the east (Russia), there was uniformity of geographical features and uniformity of history. But, in the west, there was variety of environment and history, and in the diverse south (Europe) too history was rich, studded with the efforts and achievements of Egyptians, Carthageans, Greeks, Romans, Gauls and Iberians.[13] In *Erdkunde*, he advanced the theory of the north-west movement of civilization in Europe.

The first two volumes of the *Erdkunde* were intended to be followed by a study of history. Between 1817 and 1859, he completed 19 volumes of *Erdkunde* but these volumes cover only Africa and parts of Asia. In spite of the fact that he lived long, he was not able to finish his work of Europe.

Through his writings, Ritter tried to prove that the earth is made for man: 'As body is made for soul, so is the physical globe made for mankind'. The most logical development of Ritter's work is to be found in the writings of the geographers who studied the interaction of the various phenomena – relief, climate, vegetation, and man in a particular area.

The major geographical concepts of Ritter may be summed up as follows:[14]

1. Ritter conceived geography as an empirical science rather than one based on deduction from rational principles or apriori theory.
2. There is a coherence in the spatial arrangement of terrestrial phenomena. Areal phenomena are so interrelated as to give rise to the uniqueness of the areas as individual units.[15]
3. Boundary lines, whether wet or dry (such as rivers or mountains), were instruments for understanding the real purpose of geography which is understanding the content of areas.[16]
4. According to Ritter, geography was concerned with objects on the earth as they exist together in an area. He studied areas synthetically, i.e. in their totality.
5. Ritter holds a holistic view with respect to the content and purpose of geographic study, and the whole study was focused on and culminated in man.
6. He believed that the earth was an organism made, even in its smallest details, with divine intent, to fit the needs of man to perfection. He was a teleologist in his approach.

Both Humboldt and Ritter laid great stress on the unity of nature, though one has a scientific and other a religious approach. They both believed that the ultimate aim of research was to clarify this unity and, in this respect, were in accord with the idealistic philosophies of their time. Humboldt did not pursue idealism in the same way as Ritter, for his concept of the unity of nature was more aesthetic than religious. In this respect he had more in common with Goethe than with Ritter. Unlike Ritter, he saw no reason to explain unity and order in nature as a God-given system to further humanity's development. Humboldt was very much engaged in the gradual development of natural science, and his greatest contributions lay in the field of systematic physical geography. Ritter was, on the other hand, to a considerable extent a regional geographer.

Even during the period of Humboldt and Ritter geography was still not related to a specific discipline. In fact, geography remained an umbrella concept for a variety of expeditions and other activities within the natural and social sciences, to a large extent supported by geographical societies. Some of the important geographical societies which promoted the cause of geography are given below in a chronological order:

1. The Societe de Geographie de Paris – 1821
2. The Gesellschaft fur Erdkunde zu Berlin – 1828
3. The Royal Geographical Society London – 1830
4. The Geographical Society Mexico – 1833
5. The Geographical Society Frankfurt – 1836
6. The Geographical Society Brazil – 1838
7. The Imperial Russian Geographical Society in St. Petersburg – 1845
8. The American Geographical Society – 1852

The main work of these societies was to give support for expeditions and for their publication of yearbooks and journals which included maps and other material from expeditions.

Charles Robert Darwin (1809–1882)

Born on 12 February 1809 in Shrewsbury, Shropshire (England) into a wealthy and well-connected family, Darwin was a naturalist. He is renowned for his theory of evolution and for a theory of its operation, known as Darwinism. His evolutionary theories, are propounded chiefly in two works: (i) *Origin of Species by Means of Natural Selection* (1859), and (ii) *Descent of Man, and Selection in Relation to Sex* (1871). His theories greatly influenced the scientific and religious debates of his time.

Charles Robert Darwin

Darwin's father, Robert Warning was a distinguished physician. He was brought up by his eldest sister from the age of eight. Darwin after an early life that showed little promise of his later prominence, he developed an interest in natural history. Darwin was never a model student, but he became a passionate amateur naturalist. He got his education in medicine at the University of Edinburgh. Subsequently, shifted to the Cambridge University where he obtained his degree in medicine in 1831 with no special distinction.

In 1831, Darwin sailed with an expedition as a naturalist to South America and Pacific Islands. The objective of his travel was to survey the wildlife of the west coast of South America. Darwin was particularly influenced by Humboldt and Charles Lyell. On the voyage, Darwin read Lyell's *Principles of Geology* which suggested that the fossils found in rocks were actually evidence of animals that had lived for many thousands or millions of years ago. Lyell's argument was reinforced in Darwin's own mind by the rich variety of animal life and the geological features he saw during his voyage. The breakthrough in his ideas came in Galapagos Islands 850 km west of Ecuador (South America). Darwin noticed that each of the Galapagos Islands supported its own form of finch.

On his return to England in 1836, Darwin tried to solve the riddles of these observations and the puzzle of how species evolve. Influenced by the ideas of Malthus (1798), he proposed a theory of evolution occurring by the process of natural selection. The animals and plants best suited to their environment are more likely to survive and reproduce, passing on the characteristics which helped them survive to their offspring. Gradually, the species change over time.

Darwin worked on his theory for 20 years. After learning that another naturalist, Alfred Russel Wallace, had developed similar ideas, the two made a joint statement of their discovery in 1859. In 1859 Darwin published *Origin of Species by Means of Natural Selection.*

The book was extremely controversial, because the logical extension of Darwin's theory was that *homo sapiens* were simply another form of animals. It made it seem possible than that even people might just have evolved – quite possibly from *apes* and destroyed the prevailing orthodoxy on how the world was created. Darwin was vehemently attacked, particularly by the Church. However, his ideas soon gained currency and have become the new orthodoxy.

During this five year trip he became convinced of the gradual evolution of species. Upon his return to England, he worked for 20 years refining his ideas before he started to write a definitive account of evolution in 1856 which he published in 1859 as *Origin of Species by Means of Natural selection.* His later days were spent in much physical discomfort as he suffered from 'Chagas' disease, which he had contracted while in South America.

Darwin's genius was not confined solely to questions of evolution. He explored many other natural phenomena, including the taxonomy of branches, the formation of atolls and barrier reefs, and the role of earthworms in soil fertility. His other works include *Variation in Animals and Plants under Domestication* (1868) and the *Descent of Man.* At the time of Darwin's death on 19 April 1882, there was no longer any controversy over his greatness and he was buried in Westminister Abbey, London.

Impact of Darwin on the Development of Geographical Concepts

Charles Darwin propounded the theory of evolution which revolutionized the biological, environmental and earth sciences. His theory of evolution involved commitment to common organic descent, gradualism and multiplication of species. He also spoke of natural selection, family selection, correlative variation, use inheritance and directed variation. Darwin explained how the multitude of living things in our world so finely adapted to their environment, could have come into being without recourse to a divine master plan, in a plain, causal, naturalistic way. Darwin argued that a struggle for existence must take place; it followed that those who survived were better adapted to their environments than competitors. This was essentially a theory of reproductive success in which relatively superior adaptations increase while relatively inferior ones are steadily eliminated. A similar theory was simultaneously put forward by Alfred Russel Wallace (1823–1913) who surveyed the islands of South-East Asia. Stoddart (1966) suggests that the following four main themes from Darwin's work can be traced in later geographical research:

1. Change through time or evolution – a general concept of gradual or even transition from lower to higher or more complicated forms. Darwin used the terms 'evolution' and 'development' essentially in the same sense.
2. Association and organization – humanity as part of a living ecological organism.
3. Struggle and natural selection.
4. The randomness or chance character of variation in nature.

Darwin, who rejected the teleological approach of Ritter and the theological concept in vogue about the origin of man and other species influenced significantly the growth and development of the concepts of geography both in the physical and human geography. Some of the significant influences of Darwin's theory on the development of geographical concepts, methodology and approaches have been concisely presented in the following paras.

1. Impact on Geomorphology

Darwin's theory influenced the thinking of physical geographers. In fact, it was after Darwin's work that apart from biology, geology and geomorphology became the most ambitious branches of natural sciences. Geology was important because it could with the aid of paleontology (which interprets fossils) clarify the evolution of plant and animal species. Influenced by Darwin's theory, it was Oscar Peschel, author of the book *New Problems of Comparative Geography as a Search for Morphology of the Earth's Surface* (1870), who proposed that geographers should study the *morphology* of the earth's surface. Like Ritter he was interested in the significance of landforms for the development of human beings, but he did not share Ritter's religious outlook being more concerned with cause and effects as illustrated by the methods of natural sciences.

It was after Darwin's theory that the geologists and paleontologists concentrated on the development of geological timescale, the systematic mapping of rock types and the analysis of fossils.

In physical geography, the study of landforms became the leading field of research for most of the professors appointed to geography chairs in the later half of the 19th century. It may be said as the period of 'geologification' of geography.

The science of geomorphology that analyses and describes the origin, evolution, form, classification and spatial distribution of landforms became an important branch of physical geography.

Darwin's evolutionary concept of change over time was applied by William Morris Davis in the concept of geographic cycle (cycle of erosion). Davis developed a similarity between organic life and evolution of landforms. He advocated that 'landforms evolve like the evolution of organic life'. The cycle of erosion has been defined by him as 'geographical cycle' which is the period of time during which an uplifted landmass undergoes its transformation by the process of land sculpture, ending into a low featureless plain – a peneplain. He stressed that 'landscape is a function of structure, process and time (stage)'. In the evolution of landforms, Davis identified three stages, viz., (i) youth, (ii) maturity, and (iii) old. Like organic life, each process gives rise to a distinctive landscape, so each stage of the cycle of erosion is characterized by a distinctive landform. Thus, the evolution of landforms takes place in a sequential manner like the evolution of organic life. The pioneer model of Davis became the mother of all models pertaining to the evolution of landforms. This model acquired a special position in the study of geomorphology. In fact, the entire geomorphic thought has been influenced by the concepts of Davis which he developed on the Darwin's theory of evolution of species.

2. Impact on Landschaft

The German geomorphologists, influenced by the Darwin's theory of origin of species started to define geography as 'landscape science'. Viewed in these terms, geography was fundamentally concerned with the form of landscape of particular regions and a number of schemes were proposed to classify landscapes and their elements and to provide for formal procedures of analysis. The German geomorphologists distinguished the natural landscape from the cultural landscape and in doing so recognized the importance of *Human Agency*. The Sauer's Berkeley School of cultural landscape also showed great interest in the physical features of landscape, while the British geographers considered geomorphology as the foundation of geography.

3. Impact on Human Geography

Darwin's theory about the 'origin of species' and 'descent of man' gave a new direction to the various sub-fields of human geography. The doctrine that human activities are controlled by the environment (environmental determinism) took a new turn. Darwin's notions regarding evolution were taken up by the German and American geographers to explain the man and environment relationship. Thus the environmental determinism became an important school in human geography.

Ratzel was a strong follower of Darwin. He declared in his book *Anthropogeographic* that 'similar locations lead to similar mode of life'. His American disciple Ellen Churchill Semple opened her book *Influences of Geographic Environment* (1911) with the statement 'man is the product of the earth's surface'. Huntington also asserted that 'climate controls the progress and development of human civilization'.

Reaction to the extreme generalization of environmental determinism, however, led to a counter thesis, that of possibilism, which presented the individual as an active rather than a passive agent.

4. Impact on Political Geography

Influenced by Darwin, a very important concept, namely, 'lebensraum' (living space) was coined by Ratzel. In his book on *Political Geography*, Ratzel equated a nation with a living organism, and argued that a country's search for territorial expansion was similar to a growing organism's search for space. Conflict between nations was thus seen as a contest for territory within which to expand, with the fittest surviving. The struggle and survival of the fittest notion of Darwin was thus adopted by geographers which moulded the philosophy of German political thinkers. Ratzel stressed that just as the struggle for existence in the plant and animal world always centres about a matter of space, so the conflicts of nations are in great part only struggles for territory. This fundamental concept of 'living space' helped in the development of biogeography. The concept was appropriated by the German School of Geopolitik in the 1920s and 1930s and used to justify the Nazi programme of territorial expansion.

5. Impact on Cultural Landscape

The term 'cultural landscape' was developed in American geography in 1925 by Carl O. Sauer with the publication of his article 'The Morphology of Landscape'. He developed this concept as an alternative to environmental determinism. While the environmental determinism sought to specify the causal influences of the environment on humans, the landscape approach sought to describe the interrelation between humans and the environment with primary attention given to the human impact on the environment. Sauer's emphasis was that geographers should proceed genetically and trace the development of natural landscape into a cultural landscape. The main focus of Sauer was the study of processes leading to landscape change up to the present, beginning at the prehuman stage of occupance.

Darwin's theory thus closely influenced the growth and development of geomorphology, human geography, political geography and cultural geography, and led to the development of enormous new philosophical concepts and methodologies in geography. Darwin's theory gave a new direction to the discipline of geography and it became more socially and environmentally relevant.

Notes

1. Kellner, L., 1963, *Alexander von Humboldt*, London, p. 6.
2. Humboldt, A.V., 1793, *Florae Fribergenis Subterraneans Exhibens*, Berlin, p. 30.
3. James, P.E., and Martin, G.J., 1981, *All Possible Worlds*, p. 116.
4. Ibid., p. 124.
5. Humboldt, von. A., *Kosmos*, Vol. I, p. 53.
6. James, P.E., and Martin, G.J., op. cit., p. 124.
7. Bogekamp, H., 1963, 'An Account of Prof. Ritter's Geographical Labour' in W.L. Gage, Trans. *Geographical Studies by the Late Professor Carl Ritter of Berlin*, pp. 33–51.
8. Karl, A.S., 1959, 'Carl Ritter, 1779–1859', *Scottish Geographical Magazine*, Vol. 75, pp. 152–163.
9. Dickinson, R.E., 1969, *The Makers of Modern Geography*, p. 35.
10. Ibid., p. 35.
11. Ibid.
12. Schmitthener, Carl Ritter, quoted by Dickinsons, *Regional Concept*, p. 44.
13. Freeman, *Six Hundred Years of Geography*, p. 35.
14. Dickinson, R.E., op. cit., pp. 42–45.
15. Ritter, C., *Comparative Geography*, p. 28.
16. Hartshorne, R., *The Nature of Geography*, p. 233.

7

Schools of Geography

A philosophical foundation to the subject of geography was given by Immanuel Kant, while Alexander von Humboldt (1769–1859) and Carl Ritter (1779–1859) developed the subject as an independent branch of knowledge. From the middle of the 19th century up to the contemporary period, there occurred many philosophical changes in the definition, concepts and approaches of the discipline. The major concepts and methodologies were, however, developed by the Germans, followed by the French, British, American and Soviet scientists. In the present chapter, a brief account of the different schools of geography has been given.

THE GERMAN SCHOOL OF GEOGRAPHY

German's contribution to the development of geography is enormous. In the 18th and 19th century, the Germans made great strides and put the subject on a sound footing. They gave it a philosophical and scientific base. In the post-Humboldt and Ritter period, a definite change in the role of universities occurred. During their lifetime, the major role of the universities was to train students in classical languages, theology, law and logic. It was in the middle of the 19th century when courses for various physical, biological and social sciences were standardized, and students were allowed to select their optional subjects. In Germany, the first university was established in 1809, but there were very few universities till the end of the 19th century in which geography was taught. At the initial stage, the growth of geography was very slow, and the teachers of geography did not have a proper geography background. Most of the geography teachers were pupils of Carl Ritter, and even they were not sufficiently proficient in geography as they had backgrounds of other disciplines. Some of them were trained in history, while others had the background of mathematics, biology, physical sciences, history, geology and engineering. It was during this period that a number of geographical societies were established. These societies published geographical literature and information about the earth.[1] In the second part of the last century, geographers all over the world tried to define the subject. The German scholars also gave several definitions of geography[2], and tried to delineate its scope. Some of the important concepts of geography advocated and defined by the German scholars are given in the following section.

The middle of the 19th century was a period of political turmoil in Europe. There was a great demand for maps and charts by military officers and administrators since they wished to learn about the physical and cultural conditions of different nations and regions of the world. Owing to practical utility of maps, anything that could be plotted on maps was considered geography.

Oscar Peschel (1826–1875)

Oscar Peschel, a leading German geographer, was appointed as a professor at the University of Leipzig in 1871–75. Peschel was historian of geography, and the editor of *Das Ausland* – a periodical in which articles about the geography of foreign countries used to be published.[3] Peschel started comparing different parts of the world and developed comparative geography. With the help of maps, he studied fiords of Norway and put forward the hypothesis that the fiords were fissures in the earth's crust that had been occupied and gouged out by glaciers.[4] Moreover, Peschel laid the foundation of modern physical geography. He was primarily concerned with the study of landforms. In his studies of landforms and physical geography, he, however, ignored man as an important component of the discipline of geography. He died at the age of 49 and his book on physical geography was published posthumously.[5]

Oscar Peschel

Ferdinand von Richthofen (1833–1905)

Richthofen was trained as a geologist. After field studies in China and California, he taught geography at Bonn, Leipzig and Berlin (1877–1905).

In 1871, at the end of the Franco-Prussian War, Germany was unified and the German Empire came into existence. The end of the Franco-Prussian War led to heavy demand for the teaching of geography. Under these circumstances, new geography textbooks were written, and

Ferdinand von Richthofen

geography was introduced in ten universities of the country. At this stage, Ferdinand von Richthofen who was essentially a geologist and the founder of modern geomorphology, pleaded for the cause of geography. As stated above, Richthofen was trained in geology, and as a geographer, he was primarily interested in geomorphology (study of landforms). He visited China and prepared a map of coalfields of China.[6] He also noted the loess formations of Northern China and tried to explain their genesis.

In the opinion of Richthofen, the purpose of geography is to focus attention on the diverse phenomena that occur in interrelation on the face of the earth.[7] The methodology he suggested for the study of geography was that first the physical setting (relief, climate, soil, vegetation, fauna, flora) of a region be studied and then the adjustment of man in that setting be examined. According to him, the main objective of geography is the exploration of the relationship of man to the physical earth and to the biotic features. For a substantial period, this objective and methodology remained the basic pattern of geographical studies not only in Germany but also in other parts of the world.

Richthofen was the first German scholar to differentiate between the 'general' and the 'regional' geography. He emphasized the point that regional geography must be descriptive to highlight the salient features of a region. Moreover, it should try to seek regularities of occurrence and patterns of such unique features to formulate hypothesis and to explain the observed characteristics. General geography, according to him, deals with the spatial distribution and explanation of individual phenomena in the world. For systematic regional study, he emphasized the need of fieldwork.[8] Richthofen distinguished the different methods of study in areas of different size, which he named (in order of increasing size): (i) *Erdteile* (major divisions of the world); (ii) *Lander* (major regions); (iii) *Landschaften* (landscape or the small regions); and (iv) *Ortlichkeiten* (localities).[9] He insisted that *Erdkunde* (Geography) must refer to a study of the earth where the lithosphere, the atmosphere, the biosphere and the hydrosphere are in contact with each other.

Friedrich Ratzel (1844–1904)

Ratzel was trained in Zoology and Geology at Munich. In the later part of the 19th century Friedrich Ratzel dominated the geographical scene in Germany. He was primarily concerned with the natural sciences for whom geography offered the connection between the natural sciences and the study of man. His main work was in human geography. He took his doctorate in zoology, geology and comparative anatomy. He is universally recognized as the founder of human geography.

Friedrich Ratzel

Born in 1844, Ratzel got his early education at several universities in Germany. He visited Italy in 1872, and USA and Mexico in 1874–75. He also travelled Eastern Europe widely and taught geography at the Universities of Munich and Leipzig.[10] He is known for the application of Darwin's biological concepts to human societies. Before Ratzel, the foundation of systematic geography was laid down by Alexander von Humboldt and that of regional geography by Carl Ritter. Peschel and Richthofen laid down guidelines for the systematic study of the earth's features. It was Ratzel who compared the mode of life of the different tribes and nations, and thus made a systematic study of human geography. His interest in tribes, races and nations was keen and after doing adequate fieldwork he coined the term 'anthropogeography', describing it as the major field of geographical study. Ratzel took Ritter's human geography still further, sub-dividing it into 'anthropogeography' and 'political geography'. He is best known for his organic theory of state (Lebensraum) in which he compares the evolution of the state to that of living organism. He applied the Darwin's biological concepts to human societies. This analogy suggested that groups of human beings must struggle to survive in particular environments much as plant and animal organisms must do. This is known as Social Darwinism.

Ratzel, a scholar of versatile academic interests, was a staunch German. It was because of his patriotism that at the outbreak of Franco-Prussian War in 1870, he joined the Prussian Army and was wounded twice during the war. After the unification of Germany (1871), he devoted himself to the study of the modes of life of Germans living outside Germany. For this purpose, he visited Hungary and Transylvania. Germans were in majority in these countries. He continued his mission and in 1872 crossed the Alps to visit Italy.

In 1874–75, Ratzel reached the United States and Mexico and thereby expanded his sphere of study.[11] In the United States, he started studying the economy, society, and habitat of the original inhabitants and tribes, and especially the mode of life of the Red Indians. Moreover, he focused his attention on Negroes, Africans, and Chinese living in the Central USA, the Middle West and California. On the basis of his field study, he tried to formulate some general concepts regarding the geographic patterns resulting from contact between aggressive and expanding human groups and the retreating groups.[12]

At the completion of his field study in USA and Mexico, he returned to Germany in 1875 and was appointed Professor of Geography in 1876 at the University of Leipzig. In 1878, and 1880 he published two books on 'North America' dealing with physical and cultural geography. However, the book for which Ratzel is acknowledged all over the world – *Anthropogeographie* – was completed between 1872 and 1899. The main focus of this monumental work is on the effects of different physical features and locations on the style of life of people. This first volume of *Anthropogeographie* is a study of man and earth relationship, while in the second volume the influence of man on environment has been discussed. Ratzel's work was based on the concept that physical environment controlled human activities. The *anthropogeographie* of Ratzel suggests the geography of man in terms of individuals and races (anthropological geography). Ratzel believed that society ought not to be left suspended in the air. Subsequently, he dispelled any lingering determinism by insisting that the human being joins in nature's game and that *milieu externe* was a partner, not a slave of human activity. Thus, it was Ratzel who provided the guidelines for a comparable systematic study of human geography.

In *Anthropogeographie*, Ratzel used deductive approach to present the first systematic study of the geography of man. The first volume of *Anthropogeographie* was organized in terms of physical features and their influence on human culture while in the second volume he analysed human activities and human cultures in relation to the physical environment.[13] He pointed out the great contrast in the way people make use of the land in two places that are physically very much alike. The two places are the low mountain regions on either side of the Middle Rhine Valley – the Vosges Mountains in France and the Schwarzwald in Germany. The difference to be observed in these two regions are related to the contrast between the French and German cultural traditions. This environment-dominated-man response approach was reserved for his second volume. This second approach was influenced by the chronological approach.[14]

Ratzel's deterministic approach became very popular outside Germany, especially in France, England and the United States. Ms. Allen Churchill Semple was one of his pupils and a staunch supporter of his environmental superiority philosophy.[15] Thus, Ratzel's basic philosophy was also 'survival of the fittest' in a physical environment.*

In the 1890s his active campaigning to persuade Germany to acquire overseas colonies and build up its naval fleet to challenge Britain superiority as a world power. His thinking expressed the spatial consequences of the Darwinist struggle for survival. According to the 'laws' of territorial growth, states must expand to thrive, and 'higher forms of civilization must expand at the expense of lower ones'. These laws were supposedly natural, but given the recent unification of Germany, the inter-state rivalry in Europe (General Schlieffen was already up with his plan for invasion of France) and its empires (Africa was divided into European colonies at the Berlin Conference in 1884–85). Ratzel's views accorded with Germany's territorial designs. After his death and after the First World War, German geopoliticians revived Ratzel's ideas to suit their own ambitions and, as a result, his writings were condemned by Anglo-American geographers. Had Germany won the war then perhaps Mackinder would have suffered in his stead.

---

* The phrase 'survival of the fittest' was coined by Spencer – an English philosopher – in 1876.

In 1897, Ratzel wrote *Political Geography* in which he compared 'state' to an organism.[16] He emphasized the fact that a state, like some simple organism, must either grow or die and can never stand still.[17] It was this philosophy of *Lebensraum* (living space) which created the controversy of superior and inferior races by claiming that the superior peoples (nations) have a right to expand their territory (kingdom) – 'living space' – at the expense of inferior neighbours. He expounded his views saying: when a state extends its frontiers at the expense of a weak state this is a reflection of its internal strength. The superior nations thus ruling over the inferior peoples fulfil a natural organic necessity. It was Ratzel's philosophy which governed the German policy in the thirties which finally led to the outbreak of the Second World War.

While discussing the effect of physical environment on human groups, Ratzel also explained that human societies made progress in stages.[18] These stages are: (i) hunting and fishing; (ii) hoe-culture; (iii) crops cultivation where land is ploughed; (iv) mixed farming, where farming and herding are mixed; (v) unmixed herding; and (vi) horticulture. He, however, argued that it is not necessary that all the societies have to pass through the same stages of economy.

In those days, there was enormous increase of knowledge and information, data were being accumulated on a large scale from different parts of the earth. Each region, having a different physical environment, was providing different modes of production and different styles of life. It was Ratzel who tried to build a 'fundamental unity in diversity'. The controversy regarding dichotomy between physical and human geography started during Ratzel's time. There were scholars like George Gerald, who held the view that geography is the study of earth body as a whole without reference to man. Gerald was of the opinion that exact laws can be made only if man is excluded from geography, because the behaviour of man is highly unpredictable. This radical viewpoint, declaring physical geography as the sphere of geography, was mooted by Ratzel who averred that man is an important ingredient of the discipline of geography. He stressed the principle of unity in diversity, stating that in diverse physical conditions, man always adjusted himself, and therefore, in order to have a complete view of the geographical horizon of the earth we have to synthesize diverse physical and cultural phenomena.

To summarize, we believe that Ratzel's *Anthropogeographie* was a seminal work and the amount of intellectual debate it created on both sides of the Atlantic Ocean makes it an exemplar. Ratzel's view about geography dominated for decades – a tribute to his ability as a teacher and scholar. As Miss Semple wrote:

> He grew with his work, and his work and its problems grew with him. He took a mountain view of the things, kept his eyes always on the horizons, and in the splendid sweep of his scientific conceptions sometimes overlooked the details near at hand. Herein lay his greatness and his limitations.[19]

Alfred Hettner (1859–1941)

Alfred Hettner

Alfred Hettner was a pupil of Richthofen and Ratzel. He taught at Leipzig, Tubingen, and Heidelberg (1899–1928). He was essentially a physical and regional geographer. He wrote research papers after making extensive travels. His book *Europe* was published in 1907. He supported the viewpoint of Strabo who declared geography as a chorological science or the study of regions or area. In the chorological concept of geography, the geographer is required to study many immaterial (race, ethnicity, religion, language, customs, traditions, etc.) as well as material phenomena. He should not give preference either to the material or immaterial phenomena. Thus, in the study of areal differentiation, they are on the same plane. He elaborated the importance of distribution of phenomena and stressed the significance of regional geography. He was of the opinion that geography is a field in which things have to be described in their areal context on the face of the earth, much as history was the field in which things are to be considered in their time context.

Hettner claimed that geography is idiographic (regional) rather than monothetic (systematic/general).[20] In his opinion, the distinctive subject of geography was the knowledge of the earth areas as they differ from each other. Man was included as an integral part of nature of an area and thus 'mere description has been replaced in all branches of geography by search of causes'. The concept that 'the unity of geography is in methods' was advocated by him. In his opinion, the procedure of study of relationship should be from 'man to nature, rather than from nature to man'.

Hettner rejected the view that geography could be either general or regional. Geography, like other fields of learning, must deal in both the unique things (regional geography) and with universals (general geography), but the study of regions is the main field of geography. This theory of uniqueness (regional geography) attracted the attention of German geographers for decades and is still the point of controversy in geography. After Hettner, regional studies were made with the outline of location, geology, surface features, climate, vegetation, natural resources, settlement distribution, economy, transport systems and political divisions.[21]

Hettner's approach invited criticism also. Those who criticized were of the opinion that Hettner tried to give more weight to physical environment, ignoring the cultural factors. For example, there is a direct relationship between the density of population, economy, social institutions, religious beliefs, cultural traits, and political policies. Moreover, many of the interrelations observed in regional studies were in the process of change through time.[22] Many of the nomadic herders had started settling down and adopted sedentrized life under the impact of socio-cultural and political factors.

Landschafts Kunde

The concept of *landschafts kunde* (landscape science) gained popularity in Germany at the beginning of the 20th century. According to this concept, the existing landscape of a meso or micro region is the result of exogenetic, historical and cultural forces. In the study of landscape, not only man is taken as an important agent of environment but also invisible features like wind, air, temperature and humidity are considered. Landschaft is thus a homogeneous area with a more or less uniform appearance, which is the subject matter of geography and in which biotic and abiotic things are studied in relation to each other.

The research work done in the early parts of the 19th century gave rise to the idea that general geography makes use of general concepts and is essentially analytical; while regional geography deals with a unique situation and is essentially synthetic in nature. This was an erroneous concept. In fact, the concept of chorology (examination of areal association of things of diverse origin) can be applied to general (systematic) geography as well as to the studies of segments of the whole earth surface[23] or a region.

Albrecht Penck (1858–1945)

Albrecht Penck

Albrecht Penck was a leading German geographer of the early part of the 20th century, who formulated the concept of 'geomorphology', and was trained in geology. Penck was a Professor of Physical Geography at Vienna from 1885 to 1906; and then Professor at Berlin (1906–1926). He was associated with Eduard Suess, who prepared maps of the major geological regions of the world. He founded the principles of landforms evolution and showed how the systematic study of features can be approached from the chorological (regional) point of view. In 1910, Penck postulated the hypothesis that classification of climates can be made with the help of study of landforms even where the meteorological data are not available. He was the first to point out that evaporation increases with higher temperature. He also pointed out that the effective rainfall, i.e. the difference between rainfall of a place, and run-off plus evaporation is directly dependent on the prevailing temperature. Moreover, he considered man also as an important agent who carves out the face of the earth.[24] Penck stressed the importance of accurate maps, showing relief features for a systematic study of geography. The idea of topographical maps was thus put forward by him. It was because of his suggestions that topographical maps on the large scale, i.e. one inch to one mile, showing the major relief features, water bodies, vegetations and the work of man, started being prepared. Penck introduced the term 'Gestalt', borrowed from *Psychology in Geography*. He used this to express the unified from of larger areas (whole). In his opinion, 'geography is a bridge between the natural and social sciences'.

Penck's pioneering ideas inspired many young scientists to pursue research in the field of geomorphology and climatology. Walther Penck (1888–1923), German geomorphologist was the son of Albrecht Penck, who presented a major challenge to Davis' concept of the cycle of erosion. Wladimir Koppen – the Russian-born German climatologist – started studying world climate on the lines advocated by Penck. For his climatic study, he took into consideration the observable relief features, besides the recorded temperature and rainfall data. Between 1884 and 1918, Koppen made several attempts to produce a satisfactory classification of climates. For his classification, he took into consideration temperatures, vegetation, rainfall effectiveness to temperature and seasonal and annual variations of temperature and rainfall. With the help of these indicators, Koppen arrived at certain regularities in the temperature and rainfall distribution. He established that moisture deficiencies exist throughout the year, on the western margins between 20° and 30° in both the hemispheres and similarly the continental areas on the same latitude are more cold in winter and more warm in summer than the parts lying in the vicinity of water bodies (oceans, seas).

In the field of oceanography also, Germans made a substantial contribution. Gerhard Schott studied the motion, temperature, salinity, colour, ocean deposits, configuration, ocean climates, geological structures of ocean basins, ocean organisms, the routes of ocean trade and air routes over oceans.[25] Schott prepared a world map in which ocean regions were delineated on the basis of various indicators.

During the first fifteen years of the 20th century, Germans made great strides in the field of geography. This was a period of rapid growth and increasing productivity of geographical concepts and literature. The period after the First World War was a lean period in which most of the geographers who did not agree with government policies remained silent. Many of the Jews were placed in confinement. In spite of the political problems and difficulties created by government for academics, there were notable contributions in the field of urban and agricultural geography. Walter Christaller put forward the Central Place Theory with the set objective of functional organization of space. Similarly, von Thunen advocated the Crop Intensity Theory. Thus, there was stress on landscape planning during the two wars. Again, during the period, attention of the German geographers was focused on geo-politics. Houshofer, in 1924, being inspired by the ideas of Kjellen and Ratzel, was convinced that state is a living organism which needs space to grow. Houshofer, through his writings, probably influenced Nazi policies. With the defeat of the Nazis in the Second World War in 1945, Houshofer was put on trail at Nuremburg. In 1946, he committed suicide.

After the Second World War, Germany was divided. There was political turmoil and socio-economic crisis. Universities and libraries were ruined. However, within a short period of about fifteen years (1945–1960), Germans again emerged as a force in the field of geography. In 1947, a new geographical periodical entitled *Erdkunde* appeared. The Germans started studying landscape with the help of new and sophisticated statistical tools and techniques. Now, there is more stress on variation from place to place as the function of latitude, altitude, distance from the sea and direction from the nearest coast. In the post-war period, the new trend is that of 'cultural determinism' in place of environmental and physical determinisms. The new emphasis on culture is called 'social geography'. The Germans are now trying to interpret landscape with the help of attitudes of people and their technical skill since these are felt to be the vital parts of man's culture. Moreover, an integrated approach to studying landscape is now in evidence. In this effort scholars of physical and social sciences are being involved.

German geographers and cartographers made notable contributions in the art of map-making during the 19th and 20th centuries. For many years, the von Sydow maps and atlases, with some wall maps, were widely used. The Perthes Firm published Stieler's Atlas, under the supervision of A.H. Pattermann. In 1845–47, Pattermann, while working with W & A.K. Johnston in Edinburgh, prepared the *Physical Atlas*.

THE FRENCH SCHOOL OF GEOGRAPHY

The geographical ideas and concepts, which originated with Ratzel and his disciples, spread in the neighbouring countries. Alexander von Humboldt, who wrote and published his thirty volumes of *Kosmos* in Paris, created intellectual fervour in French scholars. In the middle of the 19th century, in France as in Germany, geography was taught by historians, geologists, military personnel and engineers. Even the chair of geography in the Sorbonne (Paris) was occupied by a historian who was attached to the Faculty of Letters. The major contribution made by French scholars is given in the following section.

Phillippe Buache (1752) was the first French scholar who criticized the prevalent method of representation of population, economic and other data in administrative units.[26] He felt that the right method of representation of geographical data is in the frame of natural region. According to him, a river basin was the best kind of natural region.[27] Subsequently, Baron Coquebert (Director of French Statistical Office, 1796) proposed a division of the national territory into natural regions with a brief description of each of them. This effort created interest in regional divisions in France. But this approach was contradicted by Omalius d'Halloy in 1833 who prepared a geological map of France to establish the relation between landforms and soils and the underlying rocks.

About this time (1870) there occurred a major breakthrough in the expansion of geographical knowledge. Soon several geographical societies were established in the various universities of France. The real take off in the field of geography in France, however, started during the period of Vidal de la Blache. The French geographers were concerned with the unique character of areas.

Paul Vidal de la Blache (1845–1918)

Paul Vidal de la Blache

Paul Vidal de la Blache was the one who developed the new geography in France. He is known as the founder of Human Geography. He was essentially a scholar of classical languages and literature. His interest in geography developed in 1865 when he was studying archaeology at Athens. Later on Vidal de la Blache taught geography at the University of Nancy from 1872 to 1877, and then joined Ecole as Professor of Geography. In 1891, he founded a new professional periodical for the publication of the best geographical writings. The periodical was called *Annales de Geographie*. In 1894, Vidal de la Blache published the first edition of the *Atlas Generale Vidal Lablache*.[28] From 1896 to the time of his death (1918), he was Professor of Geography at the University of Sorbonne. During his career, he devoted himself to the cause of geography, and trained geography teachers over a period of about 26 years.

While delivering his first lecture at the Sorbonne University on 2 February 1899, he laid stress upon the relationship between man and his immediate surroundings (milieu) which could best be studied in small homogeneous areas. In France, such homogeneous areas are known as *pays*.[29] In his opinion, the concept of country is inseparable from its inhabitants.

Vidal de la Blache was a strong opponent and critique of the environmental deterministic approach. He was influenced by the writings of Ratzel, and from his second volume of *Anthropogeographie*, Vidal de la Blache advocated the concept of 'possibilism' as postulated by Febvre. His basic approach towards the study of man and environment – the two major components of geographical study – was that nature (milieu) sets limits and offers possibilities for human settlement, but the way man reacts or adjusts to these given conditions depends on his own traditional way of living.[30] La Blache insisted that human being 'joins in nature's game' and the *milieu externa* (external environment) was a partner, not a slave of human activity. He opined that 'nature is never more than an adviser'. Vidal de la Blache's belief was endorsed by the historian L. Febvre in a famous phrase: 'There are not necessities but everywhere possibilities'. And, man as a master of these possibilities is the judge of their use. Febvre, however, regarded geography as a natural science, rather than a social science. He considered the earth's surface as the terrestrial organism. He coined the concept of *genre de vie* (lifestyle). The concept of *genre de vie* (way of living) has been widely used in French geography. It refers to the inherited traits that members of a human group learn – what is called 'culture'. The term *genre de vie* stands for the complex of institutions, traditions, attitudes, purposes, and technical skills of people. Vidal de la Blache pointed out that the same environment has different meanings for people with different *genre de vie.* The lifestyle is the basic factor in determining which of the various possibilities offered by nature, a particular human group will select. He was convinced that *genre de vie* were themselves reflective of nature, even as they transformed it. He always conceived of human geography as natural, not a social science (Buttimer, 1971).

Vidal de la Blache's book *Tableau de la Geographie de la France* was a good addition to the literature of geography. In this work, Vidal de la Blache attempted a harmonious blending of physical and human features in the *Tableau* (France Plateau). He also tried a synthesis of *pays.* Vidal de la Blache's book deals with the recognizable regional units of France one by one and shows that each *pay* has its own distinctive agriculture due to its soil and water supply, and also due to the economic specialization made possible by the demands of the people living in towns. Far from reducing the individuality of each *pay*, modern trade had accentuated it by making their agriculture distinctive. Settlement showed a clear relationship to soil and water; for in some areas it was scattered and in others in the form of compact villages. Many of the *pays* had for generations been recognized as separate from, but complementary to their neighbours. These *pays* were, however, not homogeneous as in some there were local deposits such as lemon over chalk which gave sharply contrasting soils reflected in difference in land use.[31] The Tableau is a deeply human work with a firm physical base. From this time, French geographers published a series of regional monographs.

Vidal de la Blache was opposed to the idea of drainage basin as a unit of study. While criticizing the idea of taking drainage basin as the unit of study, he felt that such a unit will create many complications in understanding the reality of a region. For example, the Central Massif of France is a well demarcated natural region, but if it is divided into drainage basin units, then the culture, institutions, traditions and attitudes of the people cannot be properly understood. Regarding the method of geographical study he held the view that the basic objective of geography is to study the phenomena mutually interacting in a segment of the earth surface (*pays*).

In the opinion of Vidal de la Blache, the relatively small regions (*pays*) are the ideal units to study and to train geographers in geographical studies. The tradition of micro region study still persists in France. Many French geographers consider regional geography as best suited to doctoral work. He was, however, of the opinion that regional studies at the meso and macro levels can be of practical utility which can help in the planning of areas. It was with this objective that he prepared a scheme to study the larger regions of the world. This programme was partly carried out by Lucien Gallors after Vidal de la Blache's death.[32]

Vidal de la Blache's monumental work *Human Geography* was posthumously published in 1921 as he died suddenly in 1918. The partially completed work – *Human Geography* – was given final shape by Emmanuel de Martonne – son-in-law of Vidal de la Blache. In this book, Vidal de la Blache started with aims and objectives of human geography; the principle of terrestrial unity and the concept of man and environment (milieu); man as a geographical factor; the patterns of civilizations; circulation (means of transport); and cultural regions and cities.[33] The chapter scheme of Vidal de la Blache's book reveals the pattern on which he tried to examine the man-environment interrelationship.

Prior to Vidal de la Blache, the Germans, especially the followers of Ratzel, strongly favoured deterministic approach for the study of man-nature interaction. Vidal de la Blache had a clear insight in the weaknesses of deterministic arguments, realizing the futility of setting man's natural surroundings in opposition to his social milieu and of regarding one as dominating the other. He considered it even less useful to tackle these relationships along systematic lines in the hope of discovering general laws governing the relationship between man and nature.

According to Vidal de la Blache, it is unreasonable to draw boundaries between natural and cultural phenomena; they should be regarded as united and inseparable. In an area of human settlement, nature changes significantly because of the presence of man, and these changes are the greatest where the level of material culture of the community is the highest. The animal and plant life of France during the 19th century, for example, was quite different from what it would have been had the country not been inhabited by man for centuries. It thus becomes impossible to study the natural landscape as something separate from the cultural landscape. Each community adjusts itself to prevailing natural conditions in its own way, and the result of adjustment may reflect centuries of development. Each single small community, therefore, has characteristics which will not be found in other places, even in places where the natural conditions are practically the same. In the course of time, man and nature adapt to each other like a snail in its shell. In fact, the relationship between man and nature becomes so intimate that it is not possible to distinguish the influence of man on nature and that of nature on man.[34] The area over which an intimate relationship between man and nature has developed through centuries constitutes a region.[35] The study of such regions should be the task of a geographer. Vidal de la Blache, therefore, argued for regional geography as the core of geography. According to him:

> Human societies, like those of plants and animal world, are composed of different elements subject to the influence of environment. No one knows what winds brought them together; but they are living together side by side in a region which has gradually put its stamp upon them. Some societies have long been part of the environment, but others are in process of formation, continuing to recruit numbers and to be modified day by day.

Societies have always begun to seek ways of satisfying their needs in the immediate vicinity. Vidal de la Blache believed that population is a constantly changing phenomenon. Mankind has in common with all other forms of life the tendency toward expansion. Man is the most adaptable and mobile organism on the face of the earth. He ensured that the population did not spread like a drop of oil; at the beginning it grew in clumps like corals.[36]

Vidal de la Blache used the following illustration in order to underline the long association between the major factors governing the development of a community. While the surface of a shallow lake is being swept by a gust of wind, the water is disturbed and confused but after a few minutes the contours of the bottom of the lake can clearly be seen again.[37] Similarly, war, epidemics and civil strife can interrupt the development of a region and bring chaos for a while, but when the crisis is over the fundamental developments reassert themselves.

Vidal de la Blache's model fitted well in the agricultural societies of France and other western countries of Europe. During the medieval period, these societies were agrarian. After the industrial revolution the situation has changed in the developed countries and now in such societies 'cultural determination' seems to be more conspicuously dominant. Up to the industrial revolution, Vidal de la Blache's approach was well suited to explaining the development of European agricultural landscape. In those parts of the world where industrialization is yet to take place, his hypothesis and theory of possibilism has great utility.[38]

After the industrial revolution in France, the traditional physical setting was disturbed. The railway tracks, canals, roads and industrial complexes initiated the decline of the traditional local self-sufficient economy. Industry was developed on the basis of new cheap and rapid means of transport and large scale production for a wider market.[39] These developments reduced the value of the regional method in a growing number of areas.

In the later part of his life, Vidal de la Blache reached the conclusion that with the industrial development the best in French life was vanishing. For future, he suggested that we should study the economic interplay between a region and the city centre which dominates it rather than the interplay of natural and cultural elements.

As a result of Vidal de la Blache's efforts, by 1921, there were 16 departments of geography in France, one in each of the 16 universities.[40] Interestingly enough, all the Chairs of Geography were occupied by the pupils of Vidal de la Blache. Thus, geography in France owes much to Vidal de la Blache, and he is rightly considered as the 'father of human geography' who advocated and pleaded for 'possibilism'.

The Principle of Terrestrial Unity

Paul Vidal de la Blache developed the idea of 'terrestrial unity'. In his opinion, the dominant idea in all geographical progress is that of terrestrial unity. The concept of the earth as a whole, whose parts are coordinated, where phenomena follow a definite sequence and obey general laws to which particular cases are related, had earlier entered the field of science by way of astronomy. In the words of Ptolemy, geography is 'the sublime science which sees in the heavens the reflection of the earth'. But the conception of terrestrial unity was confined to the domain of mathematics. It did not become part of the branch of geography until the time of Vidal de la Blache. In his opinion the phenomenon of human geography is related to terrestrial unity by means of which alone can they be explained. They are everywhere related to the environment. In fact, man is the creature of a combination of physical conditions.

The idea of terrestrial unity was borrowed from the Botanical Geography which was the first to use a conception of environment. Alexander von Humboldt, with his usual foresight pointed out how important is the appearance of vegetation in determining the character of a landscape. The general appearance of vegetation is certainly the most characteristic feature of a region. Absence of it is striking. Vegetation not only accentuate landforms, but give to the landscape by their shape, colour, and manner of grouping a common, individual character. Steppe, savanna, silva, taiga, etc. are collective terms which give an idea of such an ensemble. The rivalry of plants among themselves is so active that only those best adapted to the environment are able to survive. Even so, only a state of unstable equilibrium is maintained. Adaptation finds expression in different ways, in the height, size, and position of leaves, hairy covering, fibrous tissues, root, development, etc. Not only does each plant provide as best as it can for the carrying of its own vital activity, but many different plant associations are formed so that one may profit by the proximity of other. Whatever be the variety of species living side by side, whatever be the external differences in the process of adaptation, the entire plant population has a common stamp not to be mistaken by trained eye.

Similarly, animals with their power of locomotion and man with his intelligence are better able than plants to cope with the environment. Thus environment, as a composite, is capable of grouping and holding together heterogeneous plants, animals and mankind in mutual, vital interrelationship. This idea seems to be the law governing the geography of living creatures. This law of terrestrial unity is universally applicable to peoples of indigenous origin, ephemeral and migratory character.

In the study of man and environment this perspective is quite conspicuous. Prehistoric research has shown that man has been established since time immemorial in widely diverse parts of the globe, equipped with fire and fashioning tools; and however rudimentary his industries, the modifications that the face of the earth has undergone because of them cannot be ignored. The paleolithic hunter and earliest neolithic agriculturists destroyed certain species of plants and animals and favoured others. That these hunters and agriculturists operated independently of one another, in different localities, is proved by the various methods of making fire still in use. Man has influenced the living world longer and more generally than supposed.

There are numerous races, ethnic groups, and sub-races living in different physical environments in the various parts of the world. Nevertheless, all such heterogeneous groups blend in a social organization which makes the population of a country/region a unit when looked at in its entirety. It sometimes happens that each of the element of composite whole is well established in certain modes of life; some as hunters, other agriculturists, other shepherds; if such is the case, they cooperate with and supplement with one another. It most often happens, except in some migratory Gypsies, Gitanos, Zingani, Gaddis, Bakarwals, and some of the desert tribes like Badwins. Human societies, like those of vegetation and animal world, are composed of different elements subject to the influence of environment. No one knows what winds brought them together, nor whence, nor when; but they are living side by side in a region which has gradually put its stamp on them. The lifestyle of most of the societies of the world are in adjustment to their physical environment. The principle of terrestrial unity is of vital importance and is universally applicable.

Jean Brunhes (1869–1930)

Born in 1869, Jean Brunhes was a disciple of Vidal de la Blache. After studying history and geography, he prepared himself for the conceptual framework of human geography. On the lines of his master, he tried to identify the scope and method of human geography. His main work, *Geographic Humaine: Essai de Classification Positive* was published in 1910. He limited human geography[41] to: (1) unproductive occupation of soil; (2) things connected with the conquest of plant and animal worlds; (3) the destructive economy – 'robber economy' or violent attack on nature

Jean Brunhes

which may result in poverty. Moreover, he held the view that these three sets of observable phenomena did not mark the boundary of geographical regions. Beyond these phenomena, the geography of history, and social geography – are also the important components of a geographical region. He considered regional geography 'the range of observation is well-nigh unlimited, including epidemics, physical aptitudes, moral habits and social rules, property rights, social organization, collectivization, stock companies and social anarchy in large cities'. All these may be studied by a geographer as long as he sees any relationship between them and the facts of the earth surface. He opined that nature is not mandatory but permissive. All these physical and cultural factors, therefore, need to be taken into account for undertaking a comprehensive study of a region. In his method of geographical study, he emphasized two principles: (i) principle of activity, and (ii) principle of interaction.[42]

Principle of Activity

Jean Brunhes was of the opinion that the physical and cultural phenomena are in a state of perpetual change and they must be studied in the temporal change, instead of taking them as static in the time scale. He held the view that 'everything is either growing or diminishing', 'expanding or shrinking' and nothing is 'stable and static'. For example, the heights of mountain peaks, sea level, ice sheets, glaciers, size of valleys, deltas, volcanoes, and forests are continuously changing in their shape, size and altitude. So, in order to understand the interrelationship of physical and cultural components of a meso or micro unit, the principle of activity is to be kept in mind to arrive at a just synthesis.

Principle of Interaction

The idea of principle of interaction Jean Brunhes borrowed from Vidal de la Blache, who advocated the principle of terrestrial whole. Brunhes assumed that geographical phenomena (both physical and social) are closely interrelated with one another, and must be studied in all their numerous combinations or by keeping in mind their permutations and combinations. The idea of terrestrial whole or terrestrial unity was a fundamental concept which later on inspired 'regional synthesis'. All the physical and human forces are thus closely bound together because of the endless interrelations of the conditions they bring out.

In support of his principle of interaction, Brunhes examined the relationship between animals and cultivated plants, and determined with what forms of soil exploitation, with what kinds of cultivation and with what type of economic organization, these animals are generally associated. In brief, our efforts are based fundamentally on the great geographical principle of interaction, for man is like plants and animals, and therefore, the concept of interaction should dominate every complete study of geographical facts. The forces of physical nature are bound together in their consequences, in relations and in the consequence of these relations.

Elisée Reclus (1830–1905)

Elisée Reclus

A student of Carl Ritter, Elisée Reclus was a French geographer and anarchist. Elisée Reclus was first and foremost an uncompromising idealist. He was, however, the most successful of Ritter's students. He derived his main principles and ideas concerning geography from Ritter. He was expelled from the teacher seminary at Montauban in his very first year because he supported the ideals of the 1848 revolution. When only 20 years old he went to Berlin to study theology, but began to attend the popular lectures of Carl Ritter, which awakened his interest in geography. Returning to France in the autumn of 1851, Reclus resumed his political activity. Resisting the *coup d'état* of Napoleon III in the same year, he was obliged to flee to England with his brother (Dunbar, 1981: 155). Reclus then travelled extensively in North and South America, Central America, and Colombia more in order to observe than to do research. Sustained research was, in fact, beyond his means, for Reclus had to make a living from the meagre wages he could earn as a tutor and worker en route. In 1857, he returned to France but was again banished in 1872. In 1879 he was an ardent supporter of the Anti-Marriage Movement and demonstrated his sincerity by permitting his two daughters to live with their husbands without either civil or religious sanction. He was identified by the French government as one of the leading promoters of anarchism. In fact, he befriended the leading anarchist, Mikhail Bakunin (1814–76). From that time onwards, Reclus belonged to the inner circle of the secret anarchist association Fratemité Intemtionale. In 1871, he took active part in the Paris Commune but was captured during the first days of fighting. He was held in prison for almost a year. A sentence of deportation to New Caledonia was commuted to ten years' banishment as a result of the active intercession from geographical societies and such leading personalities as Charles Darwin. Reclus chose to settle in Switzerland in his exile. Although promised a readership in geography in the Université Libre in Brussels in 1892, the university reneged on the appointment for fear of demonstrations after an outbreak of anarchist violence in France in the autumn of 1893. A support committee for Reclus started to collect money and eventually founded the New University of Brussels where Reclus was professor for the last years of his life, refusing to take any salary since his modest needs could be satisfied by the income from his books.

Elisée Reclus was a social anarchist. He got recognition as a leading French geographer with a work of systematic physical geography called *La Terre*

(1866–67). He is, however, best remembered for his 19-volume regional geography *Nouvelle Geographie Universelle* (1875–94). The clarity and accuracy of this work made it much more popular than Ritter's *Erdkunde*, which had been its exampler in many respects. Reclus' work became a model for a range of encyclopaedic studies of the geography of the world and of particular countries.

Reclus was probably the most productive geographer of all time. In the ending part of his life, he wrote *L'Homme et la Terre* (mostly published posthumously in 1905–08), which may be described as *Social Geography*. This book gives a historical account of humanity's life on earth and its use of its resources. Reclus also wrote travel handbooks and articles. Although he was the best known French geographer of his time, Reclus never held a University Chair in France and had to earn a living from his writing. Because of his political and social anarchistic activities he was obliged to live abroad in exile and hence could not directly influence the development of French academic geography.

Like Ritter whose some lectures he attended, Reclus was mainly interested in the human aspect of geography. He had a sharp eye for the inequalities of human conditions around the world, and made it a central theme of his books.

Reclus devoted his political life to social justice and social conditions were inevitably always discussed in his books. He described the poverty and relief among the poor of London. Reclus also established a connection between geography and modern town planning and sociology. He was a possibilist who also recognized that human kind could have negative influences on the environment. He influenced and had close contacts with Frederic Le Play, the French sociologist and Sir Patrick Geddes, the Scottish biologist, social scientist and planner. Geddes, although not an anarchist, became a close friend of Reclus in the last decade of Reclus' life.[43] He spread Reclus' ideas in Britain and was most interested in his ideas on social geography, which he found a suitable basis for the development of his work on applied research and planning.

Reclus was a prolific writer. He became 'the Ritter of France'. His style was lucid, detailed, systematic, and balanced. He documented his writings with 3,000 maps, giving a very accurate picture of different world societies. In his work, *L'Homme et la Terre* (*The Earth and Its Inhabitants*), he asserted that man is not the product of his environment but an important part of it. Reclus declares:

> Man may modify (his dwelling place) to suit his own purpose, he may overcome nature, as it were, and convert the energies of the earth into domesticated forces. One must seek the gradual changes in the historical importance of the configuration of the land and in studying space we must take into account another element of equal value-time.[44]

Reclus was keenly interested in the conservation of nature and natural beauty. He was concerned about the destruction of the beauties of nature, and believed that man, in developing resources and building his works, should give grace and majesty to the scenery. But, as Reclus said, through abuse of his powers 'the barbarian gives to the earth he lives on an aspect of rough brutality', and in extreme cases 'where all grace and poetry have disappeared from the landscape, imagination dies out, the mind is impoverished, and a spirit of routine and servility takes possession of the soul'. Rural life is for these reasons preferable to the life of towns.[45]

Reclus pointed out that man has destroyed natural flora and wild animals and replaced them with his own cultivated crops and domesticated animals. He has changed the balance of nature, sometimes to his disadvantage, by introducing 'ruptures in the harmony of nature'. What is needed is 'a robust education face to face with nature'. This will give us the grandest development of the 'real love of nature'. Thus, Reclus discussed man-nature interrelationship in a very scientific way.

Emmanuel de Martonne (1873–1955)

The Vidalian approach which dominated the scene of French geographical writings in the late 19th and early 20th centuries, was vigorously pursued by his disciples and successors. Emmanuel de Martonne was a pupil and son-in-law of Vidal de la Blache. Martonne, from the beginning, specialized in physical geography and his special area of concentration was Central Europe. He had received training in geology, geo-physics and biology. He studied the physical geography of Carpathians. He was also interested in studying the problem of glacial erosion of the Alps. In 1904, he met William Morris Davis in the United States on the occasion of the Eighth International Geographical Congress and was much impressed by his work on geomorphology. The most popular works of Martonne include *Traite de Geographie Physique* and *La France Physique*. He inspired many French geographers to work in the field of physical geography.[46]

Emmanuel de Martonne

Albert Demangeon (1872–1940)

Albert Demangeon was also one of the outstanding pupils of Vidal de la Blache. He started his career as a school teacher and while teaching at Picady, he produced a monograph entitled 'La Picardie et les Regions Veisines'. This dissertation was highly appreciated and he was appointed in the University of Lile where he remained until 1911.

Though Demangeon was a close friend of Martonne, he concentrated mainly on human geography.

Albert Demangeon

Demangeon devoted most of his time in editing *Annals*, and contributed to this journal 31 articles and 89 notes.[47]

Demangeon made valuable contributions to the spatial variations of farmsteads, which he pursued throughout his life. He wrote on transport geography, population and international economics. He also made a classification of lands and prepared land use maps. He advised his students to work on major population groups of the Far East, the relations of the Whites and the Negroes, irrigation in arid countries, and the growth of great cities.

Demangeon was a teacher of innumerable qualities and possessed great vision. He was a modest person of polite nature. His contribution to the Vidalian tradition is widely acknowledged.

After the weakening of the Vidalian tradition, geography was studied with a new approach during the inter-war period. With the introduction of new tools and techniques, stress was laid on analytical study. This trend was more visible in the field of agricultural geography. The French scholars also adopted sophisticated statistical techniques. This trend led to interdisciplinary approach, and to developing regional synthesis. Among the younger generation, there is greater emphasis on locational analysis of phenomena. There is increasing cognizance of the branches of climatology, botany, and sociology. The new trend shows that the spatial phenomena cannot be explained simply by correlation as expressed by Vidal de la Blache; it needs to be explained as a part of the geographical whole (*ensemble*).

In brief, recent researchers are and have been oriented less towards description and regional analysis in a regional presentation, and more towards explanation with a genetic prospective. In physical geography, there is greater emphasis on process and systems of erosion, and reconstituting the stages of morphogenetic evolution. In human geography, emphasis has been placed on economic factors of production and consumption, economic systems and regimes, zones of influence, economic regionalization, the analysis of traffic flow, far more so than on the relations of these with natural milieu.[48] Regional geography in France is now more concerned with describing and explaining the complexity of the organization of space. To achieve this goal, there is more use of detailed maps, and above all areal photographs.[49]

THE BRITISH SCHOOL OF GEOGRAPHY

Geography, one of the oldest fields of human curiosity, drew very little attention in Britain until the middle of the 19th century. Exploration of new lands, description of travels and voyages, description of the newly discovered lands and their people were considered areas of geographical studies. Geography was introduced in educational institutions at a very late stage; in schools and colleges, memorization of facts, names of places, mountains and rivers used to be taught by historians and geologists. Up to this time, geography to the British was nothing but an encyclopaedia of information.[50]

In the middle of the 19th century, academics were busy in accumulating facts and information about the newly discovered places of the world. In 1859,

Darwin published his *Origin of Species* which attracted the attention of biologists, geologists and sociologists.[51] After this evolutionary theory, the British scholars started paying attention to the earth as the home of man. It was at the end of the 19th century that geography was introduced as a discipline in the British Universities.

Halford J. Mackinder (1861–1947)

Halford J. Mackinder

Halford J. Mackinder, a scholar of dynamic personality and a person who can give simple expression to complex ideas, with an imaginative mind, is known as the founder of the British School of Geography. He was the first recorded person to climb Mt. Kenya. Before becoming known for his writings on geopolitics, Mackinder was active in lecturing to audiences around the country on his vision of geography. On the strength of his views he was appointed to a position at Oxford University becoming the first of a new generation of academic geographers and a vocal supporter of geographical education. In his opinion to become a discipline rather than a mere body of information, the subject had to bridge the natural sciences and humanities and take as its core 'the interaction of man in society and environment varies locally'. Mackinder's version of the geographical experiment held physical and human geography together in an evolutionary perspective, while stressing the utility of the subject for teachers, scientists, statesmen and merchants alike. His views have often been recalled in debates as to whether human and physical geography should be kept together. His thinking was governed by visualization, both on the map and in the mind, of the world's regional complexes as combinations of varied physical and human elements. He considered geography as a bridge between the humanities and the natural sciences, between history and geology. He applied these concepts to the interpretation of world political affairs.

With the appointment of Mackinder as Professor of Geography in the University of Oxford in 1887, geography started growing in the United Kingdom. At the initial stage, the British geographers were exclusively concentrating in the field of physical geography, in which there was hardly to be found any description of man as an agent of change in the physical surroundings. Mackinder identified geography as a discipline that traces the interaction of man with his physical environment. In 1904, he formulated the concept of 'the geographical pivot of history', which is also known as the 'Heartland Theory of Mackinder'. Heartland is a geopolitical concept which was used in the cold war discourse to denote an area of Eurasia roughly synonymous with the Siberia of USSR. In this theory, Mackinder identified a 'World Island' (Figure 7.1) consisting of the continents of Eurasia and Africa. The most inaccessible part of the world, he called heartland. This is the area of low population and difficult accessibility. He summarized his view of global strategy in the following famous lines:[52]

Figure 7.1 Mackinder's World Island

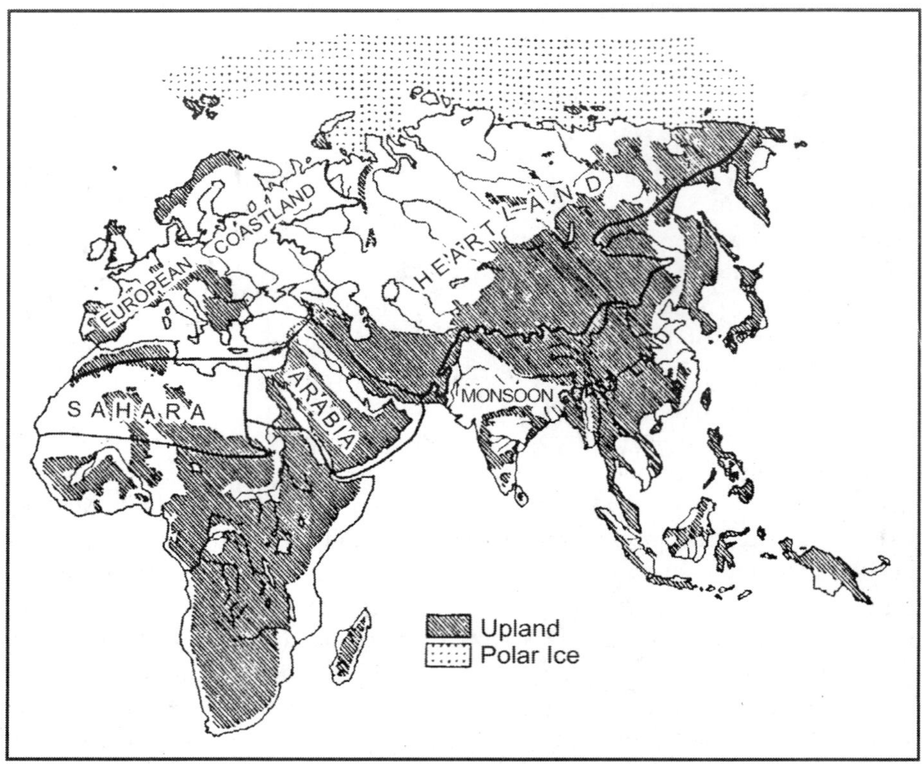

Who rules East Europe commands the Heartland;
Who rules Heartland commands the World Island;
Who rules the World Island commands the World.

Mackinder declared that throughout the history of mankind, the coastal lands had always proved vulnerable to attack from the heartland, and the heartland remained invulnerable because sea power could be denied access to it. Thus, Mackinder was mainly concerned with a global view. He repeated the same view in 1943 – a few years before his death during the Second World War. He warned of the danger of the heartland falling in its entirety under the control of the Soviet Union and of Russia's ability then to strike out to peripheral lands to the east, south and west of the 'World Island'. Despite the environmental determinism inherent in much of Mackinder's geopolitical writings, the simplicity of the heartland concept was to play an influential part in Western geopolitical thinking concerning the image of expansionist USSR. His regional concept also pervaded his interpretation of countries. According to some scholars, Mackinder's thinking was a generation ahead of his time.

Mackinder wrote *Britain and the British Seas*, which was published in 1902. This book is considered a classic in modern British literature which shows a more mature and sounder approach to a regional interpretation of Britain and its seas. His second great work *Democratic Ideals and Reality* was published in 1919. In this book, he discussed world power politics.[53]

A contemporary of Mackinder, H. Robert Mill wrote a book *Realm of Nature* in which he discussed the races of man.

The concept of 'region' was a popular theme for British geographers before the First World War. The British geographers were influenced by Vidal de la Blache, and the leading French sociologist Frederic Le Play.[54] Sir Patrick Geddes – the Scottish geographer – was a follower of Le Play, who carried on research on family lifestyles and family budgets. He recognized that family life is dependent on family lifestyle and family budgets. Le Play recognized that family life is dependent on the means of obtaining subsistence, that is, work; while the character of the latter is largely determined by the nature of the environment, that is, place. This led to the famous Le Play formula which is basic to his ideas – place, work, family – which Geddes transformed into the slogan – place, work, folk as the basic concepts in the study of cities and regions.[55]

The Geddes' scheme, given in Table 1, was widely used between the wars in teaching of geography, and by regional and city planners in their diagnostic surveys preparatory to planned action. Thus, Geddes was the founder of the *Le Play Society*. The Le Play Society was named after a 19th century engineer, Frederic Le Play, who published accounts of the places that he visited and developed a scheme (with strong overtones of environmental determinism) of place-work-family to encapsulate the major features of local societies. His ideas were taken up and promoted by Sir Patrick Geddes. The society was disbanded in 1960, having organized 71 major field surveys and published 8 major monographs during its existence.

Table 1 *The Geddes Scheme of Regional Survey, Showing the Interaction of Place, Work and Folk*

Place	*Place Work (natural) advantage*	*Place Folk (native)*
Work Place (pasture, fields, mines, workshop)	Work	Work Folk (industrial)
Folk Place (village home, etc.)	Folk Work (occupation)	Folk

Geddes put forward the idea that geography is not a mere descriptive science; it is an *applied science*, dealing with what ought to be. Geddes also influenced the study of regionalization. Andrew J. Herbertson (1865–1915) – an assistant to Geddes at Oxford – presented a scheme for a division of the world into natural regions, based on association of surface features, climate and vegetation.[56] Another British geographer – A.G. Ogilive – edited *Great Britain: Essays in Regional Geography* in 1928. This book gives a glimpse of the regional variations in the physical and cultural landscapes of Britain. Roxby suggested a scheme which explained how the regional studies should proceed. He pointed out that a systematic regional study should deal with geology, drainage, coastline, climate, vegetation to demarcate natural regions, and this should follow the 'man's relation to his physical environment'.[57]

Economic Geography

In the inter-wars period, economic geography was a very popular field of study in Britain. In all the studies of economic geography, the influence of natural factors on the site of resources and location of economic activities has been studied. Buchanan, in his book, *Pastoral Industries of New Zealand*, investigated the influence of physical environment on the economic conditions, crops and industries of the area.[58] In 1949, W. Smith produced the *Economic Geography of Britain*. This was a thorough, logical and systematic investigation of the regional variants of the economics of production in the country. It is in concept and substance an ecological approach far ahead of its contemporaries. It was based entirely on the analysis of statistical data, and showed no concern with field observation. Chisholm[59] produced *Handbook of Commercial Geography*. After Chisholm, Robert Mill wrote *General Geography*. His poor health did not permit him to conduct field studies and explorations. He, however, wrote biographies of polar explorers.[60] He was appointed as the Director of British Rainfall Organization in 1901. Under his supervision, the rainfall maps of Great Britain were prepared on the basis of 50 years average.[61] He prepared a plan to plot the land use of Britain on the ordnance survey on one inch to one mile scale. But the proposal was not approved. Later on, in 1930, L.D. Stamp started preparing the land use map of Britain.

In 1905, when Herbertson became the Director of the School of Geography, he prepared a framework of natural regions for the study of world, regional and economic geography. He divided the world into 15 natural regions. These divisions show a regularity of climate because the same regions appeared in similar positions on each of the continents.[62]

Man-Nature Interaction

From 1880 till the outbreak of the First World War, geography in Britain was dominated by man and nature relationship. Geography during this period was considered to be a record of the description of the earth surface and its influence on man.[63] The British geographers of this period delineated major natural regions of the world to ascertain how man is playing his role in these regions. The regional synthesis, advocated by the French geographers, was not acceptable to the British geographers since they believed that such a synthesis was not achievable. Consequently, they paid more attention to physical geography and brought out a large number of monographs in the field of geomorphology.[64] Forde published *Habitat, Economy and Society* on primitive societies to show the influence of environment on the occupation and mode of life of the people.[65] Forde laid stress on cultural groups in their areal pattern, that are dependent upon culture and cultural contacts rather than on direct impact of physical environment. The principle of 'environmental determinism', as a central theme of selection and interpretation, has been flatly rejected in these studies.[66]

Agricultural Geography

At the beginning of the 20th century, agricultural geography became an important field of geographical research when land use survey attracted the attention of the British geographers. After the First World War, it was realized that each unit of land should be used judiciously. In 1920, Stamp prepared the land use maps of Britain, which were used for a variety of purposes in addition to the planning of emergency crop expansion during the war. After the Second World War, they became the basis for the reconstruction of Britain.[67] In 1965, in recognition of his contribution, Stamp was knighted. The geography of rural settlement and urbanization made little progress during the inter-war period. Fleure and his followers made some studies of the rural settlement but these cannot be considered substantial contributions.[68] Virtually no attention was paid to the spatial structure of city.[69]

Historical Geography

Historical geography was also a good hunting ground for the British geographers after the First World War. The foundation of historical geography in Britain was laid by Mackinder. He stressed the point that geographers should try to rebuild past geographies and show how sequence of change has led to the present observable features; otherwise geography would become a mere description of contemporary features.[70] Historical geography evaluates through time the changing interconnections of man and nature. During this period, some of the prominent scholars who contributed to the field of historical geography were M.I. Newbegin, E.G.R. Taylor, E.W. Gilbert and H.C. Darby.[71]

Contemporary Trends in British Geography

There has been a tremendous change in the philosophy, approaches, and scope of geographical studies during the last three and a half decades. The major change was in the form of 'quantitative revolution'. In this period, scholars like Richard Chorley and Peter Haggett made enormous use of sophisticated statistical techniques, and formulated models and theories in the fields of human and social geography. In the post-war period, geography in Britain has come to be recognized as a spatial science.[72] Now geography is concerned with providing accurate, orderly, and rational description of the variable character of the earth surface.[73] The quantitative revolution started in the United States and quickly spread in Britain. The British geographers, consequently, focused their attention on quantification, multivariate analysis, statistical description of patterns, formulation and testing of hypotheses. A number of books have been written by Haggett and Chorley which explain the space relations and locational analysis with the help of models. Some of the major works are *Models in Human Geography, Locational Analysis in Human Geography* and *Frontiers in Geographical Teaching*.[74]

Some other aspects to which the British geographers tend to pay attention are inter-regional and intra-regional inequalities in society, environmental degradation, ecological crisis and environmental management. Public welfare, social amenities, medical geography and landscape ecology are some of the new areas in which the British geographers have broken fresh ground. Moreover, substantial contributions are being made in the fields of concepts, physical, regional, historical, economic, transport, agricultural and political geography.

Some of the new concepts which have received the attention of the British geographers are positivism, pragmatism, existentialism, idealism, realism, environmental causation, radicalism and dialectical materialism (see Chapter 11). Positivism is an important theory on which the British geographers have focused their attention. Positivism is a naturalistic-pragmatic trend in modern thought. It is a type of empiricism which says that science can only concern itself with empirical questions (those with a factual question), not with normative questions (questions about values and factual contents). Empirical questions are questions about how things are in reality and 'reality' is defined as the world which can be sensed. This means science is concerned with objects in the world.[75]

Positivism holds that since we cannot investigate such things as moral norms with our senses, we should keep away from normative questions. We cannot justify our tastes scientifically. Another major aspect of positivism is its emphasis on the unity of science. Scientific status is guaranteed by a common experience of reality, a common scientific language and method which ensure that observations can be repeated. Since science has unified method, there can only be one comprehensive science. In brief, the followers of positivism believe that as natural sciences discovered the laws of nature, so scientific investigations of communities would discover the laws of society. They admit that social phenomena are more complex than natural phenomena but believe strongly that the laws governing society would eventually be discovered.[76]

Some of the British geographers are concentrating on the philosophy of pragmatism to solve the problems of societies. The concept of pragmatism was borrowed by the British from America. According to this philosophy, there should be greater emphasis on 'practical problems'. A pragmatist believes that a 'concrete' or a particular situation is important to obtaining scientific knowledge and for understanding the world.

The concept of realism is a new philosophy in the British geographical literature. In the Platonic-Socratic thought, the term 'realism' was used in opposition to nominalism for the doctrine that universal and abstract entities have real objective existence. But the same term can and does take separate forms. At present, realism is in opposition to idealism. Recently, behaviouralism and humanism are the philosophical themes of human geography which are attracting the attention of British geographers (see Chapter 11).

The Royal Geographical Society (RGS) sponsors scientific expeditions and the *Geographical Journal* reports their findings. In 1995, the RGS merged with the Institute of British Geographers established in 1933 by geographers to promote academic research through conferences and publications, and the Society now has a much wider remit.

THE AMERICAN SCHOOL OF GEOGRAPHY

Geography as a subject in the American schools and colleges was diffused by the Europeans in the middle of the last century. In some of the universities like Harvard, Dartmouth, Mary, Yale, Columbia, Princeton and Pennsylvania, courses in mathematical, physical and historical geography were given as early as 1795. Arnold Guyot of Princeton, was the first Professor of Geography in America. Guyot was a follower of Ritter, a teleologist, and his teleological viewpoint, considered the continents as the abode of man, and the theatre for the action of human societies.[77] Guyot's view of the role of man on the earth had a great influence on the attitudes of his era through many of his school textbooks. Guyot was a student of the physical world. He explored the Appalachian Mountains for over thirty years, and made some 12,000 observations and collected meteorological data. His philosophical approach was in conformity with that of Carl Ritter.

In 1870, physical geography was started at Harvard in the Department of Natural History. In 1878, William Morris Davis – an eminent geologist and geomorphologist – became an Assistant Professor of Geography. Moreover, the American Geographical Society was started in 1851.

Shaler was a distinguished geologist at the Harvard University. William Morris Davis was one of his pupils. Shaler wrote *Nature and Man in America*.[78] In this book, he discussed 'the effect of physical conditions of the earth on the development of organic life in general with special emphasis on the origin, and character of the relief and climate'. The second-half of his book discusses geographic influences upon man and the way in which geographic features have controlled settlements. Virtually all thinking at this time adopted the view that the life of man on the earth was ordained by the offerings of the physical environment. It is against this background that Davis, a geologist, emerges in the first quarter of this century as the main founder of modern geography in America.

William Morris Davis (1850–1934)

William Morris Davis

William Morris Davis was born of quaker parents in Philadelphia in 1850. He graduated from Harvard in 1869. From 1870 to 1873, Davis worked as an assistant at Argentina Meteorological Observatory in Cordoba (Argentina). He returned to Harvard for further geological and geomorphological studies where he was appointed an assistant to N.S. Shaler in 1876. In 1878, he was given the designation of Assistant Professor, and became Professor of Geography in 1899. He was one of the founders of the Association of American Geographers, which was established in 1904. He introduced professional geography into the United States.

While working with Shaler, he learned the art of careful observation and made use of it in logical and impersonal argument. Moreover, he acquired the habit of seeing man and his works as part of the landscape, not separate from it. He also gained a clear appreciation of the importance of processes of change in explaining the varied features associated on the face of the earth.

In 1877, while doing observations in Montana, he developed the Theory of Cycle of Erosion which he defined as *geomorphological cycle*. Elsewhere he refers to is as *topographical cycle*. This cycle, in Davis' own words, is as under:[79]

> It is a scheme under which a mental counterpart of every land form is developed in terms of its understructure of the erosional process that has acted upon it, and of the stage reached by such action in term of the whole sequence of stages from the initiation of a cycle of erosion by upheaval or the deformation of an area of the earth crust, to its close, when the work of erosion has been completed; and the observed land is then described not in terms of its directly visible features, but in terms of its inferred mental counterpart.

Davis presented his theory in the International Geographical Congress in 1899. In this model, Davis postulated that when an initial surface is raised, rivers at once begin the work of erosion. The surface is cut by narrow V-shaped valleys that are extended headward as more and more of the initial surface is consumed. But rivers cannot cut down their valleys indefinitely. There is a base level below which rivers cannot cut – a level determined by the surface of the body of water into which a stream flows.[80]

Davis was a dedicated teacher and an engaging speaker. Mark Jafferson, Isaiah Bowman, Ellsworth Huntington, Ellen Churchill Semple, and Albert Brigham were some of his students.

In his later writings, Davis shifted his focus of study and asserted that the study of man on the earth could not be limited to the elements of physical environment. The ecological study of human groups, as of plants and animals, named ontography, demands the appraisal of adjustment to the physical earth as well as migration and segregation – a view that was basic to Ratzel's work. This change of viewpoint is abundantly evident in Davis' later writings on the nature of regional geography. He recognized that regionalization of phenomena on the earth's surface is the product of three forces – the site base, migration and association. Regional geography, he writes, seeks to describe 'the geographical elements of a given area in their totality as they exist together in their natural combinations and correlation'.

Davis was a critic of human geographers. He was of the opinion that human geographers 'fail to become all-round geographers' and that their studies are unbalanced and lack homologous treatment as they have less concern with the chronology of existing landforms than with the features of existing surfaces. He insisted that regional descriptions must be 'homologous', that is, in all aspects, land, climate, vegetation, animals and man should receive equal emphasis. He claims that the purpose of 'geographical study of man is to arrive at descriptive generalization on the basis of explanation of geographical qualities'. In the later years of his life, he became markedly ecological. The ecological approach implies that the geographer studies life forms on the earth's surface in terms of their adaptation to the site base, the migration of particular elements or ideas, and to the modes of spatial association or segregation. It is far removed from exclusive concern with land-man relationship in the Darwinian sense. It is thus clear that Davis' attitude in geography passed through both phases – deterministic and ecological. Later on, the German scholar Penck criticized the work of Davis, but there is no denying the fact that he was one of the pioneers in the field of geomorphology who tried to bridge the gap between geology and geography.

Mark Jafferson (1863–1949)

Mark Jafferson was an American geographer. He studied with William Morris Davis. He was a distinguished teacher and researcher at the Michigan State Normal College in Ypsilanti. Besides being a geographer, he was a lucid and prolific writer, and a linguist. He travelled widely and corresponded frequently and at length. In America, he was the founder of real man-oriented geography. The approach of Jafferson, in the period dominated by Davis, sought for generalizations regarding man's imprint on the land. He examined the size, extent and functions of cities, and areas of access to railroads under the succinct phrase 'the civilizing rails'.[81]

Mark Jafferson

The concepts of 'central places', the laws of 'primate city' and the 'civilizing rails' were coined and identified by him. His students of urban geography, urban centres as 'central places' were far ahead of their time.[82] He stated many times that his geographical concern was with men: 'where they are', 'what they are like' and 'why they are there'. This approach is eminently ecological, which seeks to discover casually interrelated associations of various distributions that lead to further understanding of human groups in their environmental settings.

Isaiah Bowman (1878–1950)

Isaiah Bowman

Isaiah Bowman was one of the best products of Jafferson. Born in a farmer's family, he was brought up on a farm in Michigan, where his mother stimulated his early interest in geography. After completing his early education, he reached Harvard where he got education under the supervision of Davis. In 1905, he was appointed instructor at Yale under H. E. Gregory and received his PhD.

Bowman travelled extensively in Andes, especially the Peru, Bolivia and Chile. Later on he crossed the desert of Atacam. Based upon this fieldwork, he wrote his doctoral dissertation 'The Geography of the Central Andes'.[83] Moreover, in 1931, he published *The Pioneer Fringe* in which he stated the nature of the problem, and cited examples from the United States, Canada, Australia, South Africa, Siberia, Mongolia, Manchuria, and South America. In the initial part of his career, he was an environmental determinist, but later on he gave adequate importance to the role of man in transforming his natural environment. In his own words, 'man is changing himself as well as the world as he goes along'.[84]

Ellen Churchill Semple (1863–1932)

Ellen Churchill Semple was the foremost female geographer of her time and a leading exponent of environment determinism. She was M.A. in history, who introduced Ratzel's anthropogeography to America. Both her books *Influences of Geographic Environment* (1911) and *American History and its Conditions* (1903) were the result of her admiration for the work of Friedrich Ratzel whose lectures she had attended at Leipzig in 1891–92 and again in 1895. She was a pupil and follower of Ratzel. She was born in a highly cultured family of the United States and graduated from Vassar in 1891. She went to Germany to continue her studies under the guidance of Ratzel. In Germany, she took keen interest in the study of anthropology, and was greatly stimulated by Ratzel's

Ellen Churchill Semple

new approach in the study of man and environment. She visited Siberia, Mongolia, Japan, China, Philippines, Java, India and many of the European countries. She presented her version of the first volume of Ratzel's *Anthrogeographie* in her classic work *Influences of Geographic Environment* which was published in 1911. Semple's philosophy and methodology were based on the ideas of Ratzel. A combination of geography and history can be found in her writings. About spatial relationship, Semple has aptly remarked: 'Location means climate and plant life at one end of the scale, civilization and political status at the other'. She was a determinist, which can be appreciated from the opening para of her book.[85]

> Man is a product of the earth's surface. This means not merely that he is a child of the earth, dust of her dust; but that the earth has mothered him, fed him, set him tasks, directed his thoughts, confronted him with difficulties that have strengthened his body and sharpened his wits, given him his problems of navigation or irrigation, and at the same time whispered hints for their solution. She has entered into his bone and tissue, into his mind and soul. On the mountains she has given him leg muscles of iron to climb the slope; along the coast she has left these weak and flabby, but given him instead vigorous development of chest and arm to handle his paddle or oar. In the river valley she attaches him to the fertile soil, circumscribes his ideas and ambitions by a dull round of calm, exacting duties, narrows his outlook to the cramped horizon of his farm. Up on the wind-swept plateaus, in the boundless stretch of the grasslands and the waterless tracts of the desert, where he roams with his flocks from pasture to pasture and oasis to oasis, where life knows much hardship but escapes the grind of drudgery, where the watching of grazing herd gives him leisure for contemplation, and the wide ranging life a big horizon, his ideas take on a certain gigantic simplicity; religion becomes monotheism, God becomes one, unrivalled like the sand of the desert and the grass of the steppe, stretching on and on without break or change. Chewing over and over the cud of his simple belly as the one food of his unfed mind, his faith becomes fanaticism; his big spatial ideas, born of that ceaseless regular wandering, outgrow the land that bred them and bear their legitimate fruit in wide imperial conquests. (Semple, 1911: 1–2)

Man can no more be scientifically studied apart from the ground which he tills, or the lands over which he travels, or the seas over which he trades, than polar bear or desert cactus can be understood apart from its habitat. Man's relations to his environment are infinitely more numerous and complex than those of the most highly organized plant or animal. So complex are they that they constitute a legitimate and necessary object of special study. The investigation which they receive in anthropology, ethnology, sociology, and history is piecemeal and partial, limited as to the race, cultural development, epoch, country or variety of geographic conditions taken into account. Hence all these sciences, together with history so far as history undertakes to explain the causes of events, fail to reach a satisfactory solution of their problems largely because the geographic factor which enters into them all has not been thoroughly analysed. Man has been so noisy about the way he has conquered nature and nature has been so silent in her persistent influence over man, that the geographic factor in the equation of human development has been overlooked.

In every problem of history there are two main factors, variously stated as heredity and environment, man and his geographic conditions, the internal forces of race and the external forces of habitat element in the long history of human development has been operating strongly and operating persistently. Herein lies its importance. It is stable force. It never sleeps. This natural environment, this physical basis of history, is for all intents and purposes immutable in comparison with the other factor in the problem – shifting, plastic, progressive, retrogressive man.

Geographical environment, through the persistence of its influence, acquires peculiar significance. Its effect is not restricted to a given historical event or epoch, but, except when temporarily met by some strong counteracting force, it tends to make itself felt under varying guise in all succeeding history. It is the permanent element in the shifting fate of races. Islands show certain fundamental points of agreement which can be distinguished in the economic, ethnic and historical development of England, Japan, Melanesian, Fiji, Polynesia, New Zealand and pre-historic Crete. The great belt of deserts and steppes extending across the Old World gives us a vast territory of rare historical uniformity. From time immemorial they have borne and bred tribes of wandering herdsmen; they have sent out the invading Hordes who, in successive waves of conquest, have overwhelmed the neighbouring river lowlands of Eurasia and Africa. They have given birth in turn to Scythians, Indo-Aryans, Avars, Huns, Saracens, Tartars and Turks, as to the Tuareg tribes of the Sahara, the Sudanese and Bantu folk of the African grasslands. But whether these various people have been Negroes, Hamites, Semites, Indo-Europeans or Mongolians, they have always been pastoral nomads. The description given by Herodotus of the ancient Scythians is applicable in its main features to the Kirghis and Kalmuck who inhabit the Caspian plains today. The environment of this dry grassland operates now to produce the same mode of life and social organization as it did 2,400 years ago; it stamps the cavalry tribes of Cossacks as it did the mounted Huns, energizes its sons by its dry bracing air, toughens them by its harsh conditions of life, organizes them into a mobilized army, always moving with its pastoral commissariat. Then when population presses too hard upon the meager sources of subsistence, when a summer drought burns the pastures and dries up the water holes, it sends them forth on a mission of conquest, to seek abundance in the better watered lands of their agricultural neighbours. Again and again the productive valleys of the Hoangho, Indus, Ganges, Tigris and Euphrates, Nile, Volga, Dnieper and Danube have been brought into subjection by the imperious nomads of arid Asia, just as the 'hoe-people' of the Niger and upper Nile have so often been conquered by the herdsmen of the African grasslands. Thus, regardless of race or epoch – Hyksos or Kaffir – history tends to repeat itself in these rainless tracts, and involves the better watered districts along their borders when the vast tribal movements extend into these peripheral lands.

Climatic influences are persistent, often obdurate in their control. Arid regions permit agriculture and sedentary life only through irrigation. The economic prosperity of Egypt today depends as completely upon the distribution of the Nile waters as in the days of the Pharaohs. The mantle of the ancient Egyptian priest has fallen upon the modern British engineer. Arctic explorers have succeeded only by imitating the life of the Eskimos, adopting their clothes, food, fuel, dwelling and mode of travel. Intense cold has checked both native and Russian development over that major portion of Siberia lying north of the mean annual isotherm of 0 degree C (32 degrees F); and it has had a like effect in the corresponding part of Canada. It allows these sub-arctic lands, scant resources and a density of population of less than two persons per square mile. Even with the intrusion of white colonial people, it perpetuates the savage economy of the native hunting tribes, and makes the fur trader their modern exploiter, whether he be the Cossack tribute-gatherer of the lower Lena river, or the factor of the Hudson Bay Company. The assimilation tends to be ethnic as well as economic, because the severity of the climate excludes the white woman. In the same way the Tropics are a vast melting pot. The debilitating effects of heat and humidity aided by tropical diseases, soon reduce intruding peoples to the dead level of economic inefficiency characteristic of the native races. These, as the fittest, survive and tend to absorb the newcomers, pointing to hybridization as the simplest solution of the problem of tropical colonization.

The more the comparative method is applied to the study of history – and this includes a comparison not only of different countries, but also of successive epochs in the same country – the more apparent becomes the influence of the soil in which humanity is rooted, the more permanent and necessary is that influence seen to be. Geography's claim to make scientific investigation of the physical conditions of historical events is then vindicated. 'Which was there first, geography or history?' asks Kant. And then comes his answer: 'Geography lies at the basis of history'. The two are inseparable. History takes for its field of investigation human events in various periods of time; anthropogeography studies existence in various regions of terrestrial space. But all historical development takes place on the earth's surface, and therefore is more or less molded by its geographic setting. Geography, to reach accurate conclusions, must compare the operation of its factors in different historical periods and at different stages of cultural development. It therefore regards history in no small part as a succession of geographical factors embodied in events. Back of Massachusetts' passionate abolition movement, it sees the granite soil and boulder-strewn fields of New England; back of the South's long fight for the maintenance of slavery, it sees the rich plantations of tidewater Virginia and the teeming fertility of the Mississippi bottom lands. This is the significance of Herder's saying that 'history is geography set into motion'. What is today a fact of geography becomes too narrow a factor of history. The two sciences cannot be held apart without doing violence to both, without dismembering what is a natural, vital whole. *All historical problems ought to be studied geographically and all geographic problems must be studied historically.* Every map has its date. Those in the Statistical Atlas of the United States showing the distribution of population from 1790 to 1890 embody a mass of history as well as of geography. A map of France or the Russian Empire has a long historical perspective; and on the other hand, without that map no change of ethnic or political boundary, no modification in routes of communication, no system of frontier defences or of colonization, no scheme of territorial aggrandizement can be understood.

Montesquieu ascribes the immutability of religion, manners, custom and laws in India and other oriental countries to their warm climate. Buckle attributes a highly wrought imagination and gross superstition to all people, like those of India, living in the presence of great mountains and vast plains, knowing nature only in its overpowering aspects, which excite the fancy and paralyse reason. He finds, on the other hand, an early predominance of reason in the inhabitants of a country like ancient Greece, where natural features are on a small scale, more comprehensible, nearer the measure of man himself. The scientific geographer, grown suspicious of the omnipotence of climate and cautious of predicating immediate psychological effects which are easy to assert but difficult to prove, approaches the problem more indirectly and reaches a different solution. He finds that geographic conditions have condemned India to isolation. On the land side, a great sweep of high mountains has restricted intercourse with the interior; on the seaside, the deltaic swamps of the Indus and Ganges rivers and an unbroken shoreline, backed by mountains on the west of the peninsula and by coastal marshes and lagoons on the east, have combined to reduce its accessibility from the ocean. The effect of such isolation is ignorance, superstition, and the early crystallization of thought and custom. Ignorance involves the lack of material for comparison, hence a restriction of the higher reasoning processes, and an unscientific attitude of mind which gives imagination free play. In contrast, the accessibility of Greece and its focal location in the ancient world made it an intellectual clearing house for the eastern Mediterranean. The general information gathered there afforded material for wide comparison. It fed the brilliant reason of the

Athenian philosopher and the trained imagination which produced the masterpiece of Greek art and literature.

A race or tribe develops certain characteristics in a certain region. Then moves on, leaving the old abode but not all the accretions of custom, social organization and economic method required there. These travel on with the migrant people; some are dropped, others are preserved because of utility, sentiment or habit. For centuries after the settlement of the Jews in Palestine, traces of their pastoral life in the grasslands of Mesopotamia could be discerned in their social and political organization, in their ritual and literature. Survival of their nomadic life in Asiatic steppes still persists among the Turks of Europe, after six centuries of sedentary life in the best agricultural land of Balkan Peninsula. One of these appears in their choice of meat. They eat chiefly sheep and goat, and beef very rarely and swine not at all. The first two thrive on poor pastures and travel well, so that they are admirably adapted to nomadic life in arid lands; the last two, far less so, but on the other hand are the regular concomitant of agricultural life. The Turk's taste today, therefore, is determined by the flocks and herds, which he once pastured on the Trans-Caspian plains.

Semple in the final phase of her life started writing a book on the geography of the Mediterranean which was slowly completed. Apart from being a keen researcher, Semple was a very engaging and enormously persuasive teacher. Her style of writing has exceptional literary quality which make reading a delight. She produced a large number of future geographers.

She emphasized on the methodology of comparison of typical people of all races and all stages of cultural development, living under similar geographic conditions. Such people have similar history, culture and mode of life.[86]

Albert Perry Brigham (1855–1932)

Albert Perry Brigham was a minister of a large church in Utica. He was interested in geology and geomorphology. In order to quench his thirst for study, he resigned from the post and joined the University of Harvard to work with Shaler – the well-known geologist of his period in America. Later on, he shifted his interest from geomorphology to the study of human groups and became critical of the environmental determinists.[87] Brigham was against vague and unproved assertions of climatic influence on racial character, skin colour, and man's constitution. In his lectures on the problems of geographic influence, he asserted that whether we speak of influence, or response, or adjustment, matters little. We should interpret cautiously similar human phenomena in different climates and regions of the world. There is a possibility that the same thing appears in many places, either through the unity of human spirits or the likeness of environment or from both causes.[88] Apart from attending to his academic work, he helped run the Association of American Geographers.

Carl O. Sauer (1889–1975)

Carl O. Sauer

Carl Sauer was one of the leading American geographers of the 20th century. For most of his academic life he was associated with the University of California, Berkeley. During his graduation course he attended lectures by Ellen Semple. He, however, rejected the environmental determinism of Ratzel and Miss Semple. Most of his survey was done in Latin America and the less industrialized parts of USA.

He contributed mainly in the field of origin of agriculture, the diffusion of plants and animals, and the impact of conquest upon indigenous American societies (Red Indians). The focus of his study was on the processes leading to landscape change up to the present, beginning at the prehuman stage of occupance. Thus, the human geographers should make cultural processes the base of their thinking and observation. His presentation in the symposium of Princeton (1955) on 'Man's Role in Changing the Face of the Earth' which was conceived by Thomas was highly appreciated. He specifically insisted upon the use of the historical method in geography. In his opinion, the present 'cultural landscapes' are to be studied in terms of their development from the original 'natural landscapes'. He also opined that 'geography dissociates itself from geology at the point of the introduction of man into the areal scene'. The geologists, however, do not agree with this viewpoint as changes brought by man on the earth surface may have long and serious geological consequences (e.g. landslides, lowering of underground watertable, waterlogging and soil creep).

In the opinion of Sauer geography deals with area, region, or landscape. In fact, no other subject pre-empted the study of area. He emphasized that the phenomena that make up an area are not simply assorted (independent) but are associated, or interdependent. To discover this areal 'connection of the phenomena and their order' is a scientific task to which the geography should devote its energies.

Carl Sauer opined that the task of geography is conceived as the establishment of a critical system which embraces the phenomenology of landscape. The term landscape, under the influence of the Dutch painters refers to the appearance of an area, more particularly to the representation of scenery.

The term 'landscape' was introduced into American geography by Carl Sauer in 1925 with the publication of 'The Morphology of Landscape'. This influential article drew on the concept of *Landschaft* developed by German geographers, most prominently Passarge and Schluter. Sauer put forward the concept of landscape as an alternative to environmental determinism. While environmental determinism sought to specify the causal influence of environment on humans, the landscape approach sought to describe and explain the interrelations between humans and the environment, with primary attention given to the human impact on the environment. He envisaged the study of landscape in geography as a scientific endeavour. Under this view the landscape was defined as an area made up of a distinct association of forms, both physical and cultural. He advocated: 'The geographic study of any region must begin with a study of its entire previous geography, chronologically arranged'. Landscape, in his opinion, has form, structure and functions.

Sauer further stressed that geographers should proceed generally and trace the development of a natural landscape into a cultural landscape. The difficulty with this methodology, as Sauer himself soon realized, was that it was seldom possible to construct the appearance of the natural landscape, because the human impact on the face of the earth had been pervasive for many millennia. In effect, all landscapes, irrespective of their locations, had become cultural landscapes. Thus the study of landscapes by Sauer and his students (who constituted the so-called Berkeley School) became the study of cultural history.

He emphasized the cultural landscape, and the study of its creation involved: (a) general geography, or the study of the form or elements themselves (what today is called systematic geography); (b) regional geography, or comparative morphology; and (c) historical geography, which studies the development sequence, as in sequent occupance.

Sauer argued that every landscape (area, region) has an individuality as well as a relation to other landscapes, and the same is true of the forms that make it up. No valley is quite like any other valley; no settlement pattern is similar to any other settlement; no city the exact replica of some other city. The peculiarities of different landscapes (area, regions) may be clearly illustrated from the examples of Norway and Libya. The most important geographic fact about Norway is that four-fifths of its surface is barren highland, supporting neither forests nor flocks, while over four-fifths of Libya is a desolate desert without vegetation and human habitation.

Sauer also asserted that geography is distinctly anthropocentric, in the sense of value or use of the earth to man. Geographers are thus interested in that part of the landscape which concerns us as human beings because we are part of it, live with it, and modify it. So we select those qualities of landscape in particular that are or may be of use to us. We relinquish those features of area that may be significant to geologist in earth history but are of no concern in the relation of man to his area. The physical qualities of landscape then are those that have habitat value, present or potential. The diagrammatic representation of morphology as advocated by Sauer is given in the following Figure 7.2.

The cultural landscape is fashioned out of a natural landscape by a cultural group. Culture is the agent, the natural area is the medium, the cultural landscape – the result. Under the influence of a given culture, itself changing through time, the landscape undergoes development, passing through phases, and probably reaching ultimately the end of its cycle of development. With the

Figure 7.2 Morphology as Advocated by Sauer

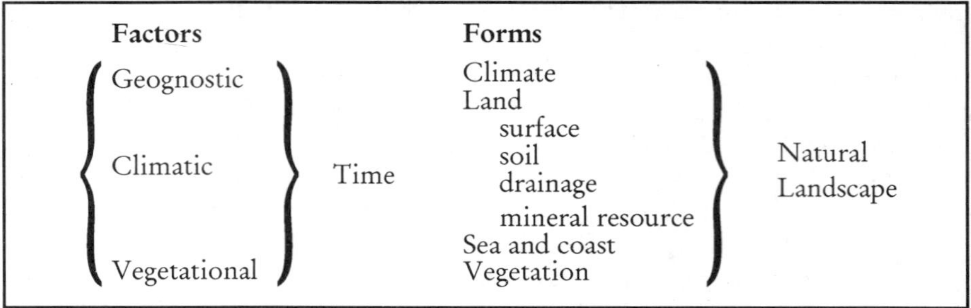

introduction of a different, i.e. alien culture, a rejuvenation of the cultural landscape sets in, or a new landscape is superimposed on remnant of an older one. The natural landscape is of course of fundamental importance, for it supplies the material out of which the cultural landscape is formed. The shaping force, however, lies in the culture itself. Within the wide limits of the physical equipment of area lie many possible choices for man. These choices are derived from the mind of man, not imposed by nature, and hence are cultural expression.

Sauer also gave weightage to the aesthetic qualities of landscape. The aesthetic qualities of landscape can, however, be described subjectively. For the aesthetics of landscape Humboldt used 'physiognomy' Banse's, 'soul', Volz's 'rhythm', Gradmann's 'harmony' which all lie beyond science.

In brief, landscape is undergoing manifold change. The contact of man with his changing environment (home) as expressed through the cultural landscape is the field of work of geographers. Geographers are concerned with the importance of the site to man, and also with his transformation of the site. Although geographers deal with the interpretation of group, or cultures, and site, as expressed in the various landscapes of the world. Thus, Carl Sauer developed geography as the study of *landscape morphology* in opposition to the environmental determinism of Ratzel and Miss Semple.

Ellsworth Huntington (1876–1947)

Ellsworth Huntington, a pupil of Davis, is considered a giant of his era. He was a creative thinker, a prolific writer, an enormously persuasive teacher and an imaginative interpreter of the effects of climate on human life. His was the search for broad interrelationships, for tendencies, for a worldview, and a seeking for the 'architectural unity' of the world structure. In order to achieve this objective, he travelled in North Africa, going as far

Ellsworth Huntington

south as the equator. He also extensively travelled in Europe and South America.[89] On the basis of his observations, he produced an enormous volume of work. He was the author or co-author of 28 books.

Huntington was an environmental determinist who tried to explain the styles of life of human groups and nations in the light of their weather and climatic conditions. He is noted for showing 'the effects of climate on human life'. In his opinion, 'a rational understanding of history requires a good knowledge of changing physical background upon which the historical events occur'. Since historians are not well equipped with such geographical knowledge and their attempts at synthesis have not been highly successful. He developed the hypothesis that the great outpouring of nomadic peoples from Central Asia, which led to the Mongols' conquest of India and China, and the invasion of Eastern Europe in the 13th century, could be explained by the drying up of pastures on which the nomads were dependent.[90] He explained this hypothesis in his book, *Pulse of Asia* (1907). The book was appreciated all over the world. In 1915, he published *Civilization and Climate* in which he asserted that civilizations could develop only in regions of stimulating weather. Contrary to this, the monotonous heat of the tropics would forbid attainment of higher levels of civilizations. In 1920, he published *Principles of Human Geography*. This book – though a textbook – became popular not only with students of geography and geographers, but was also widely read by historians, sociologists, scholars of medicine, agriculturists, ecologists, climatologists and geologists. The climatic determinism of Huntington is reflected in his writings. He writes: 'They have the kind of energy which is needed for innovations, trying new methods and carrying out reforms. In an invigorating climate, it may be easier to be *honest* and *sober* and self-controlled than in a more enervating one' (*Principles of Human Geography*, 1945).

According to Huntington, changes in the sun are a major cause of changes in the terrestrial climate, and climate profoundly influences man, his culture, other forms of life and geological processes. Weather influences man's energy, health and longevity and his attitudes and achievements. The distribution of the various levels of civilization, he suggested, corresponds roughly with the climatic regions. The climate and weather best suited to intellectual and manual activities and progress are characterized by a well-defined seasonal pattern, having frequent changes of weather and sufficient warmth and rainfall to permit extensive agricultural production. Moreover, civilization has tended to shift towards cooler climates as mankind has advanced in its culture. In his opinion the best climates for work were those in which there was variety and in which temperatures fell within a certain range. According to him a temperature of about 20°C and variable atmospheric conditions like that of temperate cyclonic weather are ideal for the high mental and physical efficiencies. Such climatic and weather conditions are found in the countries of North-West Europe and North-East USA. The advancement of Europeans in science and technology, he attributed to the weather and climatic conditions of the given countries and regions.

He also advocated that selective migration and selective survival, together with intermarriage of people of relatively homogeneous cultures, have profoundly influenced the course of history.[91] His books were immensely popular among historians, sociologists and students of biological and medical sciences.

Huntington consistently followed a quantitative approach in the measurement of civilization. This is evident in his early writings in which he describes the remarkable regional coincidence of weather conditions conducive to human, physical and mental effort and the levels of assessment of civilization. He also gave importance to heredity, stage of culture and diet. However, he gave too much weight to 'physical factors' especially to climate and too little to the autonomous development of societies.

Rollin D. Salisbury (1858–1922)

Rollin D. Salisbury

Rollin Salisbury was a major force in the development of geography in the United States during the period 1903–1919. He established and organized the department of geography in the Chicago University. He was a stimulating teacher and distinguished administrator. He wrote several valuable university level textbooks. He was keenly interested in physical geography. However, he rejected the idea of simple cause and effect between the physical earth and the human response. Physiography for Salisbury was the scientific study of stage setting on which the human drama unfolded. But, the relation of stage setting to human action was not a causal one.[92] His main work – *Physiography* – is a classic on the evolution of landforms.

Geography in the United States Between the Two World Wars

During the inter-wars period, an important change came about. A shift towards social aspect away from the physical side took place. Geographers of America started describing the unique features of places and regions and there was very little stress on the formulations of general concepts. Much attention was given to collecting information and use of concepts and models. It is interesting to note that many principles and ideas current in the 1970s had their origin in the inter-wars period. During this period, Social Darwinism was under attack. The strong environmental determinism was rejected and the simple cause and effect relationship was also not accepted.[93] The new approaches developed during this period were:

1. Human ecology
2. Chorology
3. Historical geography
4. Functional organization of space (a region defined by its functions, e.g. a city and the hinterland that uses the city as its service centre).

Thus, there occurred a shift from academic studies to the use of geographic concepts and methods in the study of practical economic, social and political questions, i.e. geography became an applied science to be used for planning purposes and to help solve socio-economic and political questions. In the following lines, a brief description of the major trends during the inter-wars period has been given.

Human Ecology

Barrows (1877–1960) was the founder of the concept of 'human ecology' or the adjustment of man to his natural environment.[94] He was of the opinion that geomorphology, climatology, and biogeography should be relinquished as the subject is too broad, and argued that geographers should work at a unifying theme, which is human ecology. Barrows insisted that the physical conditions should not only be studied in relation to man. He emphasized that human ecology should be used for the study of man and environment, not in a deterministic sense, but for man's place in the 'web of life' or the 'economy of nature'.

Chorology

M. Jafferson and Carl O. Sauer were the protagonists of the chorological theme. The main objective of chorology is to describe the visible characteristics of an area. In the opinion of Sauer, geography is concerned with the study of things associated in an area on the earth's surface and with the differences noticed from place to place – both physical and cultural. Man behaving in accordance with the norms of his culture performs work on the physical and biotic features of his natural surroundings and transforms them into the cultural landscape.[95] Thus, in his opinion, '*the landscape is considered as having organic quality*'.

During the inter-wars period, new jargon was developed including the use of symbols in geographical writings. For example, Koppen adopted letters (English alphabet) to delineate the world climatic regions. Moreover, environmental determinism was rejected. In the study of locational analysis greater importance was given to raw material, markets, power and labour than to relief, drainage, climate and soils.

Historical Geography

In the twenties, geographers in America concentrated on creative human adjustments then to a passive natural environment.[96] Many dissertations were written by non-geographers. The eminent historical geographer Brown wrote

Mirror of Americans and *Historical Geography of the United States*, in which he traced geographical changes during the course of settlement.[97] The contribution of Carl O. Sauer to geography who studied and described the prehistoric Red Indian frontier settlement on the Pacific coast of Mexico is enormous. These studies reveal that land could have quite different meanings for people with different attitudes towards their environment, different objective in making use of it, and different levels of technological skills. In agricultural areas, it was clear that slope had one meaning for the man with hoe and quite another for the man with a tractor-drawn plough. Water power sites that were useful for the location of industries before the advent of steam lost that attraction when power came from other sources. In 1929, D. Whittlesey brought up the idea of an area as sequent occupance – an antithesis of environmental determinism. It was during this period that Whittlesey and Hartshorne formulated theoretical structure to enlarge political geography.[98] Platt formulated the concept of hierarchy of central places.

Studies of Scope and Method

In its early stages, geography in America was nurtured by geologists, meteorologists and biologists. Geography, as an independent discipline, as stated at the outset, in the American universities also was started in the later parts of the 19th century. A number of definitions of the subject were given. Hartshorne defined geography 'as the study of areal differentiation'. According to him, geography is expected to provide accurate, orderly, and rational description and interpretation of the variable character of the earth surface, which involves analysis and synthesis of integration composed of interrelated phenomena of a greater degree of heterogeneity of perhaps any field of science. He stressed the fact that any phenomenon, whether of nature or man, is significant in geography to the extent and degree to which its interrelations with other phenomena in the same place or its interconnections with phenomena in other places determines the areal variations of those phenomena, and hence the totality of areal variations, measured in respect to significance to map.[99] He further emphasized that to comprehend these areal variations fully we must dip back into past relationships of the factors involved, and those whose interest it is to direct them may reach as far back into history as the availability of data may permit. Release from the necessity of focusing our attention on the relations between two particular groups of features – human and non-human – permits a wider expansion of interest and at the same time a more effective coherence of the entire field. The opportunity to develop generic studies leading to scientific principles is present in many forms of topical geography. Likewise, the unlimited number of unique places in the world, each of which is important and intellectually significant at least to those who live there, provides an inexhaustible field for those most interested in this type of research.[100]

It was during the inter-wars period that the Association of American Geographers conducted widespread discussion throughout the profession of the objectives, methods, and concepts of geography. Many of the new concepts of regional geography, population and settlement geography, resource geography, climatology, field techniques, agricultural geography, transport geography, geomorphology, medical geography, administrative geography, and military geography were defined. Moreover, the American geographers paid more attention towards the applied side of the discipline, and prepared resource inventory for the purpose of planning. Land classification, land capability, and economic surveys were conducted to give a new dimension to the discipline. The objective of these surveys was to check the destruction of land and other resources and to suggest steps for their conservation.

During the Second World War, there was an enormous demand for trained professional geographers. Geographers served in the armies as commissioned and non-commissioned officers as well as in the intelligence agencies. They were also busy with the compilation of information about various countries for the benefit of the armed forces. These efforts helped in providing proper uniforms for military use in the torrid, temperate, and frigid zones. Some of the geographers specialized in transport geography. Research in the fields of natural hazards, river floods, droughts, environmental pollution was also undertaken by geographers. Some scholars worked on marketing centres and establishment of retail centres for private firms. Thus, geographers devoted themselves to the socio-economic and political problems to improve the lot of the people.

Geography after the Second World War

After the Second World War, there was a considerable shift in the conceptual framework of geography in the United States. The Americans realized that in order to develop geography as a research discipline or to obtain recognition as being scientifically respectable among other scientists, geographers must drop the all-round synthetic approach of regional geography. It was argued that geographers should develop 'systematic field' in which formulation of universal laws is required, so that one can acquire a competence that can be applied to any area of geography. And thus, the regional method and its sequential treatment, often with a physical environmental basis, was discredited.

Now, the American geographers give emphasis on 'systematic specialization' which means the study of particular phenomenon over the earth and the process that lies behind its spatial arrangement. Thus, they are focusing attention on 'man-environment system' with an ecosystem approach.

The 1960s witnessed an enthusiastic swing by the younger generation towards the use of language of mathematics, because it could be made to express ideas more precisely than the language of literature. Thus, the Americans have laid considerable stress on the application of statistical tools, in the last sixty years, to investigate geographical distributions and arrangements. The statistical tools are applied both in the physical and the socio-cultural phenomena. Mathematical concepts, even laws of astrophysics, are used to measure and explain functions, size and spacing of urban centres, transport network, commodities flow, and crops distribution.

The new trends, besides being quantitative or statistical in nature, are concerned with the 'analysis of spatial systems' which reveal two other important facets. First, there is a shift of emphasis towards group behaviour in the assessment of the relations between man and environment, currently referred to as 'spatial perception'. Second, there is concern with current social problems and the utility of research in coping with them. Moreover, with the help of quantitative techniques, the geographers are developing 'regional science'.

Interdisciplinary approach to explain the man-environment system has been adopted by the American geographers. Emphasis is also being laid on cultural geography, which deals with the material and non-material phenomena relevant to an understanding of spatial distributions and space relation of human cultures. Cultural geographers are concerned with the earth as the home of man who, by means of culture, has become the earth's ecological master. The usual approach of a cultural geographer is to study spatial distribution of the elements or traits of a culture.

Political geography is also an area in which the American geographers are increasingly making their presence felt. The study of political geography is concerned with the interaction of geographical area and political processes; it is the study of spatial distribution and space relations of political processes. Its attention centres on the part of the earth occupied by a given political system, subsystem or systems. The general political problem areas are identified.

Locational theory studies is also an area in which the American geographers are pursuing work. Locational theory studies focus on the study of the static aspects of spatial pattern, like location, spread, density, and geometry; flow linkages between places; temporal dynamics and spatial structure and spatial systems; and normative models leading to efficiency solutions.[101] Apart from computer technology and areal photography, the American geographers are making use of remote sensing technology for the interpretation of geographical phenomena. Moreover, geography has now come to be generally recognized as a laboratory study as such large cartographical laboratories are available, for map collections and exercises, simulation models and quantitative analysis. Geography in America is essentially considered to be a social science. The field of 'physical geography' has thus been woefully neglected.

THE SOVIET SCHOOL OF GEOGRAPHY

In the 19th century, as discussed in the previous pages, geography made tremendous strides all over the European and American countries. In Russia also, many faculties, institutions, and departments of geography were established. In the later parts of the 19th century, many atlases, maps, and monographs were published in the Soviet Union.

Peter the Great encouraged Alexander von Humboldt to explore the territory, east of Urals. The objective of Humboldt's exploration was to gather geographical information to help and guide the eastward expansion of the

Russian Empire. The Imperial Geographical Society of Russia was founded in 1845 with the purpose to promote the study of geology, meteorology, hydrology, anthropology and archaeology. The diverse specialities represented in the society were known collectively as the 'geographical sciences'.[102] The eminent scholars of the pre-Great Revolution Period (1917) were Semenov, Tyan-Shanski, Voeikov, Dokuchaiev, and Anuchin. Their major contribution was to the field of physical geography.

Semenov (1827–1914) was a man of diverse interests who in 1858 explored the Dzungarian Basin, Altai and Tien-Shan Mountains. In 1888, he explored the desert of Turkistan, east of the Caspian Sea. Semenov contacted Carl Ritter for the study of geography, but like Reclus, he was not attracted by Ritter's teleological philosophy. He was more concerned with using geographical knowledge for decreasing the poverty of the rural people. He tried to emphasize the practical importance of geographical study to make the subject more relevant to society. He also wrote several regional monographs.

Before the October Revolution, the Russian geographers did not agree with the extreme forms of 'environmental determinism'. A.I. Voeikov (1842–1916) was a physical geographer and worked on the earth's heat and water balance. He tried to study climatology for the development of agriculture, and contributed appreciably to the science of snow. Tea in Georgia, cotton in Turkistan (Turkmania), and wheat in Ukraine were introduced at his suggestion.[103] Voeikov considered overgrazing of the Russian steppes as the cause of gully erosion, and asserted that irrigation facilities could increase the productivity of semi-arid lands.

Peter Kropotkin (1842–1921)

Peter Kropotkin is best known as one of the founding thinkers of anarchism. He was an idealist anarchist and a close friend of Reclus (1830–1905). He contributed appreciably to physical geography of northern Europe and Siberia. His political views led him to be goaded in Russia and France and banished from Switzerland before finding refuge in Britain.

Peter Kropotkin

Kropotkin tried to revolutionize the discipline of geography in a number of ways. He presented theories of geographical education, the relations between humanity and nature, and of decentralization, which were clearly aimed at discouraging the use of geographical research for exploitative and imperialistic purposes.

Kropotkin's political thinking was shaped by his experience in two expeditions across Siberia and the Russian Far East in 1864 and 1865. In contrast to the prevalent ideas of Social Darwinism, Kropotkin observed and stressed the role of co-operation and mutual aid in both organic evolution and peasant society. He opposed political centralization and economic concentration and favoured small-scale, self-sufficient social units living in greater harmony with their natural surroundings. His vision of geography was at odds with the prevailing emphasis on imperialism, militarism, national chauvinism and racism. Kropotkin considered school geography dull and uninspiring, yet believed it could appeal to a child's imagination concerning both local knowledge and travel to far off places. Instead of the rote learning of facts, he called for greater self-instruction and discovery. He asserted that the study of dry classics and facts should be replaced by more learning about the folklores of other countries. He returned to Russia after the Great Revolution of 1917. Fifty years after his death (1921) many radical geographers have sought inspiration from his revolutionary ideas.

Most significantly for his time, Kropotkin opposed strongly the dominant interpretation of Darwin's *Origin of Species* (1859), which saw nature as an immense battlefield upon which there is an incessant struggle for life and the extermination of the weak by the strong. He attacked the *Social Darwinist* views of the influential philosopher Herbert Spencer (1820–1903). Spencer envisaged human societies as closely resembling animal organisms which must engage in a constant struggle to survive in particular environments. Spencer believed that the 'fittest individuals' would survive best in a free-enterprise system, and thus lead civilization forward. In his 'general law of evolution', he claimed that all evolution is characterized by concentration, differentiation and determination. Kropotkin, on the other hand, wrote: 'I failed to find, although I was eagerly looking for it, that bitter struggle for the means of existence between animals of the same species'. The struggle for existence may be hard, but it is not carried out by individuals – rather, by groups of individuals co-operating with each other. In the development of civilization, as in the quest for survival, Kropotkin believed that mutual aid within small, self-contained communities was the best workable solution.

Both Reclus and Kropotkin stressed and developed the idea that geography as a discipline should encompass both humanity and nature. Kropotkin was a close friend of Scott Keltie (the secretary of the Royal Geographical Society) and strongly supported the arguments that were used to establish geography as a university discipline in Britain (the 'Keltie Report' of 1886). But for Keltie it was essential that science should be politically and socially neutral and value free – that is, based on the scientific ideals of positivism. In writing the obituary of Kropotkin, Keltie (1921: 319) expressed the view that 'this is not the place to deal in detail with Kropotkin's political actions, except to regret that his absorption in these seriously diminished services, which otherwise he might have rendered to geography'. Reclus and Kropotkin may have had little direct influence on the geographical establishments of their day but they retained alight a flame of critical social inquiry that has been resurrected today.

V. V. Dokuchaiev (1846–1903)

The first professor of geography in Russia, appointed in 1885 at St. Petersburg, was V.V. Dokuchaiev. He was, however, not so well known outside Russia as his work was available only in the Russian language. He did pioneering work in the field of soil science. He noted that different soils could be identified by looking closely at the layers of horizons, which differ because of the soil formation processes.[104] The soil, said Dokuchaiev, reflects the extraordinarily complex interaction of climate, slope, plants and animals, with the parent material derived from the underlying geological formations.[105] A soil that has been exposed to all these conditions for a long time would more closely reflect the complex of climate and vegetations than would the parent material.

V. V. Dokuchaiev

In 1887, D.N. Anuchin (1843–1923) became head of the department of geography and ethnography. He was trained in Germany and wrote several geography books for schools. His students like L. S. Berg promoted the cause of geography in Russia.

After the October Revolution, only those geographers could pursue their teaching and research who followed the principles of Marx, Engels and Lenin. Lenin was an economic determinist. For him, geography was the necessary foundation on which the design of a new kind of economy was to be based. According to Lenin, the most important product of geographical study was the identification of rational regions within which segments of the new national economy could be constructed. For him, geography was essentially an applied subject.[106] In the Stalin era, scholars who ventured to express disagreement with the Communist Party policies could be arrested and punished.[107] In the 1950s, after the death of Stalin (1953), the Soviet geographers made substantial contributions to the resource planning and territorial complexes. They followed the path of Lenin in identifying suitable places for locating various industries. These places should be so situated that raw materials could be obtained with minimal effort and expenditure. The Soviet geographers kept themselves busy with the preparation of resource inventory. M. N. Bransky stressed the importance of economic geography. He also wrote the economic geography of the USSR.

The State Planning Commission GOSPLAN (Committee for State Planning) was established in 1921 to prepare regional plans. The State Planning Commission divided the Soviet Union into 21 regions and then proceeded to a detailed study of each of them. This work was done by geographers, economists and engineers. Moreover, geographers also helped in the selection of industrial location and resource development. The Ural-Kuznetsk and the Dnieper-Basin Industrial Regions were planned. Bransky laid stress on the applied side of the discipline with a regional approach and opposed the idea of topical geography. The theory of environmental determinism was also opposed. Lenin himself took exception to the notion that the steppes north of the Black Sea could never be used for agriculture because of climatic conditions. This point of view he expressed was based on the existing level of technology and ignored the inevitability of technological changes.[108]

The dichotomy between physical geography and economic geography was also quite strong in the post-Revolution period. In the early thirties, some of the Soviet geographers were of the opinion that the laws of physical geography were different from those of economic geography, and therefore, the two sets of laws could not go together. But, during the period of Stalin, the strong centralized control of the economy made the regional geography less relevant and it seemed to negate the whole idea of economic regions being 'the major territorial production complexes'.

Physical Geography

In the post-Revolution period, physical geography made a rapid progress in the Soviet Union. The Soviet geographers developed the theoretical principles of forecasting and a typology of climatic phenomena based on dynamic methodology. They studied radiation budget and moisture cycle, and their role in the formation of climates. In hydrology, their contribution is even more significant. They worked on the theory of water budget, and the relationships between surface water and soil water. They also developed the theory of glacial processes, based on the study of heat and mass exchange in various types of glaciers. Geomorphologists studied the crustal movements and developed the morphostructural approach to geomorphology. Soil scientists identified many soil types. Biogeographers gave emphasis to ecological approaches.[109]

Philosophy of Soviet Geographers

Many of the philosophical questions smouldering in the Soviet Union were discussed in the sixties. Anuchin published the book *Theoretical Problems of Geography* in which he investigated the theoretical basis of geography.[110] Anuchin was a strong supporter of the idea of unified geography. He was against the dichotomy of physical geography and economic geography. He rejected the idea of geographical determinism which he equated with the capitalistic approach. He was also against indeterminism or 'economic determinism'. According to him, geography is to be studied with regional approach to delineate territorial complexes. In the territorial approach, physical features, history of settlements, population and economy are in balance.[111] This approach was, however, considered by some of the Soviet geographers as anti-Marxist, as it gives less weight to economic geography.

The basic philosophy of the Soviet geographers was to build socialist economy and society. To achieve this objective, they adopted quantitative techniques. Instead of considering geography to be 'regional science', the Soviet geographers considered it to be 'landscape science'. The 'landscape', according to Soviet geographers, is a dynamic concept in which matter and energy are circulating and in which there are seasonal changes of heat and water balance and biological productivity.[112] The landscape definition helps in the delineation of physical regions which can be used as the basis for identifying regions useful for planning purposes. Now they are moving towards regional synthesis.

Gerasimov (1905–1985), however, described landscape science inadequate for the purposes of constructive geography. According to him, for example, in a virgin land, after investigating the water and heat budgets there should be an attempt to improve agricultural productivity by applying various technical devices to control nature.

In the field of urban geography, hardly anything worthwhile was done. Bransky, in 1946, emphasized the need to classify cities. For this purpose data were gathered from administrative handbooks, encyclopaedias, gazetteers, atlases, maps and scholarly publications. Now tremendous work is in progress in the field of urban geography in Russia.

At present, the Russian geographers are concentrating on the management of natural resources. They are directing their efforts to the regional problems and to the management of environment. Moreover, contemporary trends in the development of geographical science in Russia are conditioned to a significant extent by the tasks which the present state of development in the socialist economy places before researchers in this field. The Communist Party demands from the Russian geographers effective theoretical answers to a whole number of topical questions as also practical recommendations with a bearing upon a whole range of man's interaction with nature. 'The party and the state will benefit from research into problems connected above all with the all-round development of production and management of production, and recommendations making it possible to enhance substantially its effectiveness'. This call was made by Leonid Brezhnev at the 25th Congress of the Communist Party of the Soviet Union. 'In-depth research into questions relating to the development trends of our society and its productive forces is highly necessary'.

'Soviet scientists should not lose sight of the problem of environment and population growth which have recently assumed such a serious aspect. Improvement of the socialist use of natural resources and the formulation of an effective demographic policy are important tasks facing a whole complex of natural and social sciences'.[113]

Geography is one of the sciences involved in the implementation of this task. Its contribution to this objective consists of the detailed elaboration of the main lines for research connected with practical requirements of the present stage of interaction between socialist society and nature.

At present, the Russian geographers are investigating a promising concept of recycling of raw materials used in the industry. Investigations into these cycles demand quantitative analysis of various types of natural resources that are initially extracted from the environment and then transformed in the processes of social production, and finally returned to the environment in the form of various waste products.

Geographical science in Russia today manifests considerable interest in the concept of territorial industrial complexes (TIC). The TIC concept is planned organization of large industrial units. Moreover, the Russian geographers are concentrating on environmental problems. They are trying to find out new methods for consolidating harmonious relationship between society and nature. Another sphere in which they are doing constructive research is the study of the theoretical foundation of recreational geography. The establishment and development of recreational zones near large towns is an important move to counter the harmful effects of urbanization. Thus, the Russian geographers are doing constructive research for the betterment of society with limited natural endowments. Constructive research has to be based on a comprehensive understanding of the fundamental laws underlying the organization and functioning of natural systems, since the introduction of new technology into these systems must not disrupt natural processes but rather serve purposefully to regulate such processes and strengthen those of them which enhance the overall quality of environment and its resources.[114]

In brief, the Russian geographers, together with experts of other sciences now count applied constructive research among society's productive forces and it has become an important factor in regional planning, town planning, management of water resources, deployment of production enterprises, organization of agricultural schemes, forecasting of weather and changes in the hydrological balance, in the fight against soil erosion, mudflows and avalanches, droughts and other such spontaneous natural phenomena. They also play a constructive role in the selection of optimal sites for the construction and routing of roads, in the development of service industries, recreational and tourist facilities, provision of resorts, health care facilities and so on and so forth. Thus, geography is playing a very vital role in the judicious utilization and conservation of natural resources to provide more amenities to the society.

Notes

1. Fischer, E.C. and Miller, E.S., 1967, *A Question of Place: The Development of Geographical Thought*, pp. 81–174.

2. Valkernberg, V., 1951, 'The German School of Geography', in G. Taylor (ed.), *Geography in Twentieth Century*, pp. 91–115.
3. Peschel, O., 1865, *Geschichte der Erdkunde bis auf A.V. Humboldt and Carl Ritter Munich.*
4. Ibid., p. 30.
5. Peschel, O., 1879, *Physische Erdkunde*, Ed. Gustav Leipoldt, Leipzig.
6. Richthofen, F. von, 1877–1912, *China: Ergebnisse eigener Reisen und drauf gegrundte*, Studien, Vol. Berlin.
7. James, P.E. and Martin G.J., 1981, *All Possible Worlds*, p. 167.
8. Hartshorne, R., 1939, *The Nature of Geography*, p. 92.
9. James, P.E. and Martin, G.J., op. cit., p. 167.
10. Wankyn, H., 1961, *Friedrich Ratzel, a Bibliographical Memoir and Bibliography*, Cambridge.
11. Sauer, C.O., 1971, 'The Formative Years of Ratzel, in the United States', *Annals of Association of American Geography*, 61, pp. 145–154.
12. James, P.E. and Martin G.J., op. cit., p. 169.
13. Speth, W.W., 1977, 'Carl Ortwin Sauer on Destructive Exploitation', *Biological Conservation*, 11, pp. 145–160.
14. Dickinson, R.E., 1969, *The Makers of Modern Geography*, pp. 64–72.
15. Spencer, H., 1864, *Principles of Biology*, p. 444.
16. Ratzel, F., 1897, *Politische Geographie, Oder die Geographie der Staanten, des Verkehrs und der Krieges*, 2nd ed., 1903, pp. 1–20.
17. James, P.E. and Martin, G.J., op. cit., p. 170.
18. Ibid., p. 171.
19. Semple, E.C., 1911, *Influences of Geographic Environment on the Basis of Ratzel's System of Anthropogeography*, New York, p. iv.
20. Harvey, D., 1969, *Explanation in Geography*, p. 50.
21. Hartshorne, R., 1958, 'The Concept of Geography as a Science of Space, from Kant and von Humboldt to Hettner', *Annals of Association of American Geography*, 48, pp. 97–108.
22. Dickinson, R.E., op. cit., p. 122.
23. Troll, C. 1950, 'Die Geographiche Landschaft und Erferschung', *Stadium Generale*, Vol. 3, pp. 163–181.
24. James, P.E. and Martin, G.J., op. cit., p. 179.
25. Fischer, E., Campbell, R.D. and Miller, E.S., 1967, op. cit., pp. 99–106.
26. James, P.E., 1936, 'The Geography of Oceans: A Review of the Work of Gerhard Schett', *Geographical Review*, Vol. 26, pp. 664–669.
27. James, P.E. and Martin, G.J., op. cit., p. 189.
28. Dickinson, R.E., 1969, op. cit., pp. 208–212.
29. James, P.E. and Martin, G.J. op. cit., p. 190.

30. Vidal de la Blache, P. and Gallois, L. (eds), 1927–48, *Geographie Universelle*, Vol. 15, Paris, p. 30.
31. Freeman, T.W., 1971, *A Hundred Years of Geography*, London, p. 84.
32. Martin, G.J., 1964, 'The Region in French Geographical Thought, 1900–1930' in *Papers of Michigan Academy of Science, Arts & Letter*, Vol. 49, pp. 325–332.
33. Vidal de la Blache, *Principles of Human Geography*, p. 32.
34. Joorg, W.L.G., 1922, 'Recent Geographical Work in Europe', *Geographical Review*, pp. 431–484.
35. Ibid.
36. Vidal de la Blache, op. cit.
37. Vidal de la Blache, 1903, *Tableau de la Geographie de la France*, Paris.
38. Wrigley, E.A., 1965, 'Changes in the Philosophy of Geography', in *Frontiers in Geographical Teaching*, ed. R.J. Chorley and P. Haggett, London, pp. 3–20.
39. Holt-Jonson, A., 1981, *Geography: Its History and Concepts*, p. 29.
40. Ibid.
41. Vidal de la Blache, 1917, *France de I' Est*, Paris.
42. Brunhes, J., 1947, *Human Geography*, p. 21.
43. Dickinson, R.E., 1969, op. cit., p. 222.
44. Reclus, E., *The Earth and Its Inhabitants*, ed. E.G. Ravenstein, New York, 1882, Vol. 1, p. 5.
45. Dickinson, R.E., op. cit., p. 224.
46. Obituary article by A. Cholley, *Annals de Geographie*, 65, p. 956.
47. Em. de Martonne, 'Albert Demangeon', *Annals de Geographie*, 65, p. 956.
48. Dickinson, R.E., op. cit., p. 264.
49. Ibid., p. 265.
50. Freeman, T.W., 1950, *The British School of Geography*, p. 12.
51. Mackinder, H.J., 1902, *Britain and British Seas*, New York.
52. Mackinder, H.J., 1919, *Democratic Ideals and Reality*, New York.
53. James, P.E. and Martin, G.J. op. cit., p. 203.
54. Holt-Jensen, A., 1981, op. cit., p. 33.
55. Ibid., p. 33.
56. Herbertson, A.J., 1905, 'The Major Natural Regions: An Essay in Systematic Geography', *Geographical Journal*, Vol. 25, pp. 300–312.
57. Dickinson, R., op. cit., p. 32.
58. Freeman, T.W., 1980, *A History of Modern British Geography*, New York, p. 10.
59. Dickinson. R., op. cit., pp. 186–189.
60. Mill, H.R., 1891, *The Realm of Nature*, London, p. 26.
61. Chisholm, M., 1962, *Rural Settlement and Land Use*, London.
62. Dickinson, R.E., op. cit., p. 45.

63. Ibid., p. 47.
64. Herbertson, A.J., 1905, 'The Major Natural Regions: An Essay in Systematic Geography', *Geographical Journal*, Vol. 25, pp. 300–312.
65. Forde, D., 1934, *Habitat, Economy and Society*, London.
66. Dickinson, R.E., op. cit., p. 136.
67. Stamp, L.D., 1947, *The Land of Britain, its Use and Misuse*, London.
68. Fleure, H.J., 1919, Human Regions, *Scott. Geog. Magazine*, Vol. 35, pp. 94–105.
69. Dickinson, R.E., op. cit., p. 139.
70. Mackinder, H.J., op. cit., p. 130.
71. Gilbert, E.W., 1933, 'The Exploration of Western America', 1800–1850, *A Historical Geography*, Cambridge.
72. Johnston, R.J., 1976, *Geography and Geographers: Anglo American Human Geography, since 1945*, New York.
73. Hartshorne, R., 1959, *Perspective on the Nature of Geography*, Chicago, p. 25.
74. Chorley, R.J. and Haggett, P. (eds), *Frontiers in Geographical Teachings*, London.
75. Hill, M.R., 'Positivism: A Hidden Philosophy in Geography, in *Themes in Geographic Thought*, ed. Milton E. Harvey and Brian P. Holly, London, pp. 38–60.
76. Ibid.
77. Guyot, A., 1863, *Earth and Man*, p. 30.
78. Shaler, N.S., 1893, *Nature and Man in America*, New York, p. 131.
79. Ibid.
80. Johnson, D.W., 1909, Twenty-Six of Davis' Papers, *Geographica Essays*, p. 12.
81. James, P.E. and Martin, G.J., op. cit., p. 290.
82. Dickinson, R.E., op. cit., pp. 209–212.
83. Bowman, I., 1911, *Forest Physiography of the United States and Principles of Soils in Relation to Forestry*, New York.
84. Ibid.
85. Semple, E.C., *Influence of Geographic Environment*, New York, p. iv.
86. Ibid.
87. Brigham, A.P., 1903, *Geographic Influences in American History*, Boston, p. 78.
88. Dickinson, R., op. cit., pp. 218–224.
89. Huntington, E., 1907, *The Pulse of Asia.*
90. Huntington, E., 1920, *Principles of Human Geography*, New York.
91. Ibid.
92. Dickinson, R.E., op. cit., pp. 232–238.
93. James, P.E. and Martin, G.J., op. cit.
94. Barrows, H.H., 1923, *Geography as Human Ecology, Annals of Association of American Geographers*, 13, p. 1.

95. Sauer, C.O., 1925, 'The Morphology of Landscape', University of California, *Publications in Geography,* 2, pp. 19–35.
96. Dickinson, R.E., op. cit., p. 274.
97. Ibid.
98. Whittlesey, D., 1929, 'Sequent Occupance', *Annals of Association of American Geographers,* Vol. 19, pp. 162–165.
99. Ibid.
100. James, P.E. and Martin, G.J., op. cit.
101. Dickinson, R.E., op. cit., pp. 345–352.
102. Hooson, D.J.M., 1968, 'The Development of Geography in Pre-Soviet Russia', *Annals of Association of American Geographers,* Vol. 58, pp. 250–272.
103. Voikov, A.I., 1901, 'De 1 influence de 1' homme sur la terre, *Annals de Geographie,* Vol. 10, pp. 97–114, 193–215.
104. James, P.E. and Martin, G.J., op. cit., p. 228.
105. Cerasimov, I. et al. (eds), 1962, *Soviet Geography: Accomplishment and Task,* English translation by L. Glinka.
106. James, P.E. and Martin, G.J., 1981, op. cit., p. 228.
107. Ibid., p. 233.
108. Nikitin, N.P., 1966, 'A History of Economic Geography in Pre-Revolutionary Russia', *Soviet Geography,* Vol. 7, pp. 3–37.
109. James, P.E. and Martin, G.J., op. cit., p. 250–272.
110. Anuchin, V.A., 1956, *Regional Study of Transcarpathian.*
111. James, P.E. and Martin, C.J., op. cit., p. 240.
112. Anuchin, V., 1981, 'The Relationship Between Society and Nature in the Geographical Environment and Philosophical Aspects of Geography', *Soviet Geography Today,* Moscow, pp. 157–176.
113. Gerasimov, I., 1981, 'A Historical Background to Russian and Soviet Geography', *Soviet Geography Today,* pp. 5–17.
114. Ibid., p. 247.

8

Dichotomy between Determinism and Possibilism

In the history of geographical concepts, there have been various approaches and schools of thought of study of man-nature interaction. The first approach adopted by the geographers to generalize the patterns of human occupations of the earth surface was deterministic. Their major initial source for explanations was the physical environment, and that theoretical position was established around the belief that the nature of human activity was controlled by the parameters of the physical world within which it was set.

Determinism is one of the most important philosophies which persisted up to the Second World War in one shape or the other. The point of view is that the physical environment controls the course of human action. In other words, the belief that variation in human behaviour around the world can be explained by the differences in the natural environment. The essence of the deterministic school of thought is that the history, culture, living style and stage of development of a social group or nation are exclusively or largely governed by the physical factors of environment. The determinists generally consider man a passive agent on which the physical factors are constantly acting and thus determining his attitude and process of decision making. In brief, determinists believe that most human activity can be explained as a response to the natural environment.

The first attempt to explain the physical features and character traits of various peoples and their culture with reference to the influence of natural conditions was made by the Greek and Roman scholars. They included the physician Hippocrates, the philosopher Aristotle, and the historians Thucydides, Xenophon, and Herodotus. In the Greco-Roman period, regional studies were closely bound up with the study of history. Thucydides and Xenophon saw Athens's natural conditions and geographical position as the factors underlying its greatness. Strabo referred to similar phenomena when explaining the might and greatness of Rome. Aristotle, for example, explained the differences between Northern Europeans and Asians in terms of climatic causes. He argued that the colder climates of Europe produced brave but unintelligent people who were able to maintain their independence but who did not have the capacity to rule others. Aristotle thought that the people inhabiting the warm climates of Asia were intelligent but lacking in spirit and therefore subject to slavery. Because humans often judge their own home as the best place, it is not surprising that Aristotle believed that the middle place, combining the best of all possible worlds, was Greece (Glacken, 1967: 93). Moreover, according to Aristotle, the inhabitants of cold countries are courageous but 'lacking in political organization and capacity to rule their neighbours' and also the people of Asia lack courage and so slavery is their natural state. The people of Greece, on the other hand, who occupy 'the middle position geographically', he sees as endowed with the finest qualities and thus destined by nature itself to rule over all.[1] The Greek scholars have referred to the easy-going ways of Asiatics living in favourable environmental conditions, while the penurious Europeans had to work hard for a little amelioration of their poor environment. They contrast the tall, gentle, brave folk of the most windy mountains with the lean, sinewy, blonde inhabitants of dry lowlands. Aristotle emphatically attributed the progress of certain nations to their favourable environmental conditions.[2]

Similarly, Strabo – the Roman geographer – attempted to explain how slope, relief, climate all were the works of God, and how these phenomena govern the lifestyles of people. Montesquieu pointed out that the people in cold climates are stronger physically, more courageous, frank, less suspicious and less cunning than those in the warm climates. The people of warm climates are timorous, weak in body, indolent and passive.

Geographical determinism continued to dominate the writings of the Arab geographers. They divided the habitable world into seven *kishwars*, or terrestrial zones (climate) and highlighted the physical and cultural characteristics of races and nations of these zones. Al-Battani, Al-Masudi, Ibn-Hauqal, Al-Idrisi, and Ibn-Khaldun attempted to correlate environment with human activities and mode of life. Al-Masudi, for example, asserted that in the land like Sham (Syria) where water is abundant, the people are gay and humorous, while the people of dry and arid lands are short-tempered. The nomads who live in the open air are marked by strength and resolution, wisdom and physical fitness.[3]

George Tathan – a leading historian of the 18th century – also explained the differences between peoples with reference to the differences between the lands in which they lived. Kant was also a determinist, who stated that the people of New Holland (East Indies) have half-closed eyes and cannot see to any distance without bending their heads back until they touch their backs. This is due to the innumerable flies which are always flying in their eyes. Kant further stressed the point that all the inhabitants of hot lands are exceptionally lazy and timid. Timidity engenders superstition and in lands ruled by kings it leads to slavery. In support of his hypothesis of the influence of climate, he stated that animals and men that migrate to other countries gradually get affected by their environment. For example, the brown squirrels which migrate to Siberia turn grey and the colour of white cows in winters turns greyish.

The environmental causation continued throughout the 19th century when geographers themselves used to regard geography above all as natural science. Carl Ritter – the leading German geographer – adopted an anthropocentric approach and introduced geographical determinism in the early 19th century. Ritter attempted to establish the cause variations in the physical constitution of body, physique and health of men living in different physical environmental conditions. He stated that the narrow eyelids of Turkoman people were an obvious effect of the desert upon the human organism. Many of his pupils considered geography 'as the study of relationship between the density of people and the nature of their land'. Many geographers of his school declared that their main task was to identify the influence exerted by geographical conditions on material culture and the political destinies of the inhabitants of a given region, both in the past and present.[4] Alexander von Humboldt, one of the founders of 'modern geography' and a contemporary of Ritter also asserted that the mode of life of the inhabitants of a mountainous country differs from that of the people of the plains.

The scientific milieu in the later half of the 19th century and early decades of the 20th century was dominated by Darwin's idea, deductive approaches and an acceptance of the Newtonian cause and effect relationships. The origin of the scientific determinism lie in the work of Charles Darwin, whose seminal book *Origin of Species* (1859) influenced many geographers. Fitting well into this intellectual environment, the theory of environmental determinism, developed mostly by geographers, was the prevailing view in American geography at the turn of the 20th century. Darwin's notions regarding evolution were taken up by William Morris Davis, in his cycle of erosion model of landforms development. The concern was with documenting the control or influence of environment upon human society.[5]

The founder of the 'new' determinism was Friedrich Ratzel. He supplemented 'classical' geographical determinism with elements of 'Social Darwinism' and developed a theory of the state as an organism which owed its life to the earth and which was ever striving to seize more and more territory.[6] In the opinion of Ratzel, 'similar locations lead to similar mode of life'. He cited the example of British Isles and Japan and asserted that both these countries have insular locations, which provide natural defence against the invaders. Consequently, the people of these countries have been making rapid progress.

Ratzel – a follower of Darwin – believed in the survival of the fittest and saw 'man' as the end product of evolution – an evolution in which the mainspring was the natural selection of types according to their capacity to adjust themselves to physical environment. He was convinced that the course of history, the mode of life of people and the stage of its development are closely influenced by the physical features and location of a place in relation to mountains and plains. In his deterministic approach, he gave more weight to location in relation to topographic features.

Historical Perspective of Scientific Determinism

The theological school of thought advocated the idea of a designed earth: one especially fitted for the human species. To a great extent, this is part of the wider concept 'teleology', i.e. the concept of an overall creation with a particular purpose which was usually divine. The deterministic school of thought is that of environmental influence on culture. This drives initially from the contrast between nature and custom in different places and came to be used in interpreting the great array of human, cultural and biological differences. Thomas Robert Malthus who was an economist and a scientific determinist (1760–1834), emphasized not only the influence of different environments but also the limitations which the earth imposed on social development. The father of this generation of offsprings seems to have been Carl Ritter (1779–1859) whose theme was that the physical environment was capable of determining the course of human development. His ideas were strengthened by the publication of Charles Darwin's *Origin of Species* in 1859, with its emphasis on the close relationship of organism and their habitats and the notion of the pressures of natural selection. Thus arose a 'scientific' type of environmental determinism which accounted for such features as migrations and the national characteristics of particular people. The names of Friedrich Ratzel (1844–1904) and Ellen Churchill Semple (1863–1932) are associated with the most outspoken expression of the idea of environmental determinism. This approach was slightly modified by Ellsworth Huntington and Griffith Taylor. Huntington tried to seek out objective evidence of the effect of physical environment, and in particular climate which he regarded as an important influence on human behaviour. Taylor (1880–1963) was even more careful to gather accurate data about environment and to relate these to his idea of human habitability, especially in Australia. He tended to play down socio-economic factor. He believed that environment sets the limit of human development. His determinism was likened to a traffic control system which determined the rate but not the direction of progress, and so it became known as 'stop-and-go determinism'.

Environmental Determinism

As stated earlier, the origin of environmental determinism lies in the work of Charles Darwin, whose seminal book *Origin of Species* (1859) influenced many scientists.

The belief that variations in human behaviour around the world can be explained by differences in the natural environment is known as environmental determinism.

At the beginning of the 20th century 'environmentalism' became particularly widespread in the United States, where its leading proponents were W.M. Davis (in his cycle of erosion model of landform development),

Ellen Churchill Semple and Ellsworth Huntington. Semple was the direct descendant of Ratzel. She preached the philosophy of her master and thus was a staunch supporter of determinism. Her books *American History and Its Geographic Conditions* (1905) and *Influences of Geographic Environment* (1911), established environmentalism in America in the early decades of the 20th century. *Influences of Geographical Environment* (1911) begins with the following paragraph:

> Man is a product of the earth's surface. This means not merely that he is a child of the earth, dust of her dust, but the earth has mothered him, set him task, directed his thought, confronted him with difficulties, that have strengthened his body and sharpened his wits, gave him his problems of navigation or irrigation and at the same time whispered hints for their solution. She has entered into his bones and tissues, into his mind and soul. On the mountain she has given him leg muscles of iron to climb the slope, along the coast she has left these weak and flabby, but given him instead vigorous development of chest and arm to handle his paddle or oar. In river valley, she attaches him to fertile soil.

Semple, in her book, distinguishes the attitudinal characteristics of the people living in different physical settings and points out that the dwellers of mountains are essentially conservative. There is little in their environment to stimulate them to change and little reaches them from the outside world. Hence, innovation is repugnant to them. As a matter of fact, the process of diffusion of new ideas and innovations in the hilly tracts of isolation and relative isolation is slow as compared to the well-connected plains of the world. This relative isolation of the hill dwellers leads to orthodoxy, conservatism and suspicious attitude towards strangers. They are extremely sensitive to their traditions and do not like criticism. They have strong religious feelings and an intense love for family. The bitter struggle for existence makes the hill men industrious, frugal, provident and honest. Contrary to this, the people of plain parts of Europe are energetic, serious, thoughtful rather than emotional, and cautious rather than impulsive. The people of the Mediterranean region where the climate is temperate and mild are gay, humorous, sporting and imaginative as life is easy.

Ellseworth Huntington – the American geographer – who wrote the monumental book, *The Principles of Human Geography* in 1945, was a protagonist of environmental determinism. Huntington's writings on climate and civilization displayed his predilection for racial typecasting and environmentalist explanations. He, however, constantly reiterated the importance of genetic constitution and threw his weight behind various genetic enterprises (Spate, 1968). He took the most decisive step since the time of Hippocrates towards something new and conclusive in environmental causation thinking. Over many years he was engaged in developing the idea of climate's leading role in the advancement of civilization. He advanced theories relating to course of civilization to climatic change. The basic philosophy of

Huntington was that the supreme achievements of civilization in any region were always bound up with a particular type of climate and variation in climate led to 'pulsations' in the history of culture. He suggested that the 'best' climates for work were those in which there was variety and in which the temperatures fell within a certain range, and wrote of the correlation between a stimulating climate and high civilization based on in the UK and New England (USA). He associated with the climatic cycles the 'Golden Age' in ancient Greece, the Renaissance in Western Europe, and cyclical fluctuations in iron production or the price of share.[7] Huntington divided the world in the mild and harsh climatic zones and established that the ancient civilizations (Egyptian, Mesopotamian, Chinese, Indus) flourished in the fertile river valleys of mild climates. He also established the hypothesis of invasion and tribal warfare. The great outpouring of nomadic people from Central Asia which led to Mongols' conquest of Iran, Iraq, Turan, Turkistan, Central Asia, China and India and the raids in Eastern Europe in the 13th century could be explained by the dying of pastures on which the nomads were dependent. According to Huntington, religion and racial character are the products of climate. A temperature of about 20°C and variable atmospheric conditions (temperate cyclonic weather) are the ideal climatic conditions for high mental and physical efficiencies. Such a climatic condition is found in the North-East USA and countries of North-West Europe. The advancement of Americans/ Europeans in the fields of science and technology has thus been attributed to cyclonic weather and temperate climatic conditions by Huntington. The underdevelopment of the tropics, he explains, is owing to the humid, hot, oppressive weather which makes the people lethargic, lazy, inefficient, suspicious and timid. Huntington thus believed that out of all the factors of natural environment, climate was the fundamental factor in the rise of civilization (1939). He concluded that his homeland, which was the north-eastern part of USA, had the best environment. He even produced a map, based primarily on the opinions of other North Americans and Europeans, which showed that temperate climates had the highest level of 'health and energy' and civilization. It is obvious that this map is highly subjective and its logic differs little from Aristotle's, except that Huntington perceived the world from a different home location.

Environmental determinism is regarded by many people as overly simplistic because it neglects the cultural factors that affect human behaviour.[8] Two societies that inhabit areas having similar climates and landforms may be very dissimilar. How could two contrasting societies like Bakarwals and Kashmiris of Jammu and Kashmir, Nepalis and Khasis of Meghalaya, Assamese and Bengalis of the Brahmaputra Valley, Tharus and Sikhs of the Tarai region of Uttar Pradesh, for example, exist in a similar environment and have different modes of life and cultural ethos, if climate dictated patterns of life?

Subsequent geographers like Mackinder, Chisholm, Davis, Bowman, Robert Mill, Geddes, Sauer, Herbertson, Taylor, etc. interpreted the progress cf societies with a deterministic approach. Many scholars made it vividly clear that climate influenced physical properties of the soil which ultimately determined the cropping patterns, on which depend the dietary habits, physique and attitudes of the inhabitants. Mac Carrison demonstrated conclusively that the greater stature, strong constitution and superior physical resistance of the Sikhs of Northern India as compared to those of the Tamils of South India are a direct result of the superior Sikh diet, particularly its greater richness in protein.[9] The Khasis of the plateau of Meghalaya have in general a poor physique because the protein intake in their diet is significantly low and the humid weather throughout the year creates breathing problems to the inhabitants of this plateau.

Lord Boyd Orr and Gilkhs observed a similar phenomenon in East Africa, where they studied the Kikuyu and the Mesai tribes of Kenya. The Kikuyus are farmers living on a diet of cereals, tubers and legumes; and the Mesais on the other hand, are cattle raisers, whose diet include meat, milk and ox blood, which they take from the animals. These two human groups living side by side in the same environment differ profoundly in their physical measurements. This difference is the direct result of their fundamentally different diets.[10] Similarly, there is no doubt that the low stature and poor physique of most of the tribals, the rural masses and the slum dwellers of India are the result of starvation, undernourishment and malnutrition. The poor physique of the Somalians, Nepalis, Bangladesis and Vietnamese may also be explained against the background of their poor diet and undernourishment.

How closely soil and vegetation influence the health and stature of peoples and animals has been explained by Karl Mackey. In the opinion of soil scientists, 'the history of civilization is the history of soil'. Roosevelt once remarked: 'If soil is gone, men must go and the process does not take long.' Thus, soil is the basis of all living organism. He cites the case of Shetland ponies:

> On the Shetland Island (North of Scotland, UK), at the northern extremity of the British Isles (60°N), are found the smallest horses in the world, only about one metre in height. Traditionally, it used to be thought that these Shetland ponies constituted a separate race of horses, stabilized by inbreeding - until some businessmen decided to supply the American market by raising these ponies in USA. To their great disappointment, the ponies born under the new conditions got bigger and bigger generation after generation until they were the same size as horses of other 'races'. The fact is, there are no separate races of ponies. Even after hundreds of generations when the ponies were taken to areas with richer soil (praries) they regained the characteristics of their ancestors.[11]

A similar example can be found among the Chinese and the Japanese who migrated to Europe and America. Their weight and height increased after a period of time. The Pygmies also lose their characteristics when transplanted to plain regions where agriculture and cattle raising provide much more varied food. Thus, the short-statured races became tall-statured ones.

Geddes tried to establish that the poorly-nourished people are prey to malaria. In support of his hypothesis, he stated that the meat-eating Muslims in India are much less subject to malaria than are the Hindus with their vegetarian diet.[12]

The influence of physical factors on food habits and the consequent effect on the rate of birth in different regions can be seen in the fact that the high birth rates (above 30) are all confined to tropical countries. The geo-ecological and socio-economic conditions of these countries are all ill-adapted to either the production or consumption of proteins of animal origin. If we compare the birth rate with the intake of animal proteins throughout the world, we find a clear correlation between the two factors, i.e. fertility going down as consumption of such proteins rises. For example, the daily intake of animal protein in Sweden and Denmark is 63 grams and 60 grams respectively and the birth rate is 15 and 18 per thousand respectively. In India and Malaysia, only about 7 grams and 8 grams of animal protein is consumed respectively and the corresponding birth rate in these countries is 35 and 33 per thousand respectively. These may be overgeneralizations as many other factors like standard of living and socio-cultural attributes also contribute to birth rate, yet there is no denying the fact that the quality of diet has much bearing on the increase, decrease and longevity of the population of a region.

There is evidence showing that terrain, topography, temperature, rainfall, humidity, vegetation and soil, individually and collectively, affect social and economic institutions and thereby the mode of life of people, yet the role of man as a transforming agent of his physical surroundings is quite significant. In fact, acts of man reveal many facts for which environmental forces alone can give no satisfactory explanation. For example, similar environment does not always invoke the same response. Eskimos differ markedly from the Tundra tribes of Siberia. Pygmy hunters share the equatorial forests of Central Africa with agricultural Negroes in a remarkable symbiosis. The Khasis, Garos and Jaintias of Meghalaya and the Lushais of Mizoram, living under almost a similar climatic and environmental conditions, have marked variations in physical traits, physique, dietary habits, standard of literacy and attitude towards life. In fact, no two cultures and various ethnic groups within a physical environment evaluate and use the resources of an environment in exactly the same way. This variation in the evaluation of resources is one of the main causes of differences in the lifestyle and stage of development of various ethnic groups and nations.

It has also been observed that the same physical conditions of land could have quite different meanings for people with different attitudes towards their environment, different objectives in making use of it and different levels of technological skill. The Gujjars and Bakarwals of Jammu and Kashmir like to settle on slopes and to utilize these slopes for pastures while the Kashmiris like to settle in levelled areas and to utilize their arable land for paddy and orchard cultivation. The former are nomads while the latter are cultivators. In agricultural areas, it is clear that slope had one meaning for the man with a hoe and quite another for a man with a tractor-drawn plough. It might be that the introduction of machinery would reduce the arable area of a country or change the kind of soil considered desirable. People of one kind of culture might concentrate in the valleys (Mesais and Kikuyus of East Africa) whereas another kind of people in the same area might concentrate their settlements on fertile uplands. Water-power sites that were useful for the location of industries before the advent of steam engine lost that attraction when power came from other sources.[13]

Environment undoubtedly influences man, man in turn changes his environment and the interaction is so intricate that it is difficult to know when one influence ceases and the other begins. Many landscapes that appear natural to us are in truth the work of man. Wheat, barley, oilseeds, olive, and vine, which dominate the Mediterranean countries, are entirely the products of human effort. Apple, apricots, walnuts, and almond orchards of Kashmir and Himachal Pradesh and Kumaun division of Uttar Pradesh are the creations of man. Similarly, cultivation of *basmati* rice (a high water requiring variety) in only 50 cm rainfall areas of Punjab and Haryana is the direct and conspicuous result of human efforts. Wheat cultivation in West Bengal, Orissa and Dimapur of Nagaland is the outcome of the use being made of the innovation of high yielding varieties (HYVs). Countless such examples from the developed and the developing countries can be cited. Thus, man and environment are intrinsically interdependent and it is difficult to say which becomes more influential and when.

After the Second World War, the philosophy of environmentalism was attacked. Many geographers in the United States, Britain, Canada and other countries drew attention to the one-sided approach adopted by the environmentalists in their interpretation of historical reality, to their exaggeration of nature's active role and to the fact that they only acknowledge man as capable of passive attempts at adaptation.[14–17] Actions of man reveal many facts for which environmental forces alone can give no satisfactory explanation.

Spate criticized the fanatic approach of environmental determinists. He, for example, states that 'environment taken by itself is a meaningless phrase; without man environment does not exist'.[18] Equally important is his indication of the need to consider the psycho-physiological influence of the geographical environment via the social structure. In the final analysis, Spate concluded that geographical environment is only one of the factors of territorial differentiation and 'it acts through society; cultural tradition has a certain autonomous influence'.[19] Recently, an Australian writer – Wolfgang Hartake – argued that while the role of physical factors might well be relatively unimportant in the fringe zone of Frankfurt, 'it is hard to imagine the extreme climatic conditions not playing a direct role in any human activity which occurs in the Sahara'.[20] Similar argument is put forward by Hartshorne.

He rejected environmentalism purely on the grounds that it separates nature from man and thus is 'disruptive of fundamental unity of the field',[21] i.e. contradicts the concept of geography as an integrated science.

The environmentalist movement started in the 1960s has, however, shown quite distinctly that there is an overall limit to certain kinds of human economic activity in terms of biophysical persistence and resilience of the planet's ecosystems and ecology. In brief, at the very largest scale we can be determinists, whereas at the more local scales we can see the virtue of possibilism or cultural and social determinism.

Possibilism

Possibilism in geography developed as a reaction to extreme generalizations of environmental determinists that led to a counter thesis, of possibilism, which presented the man as an active rather than a passive agent.

This philosophy attempts to explain man and environment relationship in a different way, taking man as an active agent in environment. This is a belief which asserts that natural environment provides options, the number of which increases as the knowledge and technology of a cultural group develop. Led by French geographers, the followers of historian Lucian Febure, possibilists presented a model of people perceiving the range of alternative uses to which they could put an environment and selecting that which best fitted their cultural dispositions. This point of view was named 'possibilism' by Lucien Febvre, who writes: 'The true and only geographical problem is that of utilization of possibilities. There are no necessities, but everywhere possibilities. The natural data (factors) are much more the material than the determinant of human development. The 'essential cause' is less nature, with its resources and its obstacles, than man himself and his own nature'.

According to Febvre, a possibilist, 'man is a geographic agent and not the least. he everywhere contributes his share towards investing the physiognomy of the earth with those changing expressions which is the special charge of geography to study'.

Vidal de la Blache refuted the concept of physical determinism and advocated possibilism. 'Nature sets limits and offers possibilities for human settlement, but the way man reacts or adjusts to these conditions depends on his own traditional way of life'.

But, the possibilists recognize the limitations imposed by physical environment. Febvre echoes this view: 'Men can never entirely rid themselves whatever they do of the hold their environment has on them'. In the similar manner, Brunhes remarks: 'The power and means which man has at his disposal are limited and he meets in nature bounds which he cannot cross. Human activity can within certain limits varies its play and its environment, but it cannot do away with its environment, it can only modify it but it can never surpass it, and will always be conditioned by it'. Brunhes further writes: 'Nature is not mandatory but permissive'.

Similarly, Vidal de la Blache says: 'There is no question of geographical determinism, nevertheless, geography is a key that cannot be dispensed with'.

Possibilism is also associated with the French School of Geography founded by Vidal de la Blache (1845–1918). The French geographers saw in the physical environment a series of possibilities for human development, but argued that the actual ways in which development took place were related to the culture of the people concerned, except perhaps in regions of extremes like deserts and tundra. The historian Lucien Febvre (1878–1956) set out to demolish the environmental deterministic argument by asserting the initiative and mobility of man as against the passivity of the environment, and regarded other humans as part of environment, of any group because they contributed to the formation of the next group's cultural surroundings, or milieu. Among those influenced by this type of thinking was H.J. Fleure (1877–1969) who tried to formulate world regions based on human characteristic rather than the traditional climatic-biotic regions. So he brought forth a scheme which included 'regions of effort', 'regions of hunger' and 'industrialized regions', to name a few. Possibilism has also been influential in the rise of the school of cultural geography associated with the name of Carl Ortwin Sauer and the University of California at Berkeley, and with the development of the idea of human ecology. The founder of this latter notion (human ecology) was H.H. Barrows (1877–1960) of the University of Chicago.

The possibilists cited numerous examples in support of their argument. There are distinct zones which are distributed symmetrically on each side of the equator, great climato-botanic frames, unequally rich in possibilities, unequally favourable to the different human races, and unequally fitted for human development; but the impossibility is never absolute – even for the races least 'adapted' to them – and all probabilities are often found to be upset by the persistent and supple will of man. The 'environmental determinist' thesis has it that these frames constitute 'a group of forces which act directly on man with sovereign and decisive power', and which govern 'every manifestation of his activity from the simplest to the most important and most complicated'. What really happens in all these frames, especially in those which are the richest in possibilities, is that these possibilities are awakened one after the other, then lie dormant, to reawaken suddenly according to the nature and initiative of the occupier. 'These possibilities of action do not constitute any sort of connected system; they do not represent in each region an inseparable whole; if they are graspable, they are not grasped by men all at once, with the same force, and at the same time'. The same regions, through the changes in value of their elements, have the most varied destinies. And it is human activity which 'governs the game'.

There are no doubts about human groups similarities – or, at least, analogies – of life which are the result of the exploitation of similar possibilities. But there is nothing fixed or rigid about them. We must avoid confusing once more necessity with possibility.

The possibilists show with great precision that society interposes practices, beliefs, and rule of life between nature and man; that man's utilization of possibilities and his exploitation of his environment are thereby hampered, so as, far example, to render his food singularly monotonous. 'Nowhere is food eaten by savages without care in the choice. There are prohibitions, restrictions, taboos on sides. But this social constraint was, no doubt, not exercised at first in its full vigour. There was great homogeneity in primitive human groups, but there were necessarily differences (age and sex) and individual contingencies, however slight. In small societies the organization was not rigid enough at the beginning to stifle initiative. It is thanks to differentiation, to the individual alone, that life has been ameliorated and that society itself has been organized.

The possibilists also argued that it is impossible to explain the difference in human society and the history of that society with reference to the influence of physical environment. They hold that man himself brings his influence to bear on that environment and changes it.

The philosophy of possibilism – the belief that people are not just the products of their environment or just pawns of natural environment – became very much popular after the First World War. For the possibilists, the works of man, not the earth and its influence, are the starting points, the most important is the freedom of man to choose.

Although the philosophy of possibilism became very much popular after the First World War, it was Vidal de la Blache who advocated and preached the philosophy of possibilism. Vidal de la Blache was such a staunch supporter of this philosophy that he developed the 'school of possibilism'. Vidal de la Blache in his studies minimized the influence of environment on the activities of man. Central to Vidal de la Blache's work were the lifestyles (*genre de vie*) that develop in different geographical environments.[22] In his opinion, lifestyles (*genre de vie*) are the products and reflections of a civilization, representing the integrated result of physical, historical and social influences surrounding man's relation to milieu in a particular place. He believed that whereas society and nature were usually represented as 'two adversaries in a duel', the human being was in fact 'part of living creation' and 'its most active collaborator'. And it was this dialectic which he subsumed in the concept of the *genre de vie*. He tried to explain differences between groups in the same or similar environment, and pointed out that these differences are not due to the dictates of physical environment but are the outcome of variations in attitudes, values and habits.[23] Variations in attitudes and habits create numerous possibilities for human communities. It is this concept which became the basic philosophy of the school of possibilism.

The possibilists emphasize the point that it is impossible to explain the difference in human society and the history of that society with reference to the influence of environment; they hold that man himself brings his influence to bear on that environment and changes it.[24]

After Vidal de la Blache, possibilism continued to grow and spread on both sides of the Atlantic. In France, Jean Brunhes was a strong supporter of possibilism. Brunhes enunciated the first explicit formulation of human geography as a systematic approach to the study of human geography. Outside France, the possibilist ideas were accepted by a large number of geographers and anthropologists. Barrows – the prominent ecologist – gave greater importance to man than to environment. A more acceptable view of possibilism was presented by Sauer. He asserted that a geographer's role is to investigate and understand the nature of the transition from the natural to the cultural landscape. From such an exercise the geographer would identify the major changes that had occurred in an area as a result of occupancy by succession of human groups. Its importance is often greater in regions where it has been acclimatized than in those where it originated and domesticated. For example, wheat does not have the largest yields in regions where it was first domesticated (South-West Asia).[25] Cultivation of rice is now done largely in USA, Canada, Australia, Pakistan and India – places where it was taken up later.

According to the possibilists, nature is never more than an adviser. There are no necessities but everywhere possibilities. This, by the reversal with it, involves man in the first place, man and no longer the earth, nor the influence of climate nor the determinant conditions of localities.[26] The range of possibilities in every region is limited more by the price man is willing to pay for what he wants than by the dictates of environment. For example, man through his technical skill can grow banana, rice and rubber in Antarctica but he has to take into consideration the input cost. The prohibitive cost of production of these crops will compel man not to grow these crops in the tundra region.

Men can never entirely rid themselves, whatever they do, of the hold their physical environment has on them. Taking this into consideration they utilize their geographical circumstances more or less according to what they are, and take advantage more or less completely of their geographical possibilities. But here, as elsewhere, there is no action of necessity.[27] The limits set by nature to man's action vary from one historical period to another. In marginal environments, such as the hot and cold deserts and tundra, and at low stages of culture man's choice may be extremely restricted. In the more favourable areas of the warm and cool temperate zones, and in periods when man's techniques are highly developed, the possibilities are more numerous. But notwithstanding the many skills man may acquire, he can never free himself entirely from nature's control. Bowman asserted: 'While the physical laws to which mankind responds are available in their application and degree of effect, yet this is also true that all men everywhere are affected to some degree by physical conditions'.

In spite of the fact that man has numerous possibilities in a given physical setting, he cannot go against the directions laid by the physical environment. The possibilistic approach has been criticized by many contemporary thinkers. Griffith Taylor, while criticizing possibilism, opined that society as a whole should make a choice, and since only an advisory role is assigned to geographer, his function 'is not that of interpreting nature's plan'. Taylor was largely right when he wrote that the task of geography is to study the natural environment and its effect on man, not all problems connected with man or the 'cultural landscape'.[28] Moreover, possibilism does not encourage study of physical environment and it promotes overanthropocentrism in geography. Geographical determinism at least obliges the geographer to turn his attention to nature, and if the question is asked as to who is setting out to destroy geography, then blame should be placed above all at the possibilists' door. Possibilism thus tended to exaggerate the role of culture and to neglect the importance of natural environment. In brief, the approach of possibilism may be as ludicrous as determinism, but possibilists generally recognized the limits to action which environment set, and avoided the great generalizations which characterized their antagonists.

Neo-Determinism (Thomas Griffith Taylor – 1880–1963)

Thomas Griffith Taylor

The concept of 'neo-determinism' was put forward by Griffith Taylor – a leading British anthropologist, world explorer and geographer. He argued that possibilists had developed their ideas in temperate environments such as North-Western Europe, which offer several viable alternative forms of human occupance. But such environments are rare: in most of the world as in Australia the environment is much more extreme and its control over human activity is enormous. He coined the term 'stop-and-go determinism' to describe his views. In the short term, people might attempt whatever they wished with regard to their environment, but in the long term, nature's plan would ensure that the environment won the battle and forced a compromise out of its human occupants. He, in the 1920s, argued that the limits of agricultural settlement in Australia had been set by factors in the physical environment such as the distribution of rainfall. Taylor's view was initially most unpopular in Australia, but it has been generally accepted since then. In his book on Australia published in 1948, Taylor reaffirmed his basic position:

> The best economic programme for a country to follow has in large part been determined by nature (environment), and it is the geographer's duty to interpret this programme. Man is able to accelerate, slow or stop the progress of a country's (region's) development. But he should not, if he is wise, depart from directions as indicated by the natural environment. He (man) is like the traffic controller in a large city who alters the rate but not the direction of progress.

Neo-determinism is also known as 'stop-and-go determinism' and Griffith Taylor's philosophy can be very vividly explained by the role of a traffic controller.[29]

Man follows nature's programme only if he is wise, presuming he can act foolishly, which admits the possible contention that within broad limits set by environment man can choose, at the very least. Taylor concedes him the choice between what is wise and what is foolish.[30] But wisdom and folly are human concepts. The natural environment knows nothing of them. In nature there is only the 'possible' and 'impossible'. Finer categories are man-made.

The possibilists admit that the opportunities offered by any environment are not all equal. Some demand little for man, others continual struggle; some yield large, other meagre returns. The ratio between effort and return can be looked upon as the price nature extracts from man for the particular choice he makes; but recognition of this inequality of opportunities gives no clue as to what nature prefers, for the wise man to follow suit.

Once possibility of alternative action is conceded, then it is difficult to see how 'stop-and-go determinism' can claim that man is not a free agent, that his liberty is curtailed. In no environment are the possibilities limitless and for every choice price must be paid, proponents of possibilism admit this, but within these limits freedom to choose exists. Man makes his choice, and man himself judges its relative wisdom or folly by reference to goals he himself has established. Limits to man's freedom beyond those generally recognized by possibilists are, according to Taylor's definition, those imposed by man's conception of wisdom. There is nothing indeed that contradicts the assertion of Febvre (founder of possibilism) that there are no necessities but everywhere possibilities and man as a master of these possibilities is the judge of their use. Thus, man chooses, but only from the range which nature presents him. In brief, people might attempt whatever they wished with regard to their environment, but in the long term, nature's plan would ensure that the environment won the battle and forced a compromise out of its human occupants.

Probabilism (O.H.K. Spate – 1911–2000)

O.H.K. Spate

Born to a German father and English mother, Spate was an English geographer, known for his role in strengthening geography as a discipline in Australia and Pacific region. He died in 2000 at the age of 89. The concept of probabilism was put forward by O.H.K. Spate in 1957. The view that although the physical environment does not uniquely determine human actions, it does nevertheless make some response more than others. The term was proposed as a midway between a stark environmental determinism of Ratzel and a radical possibilism of Febvre, Vidal de la Blache and Sauer. While the environmental determinists, influenced by the cause and effect relationship of Darwin, asserted that human activities are controlled by the physical environment, the possibilists opined that physical environment provides the opportunity for a range of possible human responses and the people have considerable discretion to choose between them. According to Spate, 'human action was represented as not so much a matter of all-or-nothing choice or compulsion, but a balance of probabilities'. For example, there is a probability that the land use intensity in the Sutlej-Ganga plain decreases away from the market centres; the population density decreases away from metropolitan centres in all directions; crop yields diminish beyond a certain walking distance from the village settlement. There may be, however, exceptions to each of these generalizations, and in many cases, there are also limits to the range of territory which they hold true. The exceptions and the limits demand explanation. After this concept, the probability theory came to be regarded as an essential component of geographical analysis since it provided 'a common mode of discourse' for 'scientific study of the landscape'.

This view, in fact, is perfectly compatible with the original Vidalian conception. The geographers started to use the probability theory to determine the man and environment relationship and also to make a scientific study of the landscape.

The probability theory was criticized on several grounds. For example, a complete knowledge about the environment (resources) may not be available; the data available about the resources and their utilization may not be reliable; the perception about resources (environment) differs from man to man, community to community, region to region and country to country. The application of probability model, owing to these constraints, may be difficult and the results thus obtained may not be authentic, close to the ground reality.

Cultural or Social Determinism

Cultural or social determinism emphasizes the human element: 'Our thoughts determine our acts, and our acts determine the previous nature of the world' (James, 1932: 318). Since human interest, desires, prejudices and group values vary across space, there is a consequent variation in the cultural landscape and levels of socio-economic development. The modification of an environment largely depends on our perceptions, ideas and decision-making processes. This philosophy, advocated by American scholars, can be summed up as the principle according to which the 'significance to man of the physical and biotic features of his habitat is a function of the attitudes, objectives and technical skills of man himself'. For example, a country that is richly endowed from the point of view of the hunters, might appear poor to the agricultural people; the importance of coal is not identical to those who can and those who cannot make use of it. All these truths are self-evident. What is also true is that as technology develops, the importance of the environment does not decrease but changes and becomes more complex.[31]

The philosophy of cultural determinism is fairly widespread among American geographers. Edward Ullman, for example, wrote that 'the environment is essentially neutral, its role being dependent on the stage of technology, type of culture and other characteristics of a changing society'. The assessment of a mountain pass, for example, will differ for those who possess horses, automobiles, aeroplanes; the assessment of soil fertility will not be identical from the point of view of a Japanese farmer, on the one hand, or an Amazonian Indian, on the other.[32] Similar natural conditions may call forth different reactions on the part of man, and within similar sets of conditions, different cultures can take place. George Carter singles out three fundamental factors in human geography. He has laid greater stress on cultural forces and writes that 'ideas remain as the primary cause of change ..., it is these ideas that determine the human use of physical world'.[33] He also emphasized the point that human will is the decisive factor.

After the Second World War, the school of social determinism became quite popular in Austria, Holland and Sweden. Social geography deals with the spatial distribution of societies. It, however, does not enable us to achieve a profound understanding of social relations or landscape. Social groups can be distinguished with reference to ethnic, religious, professional and certain other features, while social changes are only noted but rarely linked with any fundamental economic causes or the class structure of society. The study of the influence exerted by these groups on landscape is reduced to the definition of purely external factors of the cultural landscape (type and deployment of houses, land use, field patterns, etc.) right down to the morphological and functional changes within the confines of a single street. Infinitely painstaking 'micro-territorial' research of this type is usually purely empirical in character and cannot provide the basis for scientific conclusions of any real significance. Social or cultural determinism thus does not adequately assess the environmental factors, i.e. the influence of natural environment upon 'cultural geographical differences'. Social determinism is thus also rigid like environmental determinism and therefore cannot be accepted in its crude form.

The debate among geographers about whether people are free agents in their use of earth (environment) or whether there is a 'nature's plan' slowly dissolved as the antagonists realized the merits in each case.

Notes

1. Ischenko, A., 1981, 'Determinism and Indeterminism in the Work of Foreign Geographers', *Soviet Geography Today*, Moscow, pp. 201–202.
2. Taylor, G., 1957, *Geography in the Twentieth Century*, London, p. 128.
3. Al-Masudi, *Tanbin*, p. 28.
4. Tatham, C., 1957, 'A Study of Growth Fields, Techniques, Aims and Trends', *Geography in the Twentieth Century*, edited by G. Taylor, London.
5. Grossman, L., 1977, 'Man Environment Relationship in Anthropology and Geography', *Ann. Ass. Am. Geogr.*, 67, pp. 126–145.

6. Spate, O.H.K., 1957, 'How Determinism Is Possibilism', *Geographical Studies*, London, Vol. 4, No. 1.
7. Semple, E.C., 1911, *Influence of Geographical Environment*, New York.
8. Ischenko, A., 1981, op. cit., p. 210.
9. Castro, J.D., 1952, *The Geography of Hunger*, London, pp. 60–65.
10. Orr, Lord Boyd, 1950, 'The Food Problem', *Scientific American*, Vol. 183, No. 2.
11. Castro, D.J., 1952, op. cit., p. 62.
12. Ibid., p. 63.
13. Taylor, G., op. cit., p. 62.
14. Kimble, G.H.T., 'The Craft of Geographer', *The Canadian Geographical Journal*, Vol. 31, No. 5, 1945.
15. Chanman, W.J., 'Outline of Theory of Area Study', *ANNALS*, Vol. 38, No. 4, 1948.
16. James, P.E. (with the Collaboration of H.V.B. Clime Jr.), *A Geography of Man*, Waltham, 1966.
17. English, P.W., 'Landscape, Ecosystem and Environment Perception: Concepts in Cultural Geography', *The Journal of Geography*, Vol. 67, No. 4, 1968.
18. Spate, O.H.K., 1952, 'Toynbee and Huntington: A Study in Determinism', *The Geographical Journal*, Vol. 118, Pt. 4, pp. 423–424.
19. Ibid., p. 424.
20. Hartke, W., 1956, 'Die Sozialbrache als Phanomen der Geographischen. Differnzierung der Landschaft', *Erdkunde*, 10th ed., Nr.4.
21. Hartshorne, R., 1959, *Perspectives in the Nature of Geography*, Chicago.
22. Febvre, L., 1932, *A Geographical Introduction to History*, London, Kegan Paul, Trench, Trubner & Co. Ltd.
23. Ibid., p. 155.
24. Taylor, G., op. cit., p. 65.
25. Ischenko, A., op. cit., p. 219.
26. Vidal de la Blache, 1898, *Ann de Geog.*, p. 157.
27. Stoddard, R.H., et al., 1990, *Human Geography*, 2nd ed., Prentice Hall, New Jersey.
28. Ibid., p. 11.
29. Ibid., p. 159.
30. Taylor, G., *Australia*, p. 445.
31. James, P., 1965, 'Conceptual Structure of Geography', *The Journal of Geography* Vol. 64, No. 7.
32. Ullman, E., 1953, 'Human Geography and Area Research', *ANNALS*, Vol. 43, No. 1.
33. Isachenko, A., op. cit., p. 228.

9

Dualism and Dichotomies in Geography

Geographers throughout the history of geography have been confronted with the methodological problem of dualism and dichotomy. In the delineation of the sphere of geography, and the methodology to be adopted for its study, there have existed and still exist significant dichotomies like systematic versus regional geography, physical versus human geography, and deterministic versus possibilistic geography. Dualism is, however, not peculiar to geography; it is shared by all sister disciplines, though in varying degrees.

During the period of prehistory of geography, an obscure and vague dualism can be found in the writings of the Greek, Roman and Arab geographers. Herodotus laid emphasis on the description of the then major tribes and nationals and their physical surroundings; Strabo concentrated on regional description, while Ptolemy stressed on mathematical geography. Hippocrates, Aristotle, Xenophone, Aryabhatta, Al-Masudi, Al-Biruni, Al-Idrisi, and Ibn-Khaldun attempted to interpret the influence of physical environment on the lifestyle of various peoples (see Chapter 7).

Dualistic thinking in geography became more conspicuous during the post-renaissance period in Europe. Since then geography seems to have been visible into a number of mutually exclusive branches.[1] These dichotomies or branching of the subject look quite logical. In this chapter, a brief appraisal of the more important dichotomies has been given. Though new dichotomies appear on the scene with passage of time, some of the most prominent of them are as under:

1. General Geography versus Regional Geography.
2. Physical Geography versus Human Geography.
3. Historical Geography versus Contemporary Geography.
4. Study of Functional Regions versus Geography of Formal Regions.

Moreover, there is branching of the discipline in separate branches, e.g. geomorphology, climatology, hydrology, agriculture, land use and population geography.

GENERAL GEOGRAPHY VERSUS REGIONAL GEOGRAPHY

The issue of general geography versus regional geography was most probably raised by Bernhard Varen, known as Varenius, in the 17th century. This period is often termed as the classical period of the modern geographical thought. Varenius recognized the two main divisions of geography – general or universal and special or particular. This branching of the subject is known as general geography versus particular (special) geography. Systematic geography deals with one or a few aspects of the human environment or the human population and study their varying performance in the world or over a predefined geographical space. General geography, as viewed by Varenius, was concerned with the formulation of general laws, principles and generic concepts. It was believed to be the noblest of ends of scientific enquiry in the initial development of geography. Gradually, all studies of a generalistic nature acquired the status of systematic geography, while the special or particular studies were described as regional geography.[2] Systematic geography drew inspiration from the existing systematic sciences with a search for the universal and generic concepts. Regional geography, on the other hand, has not moved out of the ambit of particular studies.[3] Regional geography in the traditional sense seeks to bring together in an areal setting various matters which are treated separately in topical (systematic) geography. Regional geography is the study of the geography of regions. Regional geography is often distinguished by its interest in 'a specific situation in a particular locality' and has been hailed as 'the highest form of the geographer's art' (Hartshorne, 1982). Richard Hartshorne (1899–1992) who was a prominent American geographer declared in his book *Nature of Geography* (1939), that the essence of geography is regional geography. In the 1950s Hartshorne was part of a key geographical debate over the nature of geography. F.K. Schaefer called for the 'scientific method and universal laws in geography'. He criticized the 'old method' (regional geography) promoted by Hartshorne as the 'Hartshorian Orthodoxy'. In brief, general geography deals with the whole world as a unit. It was, however, mainly restricted to physical geography which could be understood through natural laws. On the contrary, special geography was primarily intended as a description of individual countries and world regions. It was difficult to establish laws in special geography where human beings are involved, whose behaviour is always unpredictable. Special geography, nevertheless, helped in the formulation of hypothesis and structured ideas.

After Varenius, the leading German scholar – Alexander von Humboldt – spelt out the difference between systematic and regional geography. Humboldt divided the subject matter of geography into *uranography* and geography. Uranography, according to him, is the descriptive astronomy, while geography deals with the interrelationship of phenomena that exist together in an area. He believed in inductive method and emphasized the importance of empirical method of research. In his introductory chapter to

Kosmos, while categorically stressing the value of generalization in science, Humboldt writes:

> The most important aim of all physical sciences is this, to recognize unity in diversity ... to grasp nature's essence under the cover of outward appearances ... the purpose of this introductory chapter is to indicate the manner in which natural science can be endowed with a higher purpose through which all phenomena and energies are revealed as one entity.

This embodied the logical process of transition from the particular studies of the 'single aspect' to the general aspects underlying them and revealing the 'nature's essence'. He made comparative study of different geographical regions, steppe grasslands (Russia) and the arid regions (Central Asia) to ascertain the peculiarities of the various parts of the earth's surface. Thus, Humboldt also recognized the dualism of systematic versus regional geography.

Carl Ritter – a contemporary of Humboldt – was a teleologist.[4] He stressed the need for a study of natural phenomena 'as a whole, as in parts' in order to comprehend the 'inherent plan'. Although he was convinced that there were laws, he was in no hurry to establish them.[5] He conceived geography as an empirical science rather than one based on deduction from rational principles or a priori theory. He emphasized the fact that there is a coherence in the special arrangement of terrestrial phenomena. Areal phenomena are so interrelated as to give rise to the uniqueness of the areas as individual units. In brief, according to Ritter, geography was concerned with objects on the earth as they exist together in an area. He studied areas synthetically, i.e. in their totality. He believed in the centrality of regional geography. He felt that geography must rise above a mere description of a multitude of facts about a particular phenomenon. The goal of geography should be, according to Ritter, '... to get away from mere description to the law of the thing described; to reach not a mere enumeration of facts and figures, but the connection of place with place and the laws which bind together local and general phenomena of the earth's surface'.

Ritter's ideas on the 'wholeness' of things were in accordance with the writings of W.F. Hegel (1770–1831), whose attitudes amounted to an attempt to comprehend the entire universe, to know the infinite and to see all things in God. The scientific stance of Ritter was teleological (Greek *teleo* = purpose). In brief, he studied the workings of nature in order to understand the purpose behind its order. He did not regard the shape of continents as accidental but rather as determined by God so that their form and location enabled them to play the role designed by God in the development of man. Ritter combined a basic teleological standpoint with a most critical scientific precision. 'My system builds on facts, not on philosophical arguments', he said in a letter. The collection of facts was not an end in itself; the systematization and comparison of data, region by region, would lead to a recognition of unity apparent among diversity. The plans of God, which give purpose and meaning, could only be discovered by taking into account all facts and relationships in the world as objectively as possible.

In the light of the scholarly attempts of Humboldt and Ritter, the scientific organization of knowledge completes itself in stages: first an accurate and detailed inventory of all facts about a single aspect is prepared, and secondly, these facts are integrated with a coherent and intelligible body of knowledge where facts about some particular aspects are seen not as specific facts but as parts of general interrelated system and are 'subsumed under a number of laws' which express a genetic relationship.[6] To quote Humboldt:

> In proportion as laws admit of more general application as sciences mutually enrich each other, and by their extension become connected together in more numerous and more intimate relations, the development of general truths may be given with conciseness devoid of superficiality. On being first examined, all phenomena appeared to be isolated and it is only by the result of multiplicity of observations, combined with reasons, that we are able to trace the mutual relations existing between them.[7]

The above lines from Humboldt reveal the essential character of post-renaissance view of geography. Humboldt and Ritter were inspired by an overriding concern for the universal and for the then contemporary science. The contemporary science in astronomy and physical sciences was characterized by the proliferation of universally applicable laws. Geography could not remain immune to the prevailing tendencies. Thus, Humboldt was seriously engaged in the development of systematic physical geography, while Ritter, to a considerable extent, was a regional geographer who gave weight to man as an important component of the physical surroundings.

After Carl Ritter, Ferdinand von Richthofen defined geography. In his opinion, the purpose of geography was to focus attention on the diverse phenomena that occur in interrelationship on the surface of the earth. The methodology he suggested for the study of geography was that the elements of physical setting of a region be discussed and then the adjustment of man in that setting be examined.[8] For a substantial period, it remained the basic pattern of geographical studies not only in Germany but also in other parts of the world. Richthofen also emphasized the point that regional geography must be descriptive to highlight the salient features of a region. Moreover, it should try to seek regularities of occurrence and patterns of such unique features to formulate hypotheses and to explain the observed characteristics. General geography, he felt, deals with the spatial distribution of individual phenomenon in the world.

Geography after Humboldt, Ritter and Richthofen in Europe and America was dominated by Friedrich Ratzel (1844–1904). Before Ratzel, the foundation of systematic geography was laid by Humboldt and that of regional geography by Ritter. Ratzel largely used the deductive method and compared the lifestyles of different tribes and nations. In his anthropological studies, he liked to proceed with a priori hypothesis and laws and applied them to specific cases. He was obsessed by the concept of genesis of things rather than their interdependence. He applied Darwin's concept to human societies. This analogy suggested that groups of human beings must struggle to survive in particular environments as much as plant and animal organisms do. This marked a departure from the integrative morphological approach of Humboldt.

After Ratzel, Alfred Hettner – a leading German scholar – claimed that geography is an idiographic (regional) rather than monothetic (general) science.[9] In his opinion, the distinctive subject of geography was knowledge of the earth areas as these differ from each other. He considered man as an integral part of nature of an area. His approach was, however, deductive giving more importance to elements of physical environment.

The inductive method and empirical research got revived in France. Vidal de la Blache discarded the Ratzelian deductive approach and extensively employed specific studies *(pays)* for drawing conclusions of a general nature. In actual practice his efforts led to the development of regional geography, which made the understanding of particular and unique attributes the most cherished goal of geographical enquiry.

Vidal de la Blache, in his works, attempted a harmonious blending of physical and human features and tried synthesis of *pays*. He was convinced that small regions *(pays)* are the ideal units to study and to train geographers in geographical studies. According to Vidal de la Blache, man and nature are inseparable, and it is not possible to distinguish the influence of man on nature from that of nature on man. The two influences fuse. The area on which such an intimate relationship between man and nature has developed through the centuries constitutes a region. The study of such regions, each one of which is unique, should be the task of a geographer. He, therefore, argued for regional geography and against systematic geography as the core of the discipline. Vidal de la Blache's method, which was inductive and historical, was best suited to regions which were 'local' in the sense of being somewhat isolated from the world around them and dominated by an agricultural way of life. These circumstances favoured development of local relations in architecture, agricultural practices and a general way of life; the communities lived in such close association with nature that they might be self-sufficient in majority of goods. Vidal de la Blache advised geographers to carry out research in folk museums and collections and to investigate agricultural equipment which had been used in the past in order to study the individuality of development of a region. Vidal de la Blache's work, despite the breakdown of the self-sufficient regional economy, has been and still is a great inspiration to a vital tradition in geography, that of the regional monograph. Owing to these factors, Vidal de la Blache argued for regional geography as the core of the discipline of geography. To quote Vidal de la Blache:

> Human societies like those of plants and animal world are composed of different elements subject to the influence of environment. No one knows what winds brought them together, but they are living together side by side in

> a region which has gradually put its stamp upon them. Some societies have long been part of the environment, but others are in process of formation, continuing to recruit numbers and to be modified day by day.

As is clear from the above, Vidal de la Blache advocated the principle of 'terrestrial whole'. He believed that the earth and its inhabitants stand in the closer reciprocal relations, and one cannot be truly presented in all its relationships without the other.[10]

Another French geographer Reclus, while giving a precise picture of world societies, asserted that man is not the product of his environment but an important component of it. 'Man may modify his dwelling places to suit his own purpose; he may overcome nature'.

The later part of the 19th century and the middle of the 20th century have been characterized by an overwhelming development of specific knowledge with very little or no concern for 'integrative overviews' revealing the generic relationships of universal relevance. Undoubtedly, this phase enriched the subject but it also revealed the inherent weakness of the geographical methodology for the specific and the particular, and its failure to rise above mere description of the individual aspect to a level at which cognition of the general features becomes possible. Admittedly, it did not always succeed in relieving itself from the grooves of specialized knowledge in its search for laws and generic concepts.[11]

Richard Hartshorne stressed on areal differentiation (regional geography). In the post-Second World War period, geography was essentially ideographic (regional) and was articulated through the art of geographical description, a commitment to fieldwork, and the integration of physical geography and human geography within the study of particular landscape.

The post-Second World War period is characterized by quantitative revolution in the subject. It has developed new conceptual frameworks leading to the emergence of a location theory which seeks for new order in the distribution of phenomena in space in their interlinkages.

The foregoing description gives a historical background of the dichotomies of systematic or general versus special or regional geography. The approaches adopted by the scholars of systematic and regional geography are described hereunder.

As stated at the outset, systematic geography deals with the universal laws and generic concepts. Systematic or general geography is essentially analytical and makes use of generic concepts, whereas regional geography is necessarily synthetic and deals with unique situations and their peculiarities. Systematic geography, moreover, deals with the whole world as a unit. For example, if we take the patterns of distribution of temperature, rainfall, vegetation, minerals, crops and population, and examine them at the world level or continent wise, it would be a case of systematic geography. In contrast to this, if we study landforms, climatic variables, soils, vegetation, minerals, fauna and flora, and superimpose these physical factors on the cultural landscape or on any of the elements of socio-cultural aspect, then this would be a case of regional or special geography.

To illustrate this point, Figure 9.1 has been plotted. In this figure, the rows show the approach of study of systematic geography, and columns show the approach of study of regional geography, i.e. if we study the types of soils in various continents, it is an example of systematic geography, while if we take a particular continent or a region of it and superimpose all the physical and socio-economic variables, it would highlight the peculiarities of that region. This synthetic picture, revealing the special features of the region, is a case of regional geography.

Figure 9.1 further reveals various branches of the subject. As these branches of general geography are also combined into regional geography, it can be seen that these are the two main aspects of the subject.[12] The figure clearly shows how the combinations of phenomena and parts of the earth's surface can give regional or general geography.

Figure 9.1 Dichotomy of Systematic and Regional Geography

General Geography and Special Geography							
	1	2	3	4	5	6	7
General Geography	North America	South America	Europe	Africa	Asia	Australia	Antarctica
Landforms							
Climate							
Soils							
Plants							
Animals							
Economic							
Social							
A Urban							
B Settlement							
C Population							
D Cultural							
Political							

Notes: D = Many new branches growing here

(shaded) Special (regional) geography of Africa

(brick pattern) General (Systematic) geography of population which considers its geo-regional variation throughout world

The dichotomy of systematic versus regional geography seems to be quite logical. In the opinion of some scholars, there may exist several geographies rather than one. In fact, geography has been defined by different geographers differently. These definitions range from *landscapes, places, space, location, man-nature interaction, man-earth system, human ecology* and *areal differentiation* of interrelated phenomena on the earth surface to man.

Thus, geography is multidimensional not only in the number of topics and regions of the world which can be included in one study but also in the approach of study. Geography is multivariate not only in its combination of natural sciences, social sciences, and mathematics but also in the ways different geographers may combine these elements.[13] Owing to this multivariate nature of the discipline, even regional geographers now recoil from describing all the phenomena at one place which they discover are interconnected. At a time when regional description is in backwater, it may be necessary to conceive general geography, *compage regional geography*, and full descriptive regional geography as three quite separate branches. *Compage geography* will not include phenomena which are simply characteristic of a place unless they show some logical arrangement in space and connections with other important phenomena. The term *compage* was introduced to geography by Derwent Whittlesey (1890–1956) in an attempt to give greater precision to several aspects of regional geography. The central idea of *compage* is that all the features of the physical, biotic and societal environments are functionally associated with the human occupancy of the earth. Yet, thinking of geography's wider function and obligation to educate laymen, as distinct from professional geographers, full, orderly regional description may still be required outside the profession.[14]

More stress on regional geography is also not correct, because no two places, no two groups of people are exactly alike in any place at any point of time. In the words of Berry[15], 'the regional and general geography are not different approaches, but are just the two extremes of a continuum', which he likens to a three dimensional matrix – the earth, social and geometrical. Geographical studies do not fall into systematic (topical) and regional groups but are distributed along a gradual continuum from topical studies of the most elementary integration at one end of regional studies to the most complete integration at the other.

All material objects and phenomena which exist in the real world and have been observed by us have two entities – individual or the particular and the general or the universal. They have particular characteristics which are peculiar to them and make them unique; they also have some general features which are common to other objects of the same type and are, therefore, universal in nature. It is their individuality which makes them different from other objects. These individual objects also have certain recurrent features in common which link them to a group of objects with which they have general relations.

It is, therefore, important to note that the general characteristics exist in, and are seen through, the particular and the individual characteristics and are

not independent of them. The fact is the two are mutually interpenetrative. The general becomes true in the particular and the particular becomes true in the general.[16]

> ... the interrelation between the individual, particular and universal lies in the fact that they are connected, in the fact that the individual cannot exist without the universal and that the universal cannot exist without the individual, that the individual under certain conditions may become both particular and universal.

To quote V.A. Anuchin, the Soviet scholar:

> One can trace definite cycles in the history of world science. Periods when general absorbs the particular and succeeded by those during which the particular destroys the general and a single science disintegrates into an endless number of branches. This later differentiation leads to great extensions of knowledge but results in less of integrative overviews of science which show that the *whole is greater than the sum of the parts*. Contemporary geography is a victim of such a phase of differentiation.

Thus, the dichotomy of systematic and regional therefore falls, as they do not oppose but support each other in the final analysis, as the subject matter of geography.

PHYSICAL GEOGRAPHY VERSUS HUMAN GEOGRAPHY

As regards dichotomy of physical geography versus human geography, the Greeks were probably the first who started this branching of the discipline. Hecataeus gave more weight to physical geography, while Herodotus and Strabo emphasized the human aspect. Dualism of physical versus human geography is still a characteristic of the discipline. Some writers have regarded it as essential for the justification of the role of geography, while others have argued for it as also for a division of the subject into physical and human geography on the ground that the respective methodologies of physical and human geography must be different. In studies of natural phenomena, including climatology, meteorology, hydrology, oceanography, geology and landforms, it is possible to use the methods of natural sciences and to draw conclusions with a large measure of scientific precision. The methods of natural science, however, do not lend themselves very well to the study of social and cultural phenomena. Our generalization about human groups must be limited in time and space, and must relate to statements of probability rather than certainty.[17]

Verenius, whose *Geographia Generalis* was published in Amsterdam in 1650, was one of the first scholars to suggest these essential differences in the characteristics of physical and human geography. At the beginning of the 18th century, Immanuel Kant delivered lectures on physical geography at the University of Konigsburg (Germany). He studied the deflection of wind direction resulting from earth's rotation.[18] Humboldt, who is considered as the last of the great polymaths, was primarily interested in physical geography. On the other hand, Carl Ritter, the first professor of geography at the Berlin University, was more inclined towards human geography. Humboldt and Ritter believed that the ultimate aim of research in physical geography was to clarify the unity of nature.

Reclus laid emphasis on systematic physical geography called *La Terre*. After Reclus, Darwin gave importance to physical aspect of the discipline while postulating the concepts of struggle and survival. Under these circumstances, Mare Somerville published *Physical Geography* in 1848. In the second half of the 19th century, geographers concerned themselves more and more with physical geography. They established geomorphology, the study of landforms, which later became the most substantial element in physical geography. The term 'geomorphology' was coined by Albrecht Penck – the German geographer – who was a geologist by training. After doing extensive fieldwork he formulated the principles of 'landforms evolution' and showed how the systematic study of features can be approached from the chorological (regional) point of view. He stressed the importance of relief maps for a systematic study of geography. Later on, Koppen, Davis, Martonne, Mill, Jafferson and Dokuchaive put great emphasis on landforms and climate as the major concerns of geography. In all these studies, man (the most important component of ecosystem) was ignored. It was during this period that Davis put forward the idea of the normal cycle of erosion. Ratzel and Semple also gave greater importance to physical environment which determines the lifestyle of people. Semple asserted that 'man is the product of earth surface'. Huntington, while writing about the march of civilizations, has opined that the shift in their centres was due to the climate and weather conditions. Mackinder, Chisholm and Herbertson also recognized physical geography as the main field of geographers. Th\u0254mas Henery Huxley wrote *Physiography* in 1877. Physiography had a much wider meaning; it may be defined as a description of nature; physical geography (renamed physiography after 1877) became a very popular school subject during the last three decades of the 19th century.[19] The Soviet scientists also conceived geography as the branch of science which deals with *geomorphology*, *pedology*, *hydrology* and *meteorology*. This major emphasis on physical geography may be attributed to the fact that at the initial stage of development, geography was taught by teachers who had geology background. The protagonists of physical geography declared it as the only area in which geographers should contribute. In the opinion of Wooldridge and East:

> It is futile to assert that 'human' or 'social' geography can be seen in terms of formal categories and universal principles and processes as can physical geography. This imputes to it no inferiority, it is rather to admit that it is infinitely more complex, subtler, more flexible, manifold.

Wrigley has recently commented on the methodological difficulty of 'running in harness, as it were, physical geography and social geography'. By accepting the view of that explanation in the physical sciences, Wrigley implies the existence of two radically different frameworks for explanatory thinking in geography. In physical geography law statements are of importance, but in human geography such statements are irrelevant. This geographical manifestation of the Weber-Winch thesis regarding laws in the social sciences need not be accepted. Rather, there are strong grounds for rejecting such a view. It may thus be claimed that laws can be established in both human and physical geography. Some writers dissent in general from this view and claim that laws cannot be established because of multivariate nature of the subject matter, because the number of cases about which one may generalize is often small, and because the occasional exceptional circumstance may have far-reaching consequences.[20]

The real dichotomy of physical versus human geography cannot be understood unless some light is thrown on the historical development of human geography. Ritter and Ratzel were among the first who considered man as an agent who brings change in the landscape. Febvre placed emphasis on the fact that human beings are an element of the 'landscape' – an element whose activity is incorporated in it, a modifying agent of the environment which 'humanizes' it. He also argued that the same physical factors do not always produce the same effects. In geography, according to Febvre, 'we deal with man's work, man's calculations, man's movement, the perpetual ebb and flow of humanity; man not the soil or the climate – is ever in the forefront'. It was Vidal de la Blache who founded the school of human geography. He gave relatively less importance to the elements of physical environment as the major determinants of cultural landscape of a region. Vidal de la Blache had a clear insight into the weakness of physical geography and the deterministic argument. He realized the futility of setting man's natural surroundings in opposition to his social milieu and of regarding one as dominating the other. According to Vidal de la Blache, it is unreasonable to draw boundaries between natural and cultural phenomena; they should be regarded as united and inseparable. In an area of human settlement, nature changes significantly because of the presence of man, and these changes are the greatest where the level of material culture of a community is the highest.[21] Jean Brunhes prepared himself for the conceptual framework of human geography. He developed the principles of activity and interconnection. Later, Albert Demangeon was a strong follower of Vidalian tradition.

In America, Mark Jafferson brought the idea of 'central places', 'the primate city' and 'the civilizing rails' in the field of human and urban geography. In the Soviet Union, D.N. Anuchin followed the principle of 'economic determinism'.

The basic philosophy of the followers of human geography was to establish a man-nature mutual relationship in which each of the two is dependent on the other. All geographical studies are aimed at developing an understanding of the earth surface and its physical and social phenomena both as spatially varying entities unique in their own way as well as components in a mutually interacting system. The methodology adopted to promote such an understanding differs from field to field and is essentially designed to suit the requirements of the content intended to be studied and the underlying objectives involved therein. The difference in methods employed in different branches of geography may be so vast as to make the very unity of field rather doubtful. For instance, 'the difference in methods between studies of climate and of landforms is in many respects greater than the difference between the study of natural vegetation and of cultivation of crops'.[22]

It is evident from the above discussion that the dichotomy of physical geography versus human geography is artificial and illogical. This dualism is the result of historical development of the discipline. In brief, geography does not fall into two groups, i.e. physical and human; these two are just the two extremes of a continuum. Hartshorne argues that if we divide geography into physical and human phenomena, we make the rest of the work illogical. Thus, we study the effect of physical factors on man and man's activities on land and not the physiological factor. Therefore, the division into physical and human is the cause of geography being only a partial study. In fact, all geographers realize that we could not possibly explain human choices and actions solely in terms of relationship with the natural environment. For geography, to have any value, the rift between physical and human phenomena must disappear.[23]

HISTORICAL GEOGRAPHY VERSUS CONTEMPORARY GEOGRAPHY

The dichotomy of historical geography versus contemporary geography has also attracted the attention of historians, geographers and other social scientists. Historical geography deals with the geography of an area, region or world as it had been in the past. For example, if we could ascertain the cropping patterns and settlement distributions in India during the medieval period, it would be an aspect of historical geography. S.M. Ali attempted to construct the geography of ancient India in his monumental work – *The Geography of Puranas*. Many such studies are in progress in India and abroad. Ralph Brown has attempted to show what the United States was like at the time each region was settled.[24] East has constructed human geography of Europe at several stages, while other historical geographers have dealt with topics, rather than areas in the past.[25]

Historical geography, however, does not fit within the framework of geography as we know it but stands side by side with the geography of present times. The number of books on historical geography is not as great as the number of books on contemporary geography but as historical geography embraces both general and regional works, and involve all the branches listed, it is not a part of geography, in the sense social geography is a part. It is a separate entity complete in itself.[26] Contemporary geography deals with the existing patterns of spatial differentiation of geographical phenomena. Contemporary geography will become historical geography with the passage of time. Contemporary geography and historical geography are mutually exclusive and must logically exist as supporting each other.

There are other points of views also regarding the nature and scope of historical geography. These may be listed briefly as under:[27]

1. The operation of geographical factor in history.
2. The evolution of cultural landscape.
3. The reconstruction of past geographies.
4. The study of geographical changes through time.

The Geographical Factor in History

In the second half of the 19th century, historical geography was conceived as a branch of geography which should be concerned with the interrelationship of phenomena in space in a particular period or with the geography of the past period, and the influence of geographical factors on history.[28] Whittlesey and East asserted that geographical factor in history means reconstructing the geography of a past period as a part of the necessary context within which the flow of historical events may be better understood. To some others it also means the control/effect/influence of the physical environment on man's activity in the past or on history and historical events. This leads to determinism and possibilism; and it suggests a view of causation by which it was thought that phenomena could best be understood by studying, listing and classifying the operation of particular groups of factors and their effects, each group producing its own particular pattern of determinism: social, economic, technological determinisms thus take their place side by side with geographical determinism. This type of historical geography has, however, been criticized for setting out to clarify history, not geography. Thus, it may be nothing more impressive than 'adding the missing environmental notation to the work of historians'.[29]

The Changing Cultural Landscape

According to this point of view, historical geography should be the study of the changing cultural landscape. It was this approach which led to the reconstruction of the past cultural landscapes in many of the European countries. One of the most appealing virtues of the approach to historical geography through the changing cultural landscape is its apparent symmetry with genetic geomorphology. Both are seen to be concerned with the evolution of landscape features and thus help in laying the foundation of geography. It is more profitable to regard the landscape elements as a source of invaluable evidence for the reconstruction of past geographies rather than as the phenomena which are to be explained by historical study. The distribution of settlements, house types, field patterns can help in ascertaining the cultural landscape of the past.

The Reconstruction of Past Geographies

The most orthodox view of historical geography is that it should be concerned with the reconstruction of geographies of past times. There can be numerous types of historical geographies, e.g. agricultural geography, urban geography, industrial geography, social geography and regional geography. The geographies of the past in fact are of vital importance to the present geographers. The reconstruction of the geography of the past periods is necessary to establish a relationship between the past and the present distribution of phenomena. The reconstruction of past geography of a region or country is, however, a difficult task.

Geographical Changes Through Time

Geographers are essentially concerned with places and what they are like, whether in the past or the present. The geographical factors, both physical and cultural, change in space and time. Consequently, the character of a region also changes. The study of these geographical changes through time should be the main concern of a geographer.

As stated above, historical geography and contemporary geography are one and the same. The present geography will become historical geography in due course of time. Thus, Mackinder wrote that historical geography is the study of historical present: 'The geographer has to try and to put himself back into the present that existed, let us say one thousand or two thousand years ago; he has got to try and restore it'.[30]

STUDY OF FUNCTIONAL (OR NODAL) REGIONS VERSUS GEOGRAPHY OF FORMAL (OR UNIFORM) REGIONS

The concept of space is very important in geography as it deals with locations and sites. The dichotomy of functional location and formal site means the division of geography into the studies of geography of real places on the earth surface and the geometric space on paper. The idea of formal sites and functional locations emerged to avoid the controversy of regional versus systematic and physical versus human geography.

The functional location or functional region (or nodal) is a new concept in geography. In a given location (region), there is a variety of relief, climate, natural vegetation, soil, land use, industry, transport linkages and marketing centres. These phenomena are united not by being piled up on top of one another in the same space, but by functioning and working together as part of an economic and social system. Thus, crops, animals, food, raw materials, people's messages are exchanged and moved around by air, water, pipe, wire, road and rail. The basic spatial idea is that these phenomena are spread out horizontally, side by side, and are arranged thus not by chance, but in a logical layout so that they can work together well and make the best use of space. Thus, a functional location not only affects the phenomena of that very place but also the phenomena of a place hundreds of miles away.[31] According to the followers of functional regions, 'causal relationships exist between the complex and heterogeneous phenomena at one place, and the causal connections among phenomena at different places'.[32] The homogeneous areas or habitats inhabited by societies, social groups, or nations are known as formal regions. Two or more different societies combine to form a community in a functional area, which by name and definition must be similar to formal regions. These communities, like functional regions, are organized and represented by small systems or part systems. The purpose of functional locations is to study the structure and functions of a community within a space(s).

In contrast to functional locations, the formal site study concerns with the phenomena which exist in a region or place and their interdependence on each other. The basic philosophy of the study of formal locations is that the uniform relief combined with uniform climate, and uniform soil result into uniform land use, settlements and mode of life within a region. This traditional approach of formal sites looks outdated in the present context and cannot be of much help in explaining the complex relationship of man with environment.

In the study of functional locations absolute distances and space are irrelevant. In this approach, accessibility and isolations are measured in a special way, usually in terms of cost distance, time distance or mileage through a transport network, and these distances are measured from special nodes or axes. One important aspect is that geographical features like settlement patterns, land use, diffusion processes, etc. show a location and dynamics which to a large extent are due to their relative positions in space.[33] Pip Feror observes that since distances in time, cost or even network mileage are partly artefacts of socio-economic demands and technological progress these types of spaces are naturally dynamic and truly relative. This leads him to the definition of *plastic space* – a space that is continuously changing its size and forms.[34] With the adoption of quantitative techniques and computer technology, geographers are moving from the study of formal sites to that of functional locations. In fact, formal site is complementary to the study of functional locations and therefore the two are mutually interdependent. This dichotomy therefore also looks illogical.

In the words of P.E. James,[35] acceptance of many dichotomies is a semantic trap. The dualism of topical versus regional, physical versus human, determinism versus possibilism, deductive versus inductive, idiographic versus nomethetic, formal sites versus functional locations are not mutually contradictory, such as good and evil, or reason and faith. A dichotomy does not exist when one of the alleged opposites forms a subordinate part of the other or when one is derived from the other, a dichotomy may exist for some people and not for others on certain basic attitudes of the culture.[36] All the dichotomies discussed have done particular damage to geographical thinking.

Notes

1. Minshull, R., 1970, *The Changing Nature of Geography*, London, p. 138.
2. Ahmad, A., 1978, 'Dualism and Dichotomies in Geography: A Discussion Paper', p. 6; and 1983, 'Nature of Dichotomy in Geography: General and Particular', *The Indian Geographical Journal*, Vol. 58, No. 2, pp. 97–107.
3. Haggett, P., 1965, *Locational Analysis in Geography*, Arnold, p. 35.
4. James, P.E. and Martin, G.J., 1981, *All Possible Worlds*, p. 124.
5. Dickinson, R.E., 1969, *The Makers of Modern Geography*.

6. Ahmad, A., op. cit., p. 8.
7. Humboldt, A. von. 1849, *Kosmos*. Translation by E.C. Otte, London.
8. Ahmad, A., op. cit., p. 8.
9. Hartshorne, R., 1959, *Perspectives in the Nature of Geography*, Chicago.
10. Freeman, T.W., 1971, *A Hundred Years of Geography*, London, p. 84.
11. Ahmad, A., op. cit., p. 9.
12. Minshull, R., op. cit., p. 140.
13. Ibid., p. 142.
14. Ibid.
15. Berry, B.J.L., 1973, 'A Paradigm from Modern Geography' in *Directions in Geography*, London, pp. 43–63.
16. Anuchin, V.A., 1973, 'Theory of Geography' in R.J. Chorley, (ed.) in *Directions in Geography*, London, pp. 43–63.
17. Holt-Jensen, A., 1981, *Geography: Its History and Concepts*, p. 13.
18. Hartshorne, R., 1939, 'The Nature of Geography, A Critical Survey of Current Thought in the Light of the Past', *Ann. Assoc. Am. Geographers*, 29, 173, p. 38.
19. Steddart, 1975, 'The Victorian Science-Huxley's Physiography and its Impact of Geography', *Trans. Inst. of Br. Geogr.*, 66, pp. 17–40.
20. Wooldridge, S.W. and East, W.G.,1951, *The Spirit and Purpose of Geography*, p. 30.
21. Holt-Jensen, A., op. cit., p. 27.
22. Ahmad, A., op. cit.
23. Febvre, L., 1952, *Le Terra et l'Evolution Humaine Introduction Geographique a l'Historie*, Paris; and 1932, *A Geographical Introduction to History*, London, Kegan Paul, Trench, Trubner & Co. Ltd.
24. Brown, R.H., 1948, *Historical Geography of the United States.*
25. East, W.G., 1954, *An Historical Geography of Europe.*
26. Minshull, R., op. cit., p. 140.
27. Smith, C.T., 1967, 'Historical Geography: Current Trends and Prospects' in *Frontiers in Geographical Teaching*, edited by R.J. Chorley and P. Haggett, pp. 118–143.
28. Ibid.
29. Sauer, C.O., 1941, 'Foreword to Historical Geography', *Ann. Ass. Amer. Geog.*, 31, pp. 1–20.
30. Smith, C.T., op. cit., p. 146.
31. Minshull, op. cit., p. 146.
32. Ibid.
33. Holt-Jensen, A., op. cit., p. 64.
34. Ibid.
35. James, P.E., 1972, op. cit., p. 506.
36. Ibid., p. 406.

10

Quantitative Revolution, Paradigms, System Analysis and Regional Concept

Geography for more than two hundred years was confronted with the problems of generalization and theory-building. In all other physical and social sciences theory-building has a long tradition. After the Second World War, geographers, especially those of the developed countries, realized the significance of using mathematical language rather than the language of literature in the study of geography. Consequently, empirical descriptive geography was discarded and greater stress was laid on the formulation of abstract models. Mathematical and abstract models need rigorous thinking and use of sophisticated statistical techniques. The diffusion of statistical techniques in geography to make the subject and its theories more precise is known as the 'quantitative revolution' in geography.

Traditionally, geography was considered to be a description of the earth surface, but in due course of time its definition and nature changed. Now, it is concerned with providing accurate, orderly, and rational descriptions and interpretations of the variable character of the earth surface.[1] In the words of Yeates, 'geography can be regarded as a science concerned with the rational development, and testing of theories that explain and predict the spatial distribution and location of various characteristics on the surface of the earth'. In order to achieve this objective and to obtain the real picture of a region, geographers began to use and apply quantitative tools and techniques to which qualitative geography was opposed, especially till the 1960s. Thus, the most obvious change brought about by the quantitative revolution is the change of methods and techniques. After this revolution, quantitative techniques and general system theory have been used quite extensively in geography. The new electronic devices have made possible the use of complex mathematical computations never before attempted.

QUANTITATIVE REVOLUTION

The application of statistical and mathematical techniques, theorems and proofs in understanding geographical systems is known as the 'quantitative revolution' in geography. Statistical methods were first introduced into geography in the early 1950s (Burton, 1963). Consisting mainly of descriptive statistics, there was also some attempt at hypotheses testing using, for example, chi-square. Bivariate Regression Analysis followed shortly but it was not until the 1960s that the General Linear Model was fully explored. It was I. Burton who published a research paper, 'The Quantitative Revolution and Theoretical Geography' in the *Canadian Geographer* (7: 151–162) in 1963.

The statistical methods are employed in geography for generating and testing hypotheses using empirical data, whereas the mathematical techniques and theorems are used for deriving models from a set of initial abstract assumptions. In other words, statistical methods are used to estimate, and test the significance of, various parameters associated with a given mathematical model such as the distance decay and gravity models.

There has been confusion among the geographers and the public mind about the nature and social relevance of geography, especially after the Second World War. The status of geography as a university discipline was under discussion. It was also a topic of debate that what should be taught as geography at various stages of the educational processes. In 1948, James Conant, President of the Harvard University, had reportedly come to the conclusion that 'geography is not a university subject'. The Department of Geography of Harvard University was closed soon after and the discipline of geography was gradually eased out in many of the private universities of USA. The continual threat of departmental closure or staff reduction also led to frantic search in American universities for new ideas and research programmes. This resulted into the development of the 'spatial sciences school', also called 'quantitative revolution' in geography.

The last three decades have been characterized by an almost continuous debate among human geographers concerning the philosophy, nature and methodology of geography. Moreover, the geographers of the post-Second World War suffered from a complex that they did not have standard theories, models and laws like that of other social and biological sciences. Consequently, their efforts and researches were not considered of much social relevance. In order to overcome these complexes and to put the subject on a sound theoretical footing, geographers started using quantitative techniques to interpret the organization of space, to generalize and to formulate their own theories and models about the man and environment relationship.

The main objectives of the quantitative revolution in geography were as under:

1. to change the descriptive character of the subject (geo+graphy) and to make it a scientific discipline;
2. to explain and interpret the spatial patterns of geographical phenomena in a rational, objective and cogent manner;
3. to use mathematical language instead of the language of literature, like 'Af' in the Koppen's classification of climate which stands for the 'tropical rainforests';

4. to make precise statements (generalizations) about locational order;
5. to test hypotheses and formulate models, theories and laws for estimations and predictions;
6. to identify the ideal locations for the various economic activities so that the profit may be maximized by the resource users; and
7. to provide geography a sound philosophical and theoretical base, and to make its methodology objective and scientific.

In order to achieve these objectives, the preachers of quantitative techniques stressed on field surveys for the collection of data and empirical observations. In the formulation of models and theories they assumed:

1. Man is a rational (economic) person who always tries to optimize his profits.
2. Man has infinite knowledge of his space (environment and resources).
3. They assumed 'space' as an isotropic surface.
4. There is no place for the normative questions (questions about social values) in scientific research and objective interpretation of the geographical reality.
5. They assumed that normative questions, like cultural values, beliefs, attitudes, customs, traditions, likes and dislikes, prejudice, and aesthetic values have no place in geographical research and scientific explanation of geographical patterns.

Historical Perspective of Quantitative Revolution

As stated at the outset, statistical methods were first introduced into the discipline in the early 1950s. Since then, much attention has been paid to a set of very sophisticated dynamic linear (e.g. Space-Time Forecasting Model) and non-linear (e.g. spectral analysis) statistical techniques, including those that bear peculiarly upon geographical problems (e.g. spatial autocorrelation).

The inspiration for mathematical modelling came from at least two sources: first *Social Physics*, which focused initially on the 'Gravity Model' and later 'Entropy Maximization' and, second *Neoclassical Economics* which influenced geography principally through the regional science movement and 'Location Theory'. Associated with each were often different questions, and hence a different branch of mathematics. The typical pre-occupation of social physics is the spatial interaction among a set of discrete geographical points (frequently, but not always, leading to the use of matrix algebra), while for neoclassical economics it is with optimization over continuous space (usually resulting in the use of differential calculus).

Although geographers borrowed several models from economics and sociology, e.g. the 'Crop Intensity Model' of J.H. Von Thunen (1826), Alfred Weber's model of 'Industrial Location' (1909), Christaller (1893–1969) was the first geographer who made a major contribution to location theory in his study of *Central Places in Southern Germany*. Subsequently, American urban geographers developed theoretical models of urban places. It was during the post-Second World War period that A. Ackerman (1958) encouraged his pupils to concentrate on quantification cultural processes and systematic geography. Weaver, another American geographer, delineated crop combination regions in Middle West (1954) by applying the standard deviation technique which brought quantitative revolution in agricultural geography. Hägerstrand became interested in the possibilities of investigating the process of innovation with the help of mathematical and statistical methods. He was able to construct a general 'Stochastic Model' of the process of diffusion. Stochastic literally means at random. Stochastic or probability models are based on mathematical probability theory and build random variables into their structures.

Empirical studies indicated that the movement of persons between two urban centres was proportional to the product of their populations and inversely proportional to the square of the distance between them. Stewart pointed out the isomorphic equal form or structure relationship between this empirical generalization and Newton's law of gravitation. Thereafter, this concept became known as the 'Gravity Model'. In other branches of the discipline, e.g. population, regional, economic, cultural and political geography a number of statistical techniques were gradually diffused. Thus, the diffusion of quantitative techniques steadily took place in the 1960s. In Britain Richard Chorley and Peter Haggett, both from Cambridge University, applied quantitative techniques vigorously and inspired the new generation to adopt sophisticated statistical and mathematical tools and techniques to explain and interpret geographical patterns and spatial relations.

Merits of Quantitative Methods

The advantages of the application of quantitative techniques in geographical studies are many. The quantitative techniques are firmly based on empirical observations and are readily verifiable. The statistical techniques help in reducing a multitude of observations, data and facts to a manageable number of factors. These techniques help in the estimation, interpolation, simulation of data which are necessary for forecasting. They also help in describing, analysing and simplifying a geographical system. Locational theories of industries, agricultural land use intensity, and stages of development of landforms can be easily understood and predicted with the help of quantitative techniques. Moreover, these techniques provide framework within which theoretical statements can be formally presented. They also provide linguistic economy as the scientific mathematic language is based on the principles of brevity and clarity. The use of certain factorial designs allows the objective measurement of data which has to be sacrificed at successive levels of regional generalization. The models formulated with the help of quantitative techniques are generally free from biasness and they are of immense use in the building of theories, general and special laws. Finally, quantitative revolution based on positivism distinguishes science from metaphysics and religion and thus provides the subject of geography a sound philosophical, scientific and methodological base.

Demerits of Quantitative Methods

The application of quantitative techniques in geographical studies have been criticized on several counts. Some of the weaknesses and limitations of quantitative techniques have been given as under:

1. The quantitative revolution was based on the philosophy of positivism (originally proposed by August Comte in 1820) which distinguishes science from religion and metaphysics. It followed the methodology of spatial science and thus reduced the subject to space geometry. The man and environment relationship cannot be properly established by the mechanistic models designed with the help of quantitative techniques.
2. The advocates of quantitative revolution pleaded for the language of geometry. Geometry is not an acceptable language to explain the man and environment relationship – the main theme of human geography.
3. The models and theories developed on the basis of empirical data exclude the normative questions like beliefs, taboos, emotions, attitudes, desires, hopes, fears, likes and dislikes, prejudices and aesthetic values. This is mainly being done to make the study objective and scientific. In the real world, in the interrelationships of man and environment, and decision-making processes, the normative questions and social, moral, religious and ethical values have a close bearing. In fact, in any economic activity and in the decision-making process about the utilization of resources, people are largely governed by their religious, taboos, likes, dislikes, emotions, attitudes, moral, cultural and social values. It is because of these values that dairying is not developing among the Khasis (Meghalaya) and Lushais (Mizoram) of the North-East India. In fact, taking milk is a taboo in these tribes. The Muslims, all over the world hate piggery, and the Sikhs dislike the cultivation of tobacco. By excluding the normative questions the study may be objective, but it gives only a parochial picture of the man and environment relationship.
4. The advocates of quantitative techniques in geography focused on 'locational analysis'. The main weakness of the locational analysis is that it promotes capitalism. In a capitalistic society, there is exploitation of human and environmental resources (land, water, forest and minerals) which makes the rich richer and the poor poorer.
5. With the development of sophisticated machinery and automation, there is less scope of employment. Thus, it leads to unemployment and it is a system of wasteful production.
6. The assumption that man is a 'rational person' who always tries to optimize his profit has also been criticized. In the real world location decisions are seldom if ever optimal in the sense of maximizing profits or minimizing resources used. In the opinion of Simon, man, in a limited number of alternatives, chooses one that is broadly satisfactory rather than optimal. In most of the cases the satisfying model applies and man takes decisions about the utilization of his resources to satisfy his aspirations and desires.

7. The assumption that man has 'infinite knowledge' of his space or environment (resources) has also been criticized. The knowledge about a resource changes as the new technology develops. Hence, no one can say that he has full knowledge about his environment.
8. Models developed with the help of quantitative techniques reduced people (decision makers, workers) to passive agents. To a large extent such models may be seen as one of economic determinism.
9. Application of quantitative techniques demands not only considerable mathematical power, they also demand reliable data which is rarely available in the developing countries like ours. In fact, the data collected in the developing countries has many pitfalls and shortcomings. The models or theories developed on the basis of unreliable data is bound to give only a distorted and faulty picture of the geographical reality.
10. The overenthusiastic preachers of quantitative techniques have sacrificed many good qualitative statements which were quite useful in the interpretation of regional personalities.
11. The estimations and predictions made with the help of sophisticated quantitative techniques proved erroneous many a time and there remains a danger of overgeneralization.
12. The models developed with the help of statistical techniques give more prominence to some features and distort some others.
13. Making reliable models and universal laws in human geography like other social sciences with the help of quantitative techniques is, however, not possible. According to one school of thought of physics, the probabilities can be calculated but definite predictions are not possible even in pure sciences like physics. In the opinion of Stephen Hawking, 'the laws of science cannot completely determine the future of the universe'. God (God as a metaphor for the laws of nature) plays dice and God may turn out to be 'an inveterate gambler'.

Despite all these merits and demerits of quantitative revolution, it may be summarized that 'spatial science' was inaugurated in North America. By the end of 1960s it was dominating many of the journals published throughout the English-speaking world. Most research was positivist in its tone. Most of the researchers used quantitative methods, and thus contributed to the development of theories and models. But these theories and models presented only a partial picture of the man-environment relationship. This methodology was criticized and as a reaction to this behavioural and humanistic approaches were introduced in human geography. In some cases, even in social science, a purely quantitative approach is necessary and in others requires a purely qualitative approach, and often a combination of the two is more satisfactory for making estimations and predictions in geography.

Whatever were the merits and demerits of the quantitative revolution from American centres in Washington, Wisconsin and Iowa, it spread to Europe, especially in Britain and Sweden. In Sweden, the department of geography at Lund University soon became renowned as a centre of theoretical geography, attracting scholars from many countries.[2] The major advances towards a unifying methodological and philosophical basis for the quantitative schools were made in the 1960s by British geographers, notably Peter Haggett, Richard Chorley and David Harvey.[3] These scholars suggested that geography should adopt quantitative methods and the use of computer to handle data to develop geographical paradigms and models.[4] A model was defined as an idealized or simplified representation of reality which seeks to illumine particular characteristics.[5] According to Chorley and Haggett, a model was either a theory or a law or a hypothesis of structured idea.[6]

The overenthusiasm of the preachers of quantitative revolution has, however, given way to the present phase in which mathematical and statistical methods are just one of the many tools for approaching geographical problems. In the 1970s, even Harvey – a staunch supporter of the quantification philosophy – became an apostate, and declared that quantitative revolution has run its course and diminishing marginal returns are setting in.[7] L.D. Stamp vehemently opposed the quantitative revolution and preferred to term quantitative revolution a 'civil war' and noted that quantification had many points in common with a political ideology; it was more or less a religion to its followers, 'its golden calf is the computer'. Stamp pointed out that there are many fields of enquiry in which quantification may stultify rather than aid progress, because there will be temptation to discard information which cannot be punched on a card or fed onto a magnetic tape; there is also a danger that ethical and aesthetic values will be ignored.[8] Minshull observed that the landscape was becoming a nuisance to some geographers, that many of the models will only apply to a flat, featureless surface, and warned that there is a real danger that these ideal generalizations about spatial relationships could be mistaken for statements about reality itself.[9] Minshull also sounded a note of warning that scholars would try to justify their models or hypotheses many a time in a subjective way which could give a distorted picture of the geographical reality.

The quantitative revolution, as stated above, began in the developed nations of the west where theories and models were constructed on the basis of data collected. There is certainly a danger that the models developed in Europe and America may be elevated to general truth and universal models. In reality we do not have universal urban geography and universal agricultural geography. There are different urban and agrarian processes which are working in different parts of the world and leading to different cultural landscapes. Owing to this factor, generalization on the basis of quantitative techniques may be misleading and negative instead of being positive. Apart from the above fact, the data used by the western experts hardly refer to a period of about one hundred years. Further, it reflects the modes of production and distribution of the developed capitalist societies. The processes which operate in the rigidly planned economies of the socialist countries of

Europe and in the East European countries are altogether different. The processes of urbanization and development which change in space and time are different in different economic and political systems. In brief, quantitative revolution also could not enable the geographers to formulate universal laws and paradigms.

Leading Advocates of Quantitative Revolution in Geography

Peter Haggett (1933–)

A prolific writer, Peter Haggett was associated in teaching and research with the universities around the world. He has written in three scientific areas: The first area includes (i) *Nature of Geography as a Discipline* (1967), *Geography: A Modern Synthesis* (1972), *The Geographer's Art* (1990), and *Geography: A Global Synthesis* (2001). He also established two journals reviewing developments in the field: 'Progress in Physical Geography', and 'Progress in Human Geography'.

Peter Haggett

(ii) The second area is on quantitative methods in Human Geography – *Locational Analysis in Human Geography* (1965) was followed by five books: *Network Analysis in Geography* (1969), *Regional Forecasting* (ed.) (1971), *Elements of Spatial Structure* (1975), *Locational Models* (1977) and *Locational Methods* (1977).

(iii) The third area has been on applying geographical ideas to understanding the changing geography of infectious diseases. His monographs on the geography of disease include *Spatial Diffusion* (1979), *Spatial Aspects of Influenza Epidemics* (1986), *Atlas of Disease Distribution* (1988), *Atlas of AIDS* (1992), *Measles: A Historical Geography* (1993), *Deciphering Global Epidemics* (1988), *Island Epidemics*, and *World Atlas of Epidemic Diseases* (2004). Busy in social and academic activities, he is currently passing a retired life.

Richard John Chorley (1927–2002)

A versatile geographer and a great advocate of application of quantitative techniques in geographical research Chorley was a great British geographer of the contemporary period. His seminal insight placed British geomorphology for several decades at the very high centre of world stage. Appointed as a Demonstrator at Cambridge in 1957, he proceeded to move rapidly up the university hierarchy with Readership in 1970 and 'adhominem' Chair in 1974. Cambridge provided the launching pad for Chorley's revolutionary ideas.

Rejecting the prevailing paradigm of Division Cycle of Erosion, he sought to replace these with a quantitative model-based paradigm with an emphasis on general system theory and numerical modelling.

Richard John Chorley

Chorley added six volumes in physical geography that were to codify his approach and ask new questions about earth surface processes and the way they can be studied. Central to these was the concept of system dynamics, and *Physical Geography: A System Approach* (1971), and *Environmental System* (1978) were to influence a generation of young scholars. Chorley's studies ranged into Cimatology and Hydrology. He cooperated with the Colorado Meteorologist, Roger Barry on *Atmosphere, Weather and Climate* (1968). He was generous in sharing his ideas, and most of the volumes were jointly authored or edited, including *Water, Earth and Man* (1969). In addition to his contemporary scientific work, Chorley launched in 1964 the first of series of Magisterial Volumes on the history of the study of landforms. Two further volumes were published in 1973 and 1991 and the time of Chorley's death, volume 4 was nearing completion.

William Bunge (1928–)

An American geographer, self-described as quantitative analyst, spatial theorist, radical humanist and Marxist geographer. He was a versatile geographer and prolific writer who contributed exceptionally in theoretical and applied branches of geography. He made major contributions to theoretical, quantitative spatial analysis earlier in his career (1962). Subsequently, he became an urban radical, supporting applied geography of social change and justice in inner city of America and Canada. He formed the Detroit Geographical Expedition in partnership with Gwendoln Warren in 1968 and the Society for Human Exploration in 1971. His cartographic representations of spatial patterns, particularly in Theoretical Geography, were also understood to be innovative.

William Bunge

Torsten Hägerstrand (1916–2004)

A pride product of Lund University (Sweden), Hägerstrand obtained his Doctorate in Geography in 1953. He was awarded the honorary doctorates from the Universities of Bristol, Glasgow and Edinburg. He was a Professor of Geography at the University of Lund and was applauded for his influence on population geography with introduction to the diffusion models. His key works were *Innovation Diffusion* (1967); *Definition of Migration* (1973); *The Domain of Human Geography* (1973); *Impact of Transport on the Quality of Life* (1974); and *Space, Time and Human Conditions* (1975).

Torsten Hägerstrand

PARADIGMS IN GEOGRAPHY

Like psychology and economics, geography has passed from the descriptive to model formulation stage. This process started in the 19th century and got accelerated during the sixties and seventies of the 20th century. Some authorities state that one change in geography is its progress from the descriptive, through the classificatory, to the law-making stage.[10] The processes of theory-building and law-making have became quite popular among geographers in the recent past. Geographical laws are not identical to natural laws which are universal in character and can be applied with confidence in any part of the world. The law of gravity – a natural law – can be tested in India, in Europe or in America, with equal validity. A natural law can be tested *ad infinitum*. In a million experiments, if the law holds 9,99,999 times but does not hold on the millionth occasion, then the scientific law has been disproved and a new law must be formulated to fit the observed facts.[11] In the present century, natural scientists and experts of social sciences have realized that while they can formulate general behaviour of any one molecule or people conform, the behaviour of any molecule or person seems to be completely random. Thus, we have the principle of stochastic laws which say what the majority of the phenomena in question will probably do.[12] Golledge and Amadee raised a very interesting point when they said that there was not a single generally accepted definition of law. They asserted that there are probabilistic laws, cross-section laws, equilibrium laws, historical laws, developmental laws, statistical laws, mathematical laws, stochastic laws and composition rules.[13] The laws of geography are not like the laws of natural sciences as geography is not an experimental science. In other words, controlled experiments in geography like those in physics are not possible. Geographical laws can be tested only in the fields where conditions are not controlled. Moreover, like the laws of economics, geographical laws can be tested only if 'all other things are being equal'. The principle of activity applies in geography which means that every phenomenon is changing in time and space. Geography, being idiographic (regional) and nomethetic (systematic), has different types of laws and models which differ from the laws of physical sciences and in many cases even from those of the social sciences. In the following paragraphs, some light has been thrown on geographical laws and paradigms. Geography, like other disciplines, is passing through phases of chaos and tranquillity, as pointed out by Kuhn.

Kuhn's Paradigm

The American historian of science – S. Thomas Kuhn – postulated a very important theory about the growth and development of science. According to Kuhn, science is not a well-regulated activity where each generation automatically builds upon the results achieved by earlier workers. It is a process of varying tension in which tranquil periods characterized by steady accretion of knowledge are separated by crises which can lead to upheaval within subject disciplines and breaks in continuity.[14]

In order to elucidate this process of development of science, Kuhn prepared a model which he termed as the 'paradigm of science'. He defines paradigm as 'universally recognized scientific achievements that for a time provide model problems and solutions to a community of practitioners'. Haggett defines them as a kind of super model.[15] In other words, a paradigm is a theory of scientific tasks and methods which regulates the research of most geographers, for example, or, where there is conflict between paradigms, of a group of geographers. The paradigm tells researchers what they should be looking for and which methods are, in this particular case, 'geographic'.[16]

Kuhn, in his postulate, advocated that the development of science consists of pre-paradigm phase, professionalization, paradigm phase 1, crisis phase with revolution, paradigm phase 2, crisis phase, paradigm phase, crisis phase with revolution, paradigm phase 3, and so on and so forth. The concept, geographically plotted by Henriksen and depicted in Figure 10.1, shows that scientific knowledge progresses and develops like a plateau. There are sudden upheavals, and then abrupt rise which is followed by smooth and slow progress. The first phase, i.e. the pre-paradigm period, is marked by conflicts among several distinct schools which grow around individual scientists. This period is also characterized by a rather indiscriminate collection of data over a very wide field and by a low level of specialization. This period is full of communication among various schools of thought and with other scientists and laymen. One school of thought does not consider itself to be any more 'scientific' than the other.[17]

From the pre-paradigm phase, scientific development marches and enters into professionalization. Professionalization takes place when one of the

Figure 10.1 A Graphical Interpretation of Kuhn's Theory of the Development of Science (After Henriksen, 1973)

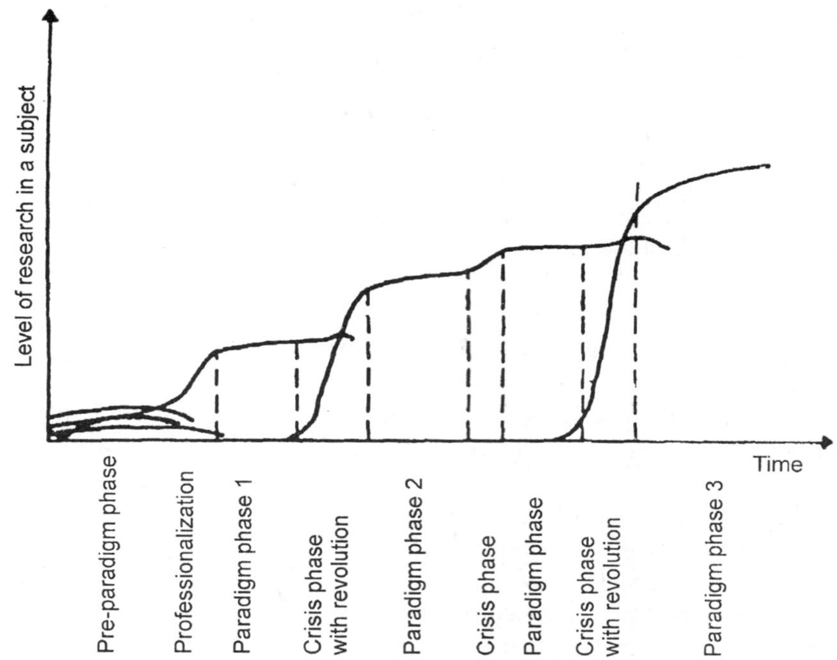

conflicting schools of thought begins to dominate the others and thus a clear answer to the questions raised is given. A particular school of thought may become dominant because it develops new methods or puts questions which come to be regarded as more interesting or significant. New researches are thereby undertaken and research makes progress. Kuhn argues that mathematics and astronomy left the pre-paradigm phase in antiquity, whereas in parts of the social sciences the transition may well be occurring today.

The third phase is the paradigm phase. This phase is characterized by a dominating school of thought which has, often in quite a short space of time, supplanted others. A paradigm is established which leads to concentrated research within a clearly distinguishable problem area – an activity described as 'normal science'.

After the 'normal science' phase, there occurs stagnation in research which leads to chaos and turmoil. This period may be termed as 'temporary dark phase' in the development of scientific knowledge. This crisis phase with revolution is the starting point for the paradigm phase 2, which, in turn, is followed by crisis phase and leads to paradigm phase. This sequence of crisis, revolution and paradigm continues throughout the history of science and helps in the advancement and decline of societies.

In this dynamic world, the period of 'normal science' also does not last infinitely. A period of 'normal science' is sooner or later replaced by a crisis phase. This occurs because more and more problems get accumulated which cannot be solved within the framework of the ruling paradigm. Either more observations shake the underlying theory or a new theory is developed which does not accord with the stipulation of the ruling paradigm. The crisis phase is characterized by a reassessment of former observational data, new theoretical thinking and free speculation. This involves basic philosophical debates and a thoroughgoing discussion of methodological questions. The crisis phase ends when it appears either that the old paradigm can solve the critical problems after all, allowing a period of normal science to be resumed, or that no significantly better theory to solve the problems can be developed and thus, consequently, research must continue for a further period within the old paradigm. Otherwise, the crisis phase ends when a new paradigm attracts a growing number of researchers.[18]

In case the crisis phase terminates owing to the acceptance of a new paradigm, it becomes the inaugural point of the revolutionary phase. This involves a complex break in the continuity of research, with a comprehensive reconstruction of the theoretical structure of a research field rather than a steady development and accumulation of knowledge. The understanding of the truth itself and the scientists' perception of the world can take on a new dimension. The acceptance of the new paradigm gives recognition to the new and younger scientists. The new researchers start competing with the old established scientists. The new scientists generally cannot convince the old scientists but nevertheless the former prevail because the elderly scientists soon pass away and their following becomes weak.

The exchange of one paradigm for another is not a wholly rational transaction. The new paradigm will generally provide solutions to the problems which the old one found difficult to resolve but may not answer all the questions which were fairly easy to solve before. It is seldom possible to argue logically that the new paradigm is better than the old. The positive approach becomes doubtful because many of the normative values and aesthetic considerations may influence the new paradigm to make it simple and more beautiful. Many a time, the younger research workers have vested interest in changing the existing scientific ideology, namely, the anxiety to take over form their elders.[19]

Kuhn's paradigm gives a very scientific explanation of the growth phases of scientific knowledge. This model, like all the other paradigms, has also its merits and demerits. Kuhn's paradigm has provided an opportunity to the younger research workers to postulate new theories without justifying their researches objectively. It is considered adequate to declare their paradigm objective. Such a paradigm cannot be free from value judgement and thus can be against the positivist approach or scientific research. In spite of negativism tendency in the new paradigm, Kuhn's theories have had a positive influence on modern science in that they facilitated the acceptance of new theories and frameworks of understanding which may widen our knowledge and perception, but may have negative influence in giving well-organized groups of poorly-qualified people a legitimate entry into research. Moreover, Kuhn's model provided an impetus to students to go in for disciplines which offered simple and popular theories.

Whatever the merits and demerits of this paradigm, Kuhn's efforts have developed a long-awaited new paradigm for the philosophy of science. This paradigm gives useful guidelines for the understanding of the historical development of a subject, but does not offer a complete explanation. The history of geography in the light of Kuhn's paradigm can be easily understood by pursuing the following description.

Geographical Paradigms: A Historical Perspective

As discussed earlier in Chapters 8 and 9, geography had to confront many evolutionary and methodological problems. It passed from the descriptive and teleological phase to the quantitative, radical and dialectical materialism stage. Various methodologies have been adopted to give precise and reliable description of places in literary as well as mathematical languages. Yet, a consensus has not been reached about the nature of the discipline and its laws and paradigms.

Geographical laws are not like the precise laws of the natural sciences. A natural law as, defined by Braithwaite, is 'a generalization of unrestricted range in time and space'; in other words, a generalization with universal validity.[20] The laws of universal validity are, however, the laws of physics and chemistry only. Nevertheless, in physics also, there are elements of uncertainty which make probability calculations necessary. In contrast to this, most of the geographical laws are empirical in nature and therefore cannot be placed in the category of laws of the natural sciences. All the empirical laws, formulated mainly in the social sciences, are valid for a specific place and specific time and are therefore termed as models, structured ideas of paradigms. Looking at the variation in the nature of laws, Harvey gives the concept of law a much wider significance when he postulates a three-fold hierarchy of scientific statements from factual statements (systematized descriptions), through a middle tier of empirical generalizations or laws, to general or theoretical laws.[21] Against this background of the classification of laws it will be interesting to know the types of models and paradigms that developed in geography during the last one hundred and fifty years.

If we begin with the period of Carl Ritter who is considered as one of the founders of modern geographical thought and an advocate of empiricism in the discipline, it may be said that he used inductive method as a framework for his presentation of data and as a means to arrive at some simple empirical generalizations. Being a teleologist, Ritter asserted that all phenomena are spatially distributed according to the plan of God for mankind. The major problem of the teleological philosophy is that such a philosophy cannot be tested empirically and therefore does not qualify as scientific explanation.

Nevertheless, it has the characteristics of a paradigm. Ritter's teleological approach is generally taken to mean that a phenomenon is explained in relation to the purpose it is believed to serve. The 'holistic synthesis' of organic relationships is strongly related to teleological explanatory models. This approach is reflected in most of the semantic religions and their philosophies.

The post-Ritter period was dominated by Darwin who revolutionized the entire philosophy of science and brought a cause and effect approach in explaining spatial distribution of phenomena. It was during this period that geographers and scientists started thinking seriously about the nature of geography, and concentrated on the issue, whether geography can be regarded as a science. Darwin laid the foundation of the deterministic approach in geography. In his opinion, the natural conditions determine the socio-cultural development of a society. After Darwin, scientists were looking for the controlling laws of nature (and the materially conditioned social laws) and to a considerable extent adopted a nomethetic (general law-making) approach. At this stage, inductive arguments were increasingly replaced by hypothetic deductive methods. Researchers, starting from inductive arrangements of their observations or from intuitive insight, tried to devise for themselves priori models of the structure of reality. These were used to postulate a set of hypotheses which could be confirmed, corroborated or rejected by testing empirical data through experiment. The theories postulated about the evolution of landforms, normal cycle of erosion, etc. The heartland theory by William Morris Davis and Mackinder respectively, fell under this category of geographical models. As a result of these paradigms, geography acquired recognition and respect in the community of sciences. But human geography showed a stunted growth. At this stage, Vidal de la Blache and his followers laid stress on possibilism and declared that man is not a passive agent ruled by the forces of nature which play their role and determine man's destiny and shape human society. For this purpose, a large number of micro-level studies were conducted which was a regional approach and thus geography became an 'idiographic' or a 'regional' science. In the terminology of Kuhn, geomorphology and determinism represented the first paradigm phase in geography (Figure 10.1). Determinism, however, had a short span of life and it was replaced by possibilism and the French School of Regional Geography (Chapter 8). The possibilists developed the approach that to understand a society and its habitat field study is most important.

Although possibilism and regional geographical school developed new paradigms and became very popular, these could not remove the deterministic model in totality. Thus, the deterministic explanatory model survived side by side with possibilism. Kuhn has termed this period as 'revolutionary phase'.

After the Vidalian tradition, the major concern of geographers became to study regions. George Chabot went to the extent of saying that 'regional geography is the centre around which everything converges'.[22] Regional geography flourished in France and got diffused in the neighbouring countries. But later on this approach also became inadequate to explain the regional personality, and therefore, a period of crisis in the discipline emerged. It brought about quantitative revolution and functional approach in geography. Now geographers began using more models, especially in the field of human geography. Many of them have been strongly pleading for system analysis which has been discussed towards the end of this chapter.

From the description given above about the development of geographical paradigms, models, laws and theories, it may be inferred that complete revolutions in geography have not taken place. Numerous schools of thought are marching side by side in search of new paradigms which can help in ascertaining the geographical personality of a region. Geographers are dividing themselves in the category of positivists, pragmatists, phenomenologists, existentialists, idealists, realists and dialectical materialists. This is a crisis phase with revolution which shall lead to new paradigm phase.

Areal Differentiation

The study of areal variation of human and physical phenomena as they relate to other spatially proximate and causally linked phenomena is known as areal differentiation. The term 'areal differentiation' was coined and used by Hartshorne in his classic work, *The Nature of Geography*, published in 1939. Drawing from Hettner, Hartshorne's central claim about geography is its integrative or synthetic purpose. The areal differentiation is also known as 'chorology' or 'chorography'. Chorology is the study of the areal differentiation of the earth's surface. Geography on this definition is solely concerned with the unique character of different areas of the earth's surface. Areal differentiation may be termed as 'idiographic as it is concerned with the unique and particular'.

Areal differentiation represents the oldest tradition of western geographical inquiry. It was first set forth by Hecataeus of Miletus in the 6th century BC, and codified in the form of chorology by Strabo in the 17 books of geography written by him sometime between 8 BC and 18 AD. The geographer, he declared, is the person who attempts to describe the parts of the earth. The two key words were 'describe' and 'parts'. In effect, Strabo was recommending what could now be called 'regional geography' as the core of geographical study.

In the 1980s areal differentiation has been reinstated as a central perspective in human geography. Intellectual inspiration has come from three general directions, none of which is directly connected to older positions in debates about areal differentiation or uses the same terminology as the others.

The first is from the streams of thought referred to collectively as humanistic geography which gives central and active role to human awareness and human agency, human consciousness and human creativity. The humanistic method (iconographic technique) seeks to explore the composition of landscapes, interpreting their symbolic content and reimbursing landscaping in their social and historical context. Successful iconographic interpretation allows us to see human landscapes as both shaped by and themselves shaping broader social and cultural processes, and thus having ideological significance.

The second focus of revival has been the analysis of uneven development and changing spatial division of labour. Some of the geographers tried to explain spatial variation in economic activity and well-being with a Marxist approach.

The third influence has come from attempts in geography to create Contextual Theory, in which the place or region is viewed as geographically mediating the interpellation of human agency and social structure and is thereby implicated directly in the production of geographical sameness and differences. The third direction could be seen as potentially integrating the other two, but this would be a superficial view. There are important philosophical differences between all the three. For example, the first direction tends either to privilege or to emphasize the human objective experience of place – more often than not that of the scholar engaged in writing about it – whereas the second and third view the division of space in terms of objective socio-spatial processes with, for the third view, sense of place arising out of the conditions created by such processes.

In the areal differentiation the most challenging dilemma is the difficulty of neat boundary delimitation between places and regions when the territoriality of social groups is dynamic and irreducible to a singular and temporally fixed set of spatial units. Areal differentiation has also been criticized that it is incapable of contributing towards effective generalization.

Areal differentiation helped in the reconstruction of regional geography. It emphasized, however, that regions must not be studied solely as separate entities. Thus regional geography must focus on the unique characteristics of the place being studied, but must not express them as if they were singular.

Despite all these criticisms, areal differentiation is being considered the rational and scientific definition of the discipline of geography.

Exceptionalism in Geography

The father of exceptionalism is Immanuel Kant. Kant claimed exceptionalism not only for geography but also for history. According to him, history and geography find themselves in an exceptional position different from that of the so-called systematic sciences. This grouping of geography with history has tempted many subsequent writers to elaborate the alleged similarity in order to obtain some insight into nature of geography. This is one of the roots of historical variant of the claim to uniqueness. It is in Kant's work that one finds the statement on geography and history that has been quoted so reverently again and again by those who make it the cornerstone of geographical method. Humboldt and Ritter used it, so did Hettner and eventually Hartshorne.

The term exceptionalism in geography is, however, usually identified with Schaefer. Schaefer was originally an economist: he joined the group of geographers teaching in the economics department at the University of Iowa after his escape from Nazi Germany. He made a critical analysis of the book of Hartshorne, *The Nature of Geography* and published a paper 'Exceptionalism in

Geography: A Methodological Examination' in the *Annals of Association of American Geographers* (1953, Vol. 43). Schaefer's essay was widely regarded as a rallying point for a new generation of human geographers.

Schaefer claimed that his paper was the first to challenge Hartshorne's interpretation of the works of Hettner and others. He criticized Hartshorne's exceptionalist claims for regional geography, and presented an alternative case for geography adopting the philosophy and methods of the positivist school of science. He first outlined the nature of science and then defined the peculiar characteristics of geography as a social science. He argued that a claim for geography as the integrating science which put together the findings of individual systematic sciences was arrogant, and that in any case its products were somewhat lacking in 'starting new and deeper insights'. A science is characterized by its explanations, and explanations require laws. 'To explain the phenomena one has described means always to recognize them as instances of laws'. In geography, in the opinion of Schaefer, the major regularities which are described refer to spatial patterns. 'Hence, geography has to be conceived as the science concerned with the formulation of laws governing the spatial distribution of certain features on the surface of the earth', and these spatial arrangements of phenomena, not the phenomena themselves, should be the subject of geographers' search for law like statements. Geographical procedures would then not differ from those employed in other sciences, both natural and social: observation would lead to hypothesis about the interrelationship between two spatial patterns, for example, and this would be tested against larger number of cases, to provide the material for a law if it were thereby verified.

The argument against this definition of geography as the science of spatial arrangements was termed 'exceptionalism'. It claims that geography does not share the methodology of other sciences because of the peculiar nature of its subject matter – the study of unique places, or regions and compares geography with history, which studies unique periods of time. Using analogies from physics and economics, Schaefer argued that geography is not peculiar in focusing on unique phenomena, all sciences deal with unique events which can only be accounted for by an integration of laws from various systematic sciences, but this does not prevent (although undoubtedly makes more difficult) the development of laws. It is, therefore, absurd to maintain that geographers are distinguished among the scientists through the integration of heterogeneous phenomena which they achieve. There is nothing extraordinary about geography in that respect.

Thus, Schaefer argued that there is nothing exceptional in the philosophy (nature) and methodology of geography as advocated by Hartshorne. As stated above, he positioned human geography in the category of social sciences rather than in humanities or natural sciences. Geography, he argued, must be like other social sciences and not 'exceptional'.

The belief that geography and history are methodologically distinct from other fields of enquiry is because they are peculiar concerned with the study of unique and particular. Thus, Schaefer rejected the idiographic orthodoxy enshrined in Hartshorne's *The Nature of Geography*. In other words, he argued for *nomethetic geography* declaring geography as general or systematic which aims to furnish general and universal 'morphological laws' about spatial patterns. He declared geography as the science of earth surface in which the general and universal laws about spatial patterns are to be formulated instead of the regional and local laws.

Earlier, Humboldt and Ritter recognized that the major concern of geography is to examine the manner in which the natural phenomena, including man, are distributed in space. This implies that geographers must describe and explain the manner in which things combine 'to fill an area'. These combinations change, of course, from area to area. These differences either in the combination of factors or in their arrangement from place to place underlie the common sense notion that areas differ. Following the Greek geographers, this viewpoint is called choreographic or chorological one, depending on the level of abstraction. Geography, thus, must pay attention to the spatial arrangement of the phenomena in an area and not so much to the phenomena themselves. Spatial patterns are the ones that matter in geography, and no others. Non-spatial relations found among the phenomena in an area are the subject matter of other specialists such as ecologist, anthropologist or economist.

Subsequently, Kraft, while discussing Humboldt and Ritter, agrees with them that geography is, a science trying to discover laws; that it is limited to the earth's surface; and that it is essentially chorological. Incidentally, he also feels that this suffices to set geography logically apart as an exceptional discipline.

Hettner, one of the leading German geographers declared that 'both history and geography are essentially chronological'. History arranges phenomena in time, geography in space. Both, in contrast to other disciplines, integrate phenomena heterogeneous among themselves. Also these phenomena are unique. No historical event and historical period is like any other. In geography no two phenomena and no two regions are alike. Thus, both fields face the task of explaining the unique (exceptional). Hettner calls history 'time-Wissenschaft' and geography 'space-Wissenschaft'. Hartshorne translated them into 'time science' and 'space science'.

Hartshorne's views were in fact more naunced than Schaefer maintained, and he never accepted any clear division between the idiographic (regional or particular) and the nomothetic (general or systematic) because both are 'present in all branches of science'. But, he did insist that any general concept used in geography should be directed towards the analysis of specific 'regions', and that its essential task was to study areal differentiation rather than (as Schaefer preferred) the elucidation of laws of location that were supposed to underpin these regional configurations.

It may be summarized that both Hartshorne and Schaefer relied upon mainly German sources such as Kant, Humboldt and Hettner, and both directed geographers towards spatial patterns. Even so Schaefer was among the first geographers to teach the ideas of Christaller, Von Thunen and Losch, which Hartshorne has not acknowledged but which were to become the mainstay of 1960s human geography. He also positioned human geography in the social sciences rather than the humanities or natural sciences. Geography, he argues, must be like other social sciences and not exceptional.

Spatial Analysis

The quantitative techniques employed in locational analysis is known as 'spatial analysis'. It is sometimes used as a synonym for the locational analysis. Unwin (1981) presents spatial analysis as the study of the arrangements of points, lines, areas and surfaces on a map.

The followers and advocates of spatial science consider human geography as that component of social sciences which focuses on the role of space as a fundamental variable, influencing both society's organization and operation and the behaviour of its individual members. Spatial analysis got popularity during the period of quantitative revolution. It is closely associated with the philosophy of positivism.

The goal of spatial analysis was 'building accurate generalizations with predictive power by precise quantitative description of spatial distribution, spatial structure and organization and spatial relationships'.

The generalizations arrived at with the application of spatial analysis could be based on just three fundamental spatial concepts: (i) direction, (ii) distance, and (iii) connection (or relative position).

In the spatial analysis some of the geographers merely apply techniques derived from the General Linear Model to geographical example, others have argued that spatial data analysis poses particular statistical problems (such as spatial autocorrelation) which means developing procedures specifically designed to counter them.

Spatial analysis for man and environment relationship has been criticized on several counts. One of the main criticisms is that spatial analysis focused on spatial determinism, and the logical impossibility of defining spatial variables independent of the context within which they were supposed to operate.

The second weakness of spatial analysis is that it does not take into consideration the cultural values and normative questions while attempting to establish the man and environment relationship.

The Marxist and radicalists consider spatial analysis as a device to promote capitalism and exploit workers and environment (see quantitative revolution).

Locational Analysis

Locational analysis is an approach to human geography which focuses on the spatial arrangement of phenomena. Its usual methodology is that of spatial science. The main objective of locational analysis was expressed as building accurate generalization, models and theories with productive power (Berry and Marble, 1968).

Locational analysis is based on the philosophy of positivism. The philosophy of positivism underpins the approach, which concentrates on the identification of theories of spatial arrangements and so is closely linked to the discipline's quantitative revolution.

A number of geographers in USA advocated the cause of locational analysis in the 1950s, although it has much deeper roots in the work of pioneers that were later adopted by geographers. Bunge (1966), for example, wrote a thesis on 'Theoretical Geography' based on the premises that stated geography is the 'science of locations'. Others such as McCarty, were strongly influenced by developments in the field of economics, to which they introduced the spatial variable. These links led to the close interrelationship between geographers and regional scientists (economists) in the 1960s and 1970, and illustrated by attempts to build economic geography theories of spatial arrangements (Smith, 1981).

Locational analysis is based on empiricism. Empiricism is a philosophy which accords special privilege to empirical observations over theoretical statements. Specifically, it assumes that observational statements are the only ones which make direct reference to phenomena in the real world, and that they can be declared true or false without reference to the truth or falsity of the theoretical statements. In empirical inquiry, it is assumed that its facts 'speak for themselves'. They presented a strong case for using geometry as the language for the study of spatial form.

Haggett, in his book *Locational Analysis in Human Geography* (1965), appealed to adopt the geometrical tradition to explain order, location order and patterns in human geography. Such a focus needed: (1) to adopt a system approach which concentrates on the patterns and linkages within a whole assemblage; (2) to employ models as to understand man and environment relationship; and (3) to use quantitative techniques to make precise statements (generalizations) about locational order. For the spatial analysis they suggested to adopt 'linear model', spatial autocorrelation and regression.

Other geographers who contributed substantially to the field of locational analysis are Morril, Col, Chorley, Cox, Harvey, Johnston, Pooler, Sack and Smith.

Morril was strongly influenced by the geometrical traditions adopted by Bunge and Haggett. In his book, *The Spatial Organisation of Society*, he argued that people seek to maximize spatial interaction at minimum cost and so bring related activities into proximity – the result is that human society is surprisingly alike from place to place ... [because of] the predictable, organized pattern of locations and interactions.

The locational approach in human geography has been criticized on philosophical and methodological grounds by the behaviouralists and humanists. Some of the main criticisms against locational analysis are as under:

1. The locational analysis based on positivism ignores the normative questions to explain the man and environment (resource) relationship. It was their mistaken belief that 'positive theory would lead to normative insight'. The cultural values are quite important in any decision-making process. The ideal location for any economic activity may not be acceptable to individuals and the society (see quantitative revolution).
2. The locational analysis did not reflect the reality of decision-making processes and so was of little value in predicting locational arrangement.
3. The models developed with the help of locational analysis conceal the complexities of the real world.
4. At present, there is economic interdependence of societies at the global level, which means that spatial interdependence has become much more important and 'locally experienced environmental dependencies lost their rationale'.
5. Locational analysis has also been criticized on the ground that it encourages the social order of capitalism in which the owners of the means of production become rich and the poor workers become poorer.
6. The locational analysis has given a chance to the capitalists to optimize their profits. It gives an uncontrolled liberty and licence for plunder and miscalled profit.
7. Owing to locational analysis, there is over production and the economy enters the era of overindustrialization.
8. It is mainly because of the locational analysis and capitalism that there is a total newness – new technology, new means of transportation, new education, new art, new morals, new media, new amusement, new weapons, new violence, new terrorism, new war and new mode of exploitation.
9. The followers of spatial science (positivists) treat people as dots on a map, statistics (data) on a graph, or numbers in an equation. They consider humans as non-living and others as living (plants and animals).

It is because of the inadequacies of the locational analysis that the 'behaviouralism' and 'humanism' achieved much significance in human geography.

Whatever the reason for its origin, there is little doubt that locational analysis substantially changed the nature of human geography from the mid-1960s, although there is some doubt that it ever dominated the discipline (Mikesell, 1984). It presented geography as a positivist social science, concerned to develop precise, quantitatively stated generalization about pattern of spatial organization, thereby enriching and being enriched by Location Theory, and to offer models and procedures which could be used in physical planning. By 1978, therefore, Haggett could write that:

> the spatial economy is more carefully defined than before, we know a little more about its organisation, the way it responds to shocks, and the way some regional sections are tied into others. There now exist theoretical bridges, albeit incomplete and shaky, which span from pure spaceless economics to a more spatial reality.

Twelve years later, he continued to promote the search for 'scientific generalization' (Haggett, 1990), while accepting, that in the search for spatial order 'the answer largely depends on what we are prepared to look for and what we accept as order': for only a minority of geographers can now claim that order is the focus of their quest.

Geography as a Chorographic or Chorologic (Regional) Science

As stated earlier, chorography or chorology is the study of the areal differentiation of the earth's surface. It represents the oldest tradition of geographical inquiry. It was first set forth and codified most elegantly by Strabo in his 17 books on geography written sometime between 8 BC and 18 AD. The geographer, he declared, is the person who attempts to describe the parts of the earth (in Greek, *chorographein*). The two keywords were 'describe' and 'parts': in effect, Strabo was recommending what would now be called regional geography as the core of geographical study. For Strabo, geography described those worthwhile and unique things which one could learn about truth, nobility and virtue. Strabo's geography was fundamentally concerned with human activities. It was directed towards social, political and military ends. The modern case for geography as a 'chorographic' science was argued most forcefully by Hartshorne in the *Nature of Geography* (1939).

Geography, as a chorographic study, has always found justification in the widespread desire of many people to know what other parts of the world are like. Further, since the ordinary person actually knows but partially the area of what is happening in the world in his lifetime, 'home geography' is as necessary as history.

In the opinion of Volz, 'the purpose and significance of geography is that it teaches us to know the space (area/region) – the surface of the earth'. According to Sauer, geography assumes the responsibility for the study of the areas (regions), because there exists a common curiosity about that subject. The fact that every school child knows that geography provides information about different countries is enough to establish the validity of regional geography. No other subject attempts the study of areas.

From the point of view of each specialized science, while each subject deals with a restricted class of phenomena, geography deals with heterogeneity of phenomena. The study of numerous types of phenomena helps to break the field up into separate specialized parts (economic geography, social geography, cultural geography, political geography, urban geography, agricultural geography, industrial geography, transport geography, settlement geography, etc.).

Vidal de la Blache also justified the regional geography as the core of geography. Vidal de la Blache observed, that man is interested above all in the totality of conditions in an area. There are many different ways of studying the world, but since men as individuals, and as individual groups, do not in any sense live in the entire world, but each in relatively restricted area of the world, one of the most significant methods of studying the world is to study it by areas (regions). Geography, thus, attempts to provide complete, accurate and organized knowledge to satisfy man's curiosity about how things differ in the different parts of the world. In order to achieve this objective, geography must consider the world in terms of limited areas (regions) within which things are closely associated.

Geography as a Science of Relationship

The concept of geography as the study of man and environment relationship is quite old. The Greek, Roman, Indian, Chinese and Arab geographers attempted to establish a relationship between man and natural environment. Kant, in the concluding part of the 18th century, advocated the impact of environment on the lifestyle and physical constitution, equatorial, hot deserts, Mediterranean, coastal and mountainous regions. According to Kant, the inhabitants of torrid zone are exceptionally lazy and timid, while the people of the Mediterranean region living in the mild temperature conditions are industrious, hard working and progressive.

The environmental causation continued throughout the 19th century. Humboldt asserted that the mode of life of the inhabitants of the mountainous countries of the Andes mountains differ from that of the people of Amazon basin, coastal plains and islands like Cuba and West Indies. Ritter attempted to establish the cause variations in the physical constitution of body, physique and health of people living in the different physical environmental conditions.

The idea of defining geography in terms of man and environment relationship developed on scientific lines in the later part of the 19th century after the publication of *Origin of Species* (1859) by Charles Darwin. This seminal work gave a new direction to the discipline of geography. The theory of evolution held that all living species have evolved from pre-existing forms. His geological observations and theories had one thing in common: the idea that things in nature change with time. He also believed that the face of the earth also changes with the change in environment over the period of time. In this book, *Origin of Species*, Darwin presented his idea that species evolve from more primitive species through the process of natural selection. In his account of natural selection, known as Darwinism, he pointed out that not all individuals of a species are exactly the same but have variations and some of these variations make their bearers better adapted to particular ecological conditions. He theorized that well-adapted individuals of a species have more chance of surviving and producing young than do the less adapted, and that over the passage of time the latter are slowly weeded out. Through his theory

Darwin showed how the multitude of living things in our world could have come into being without any recourse to a divine master plan, in a plain, causal, naturalistic way. Darwin argued that a struggle for existence must take place; it followed that those who survived were better adapted to their environment than competitors. This means that relatively superior adaptations increase while relatively inferior ones are steadily eliminated. The Darwin's concept of man and environment relationship may be summarized as under:

1. Organisms vary, and these variations are inherited (at least in part) by their offspring.
2. Organisms produce more offspring than can possibly survive.
3. On an average, offspring that vary most strongly in direction favoured by the environment will survive and propagate.

The Darwin's theory had a far-reaching impact on the growth and development of geography. It assumed that variations in animals were random. In this way, the older teleological conception (the religious belief that God has a plan and every phenomena of the earth has been created to perform certain functions for man) of nature was profoundly challenged.

Darwin's book upset many established patterns of thought, contradicted firmly held religious tenets (teleological concept) and brought in focus the concept that humans are one species among many that have evolved from a more primitive one. In his subsequent book, *The Descent of Man and Selection in Relation to Sex* (1871), Darwin provided evidence of human evolution from one primitive species and discussed the role of sexual selection in evolution.

The concept of defining geography in terms of relationship became quite popular in Germany. The work of Darwin influenced Friedrich Ratzel, who published *Anthropogeography* in two volumes in 1882 and 1891 respectively. In the first volume, he organized the material to show the influence of physical environment on history, culture and the mode of life of the people, while the second volume deals with the geographical distribution of men in the world. It was because of this book in which he discussed the man and environment relationship of the different tribes of the world that he is considered as the 'founder of human geography'.

Ratzel, by applying the organic theory to political geography, developed the concept of *lebensraum* (literally living space or the geographical area within which an organism develops). While developing the man and environment relationship, Ratzel, in his book *Political Geography* (1897), equated a nation with a living organism, and argued that a country's search for territorial expansion was similar to a growing organism's search for space. Conflict between nations was thus seen as a contest for territory within which to expand, with the fittest surviving. The concept was appropriated by the German School of Geopolitik in the 1920s and 1930s and used to justify the Nazi programme of territorial expansion. In the opinion of Dickinson and Gumplovicz, 'Ratzel's work contains more and more important knowledge concerning the state, than the entire theoretical political science literature of the last 100 years'.

Ratzel's book *Anthropogeography* had a great influence on the geographic thought of America, France, Britain, Russia and Sweden. Ratzel's most important disciple was Ellen Churchill Semple. Semple, in the introduction of her book, *Influences of Geographic Environment* declared 'man as the product of earth's surface'. The influence of physical environment on the history of people of the region in any part of the world can be found in her writings. The main cause of variations in the history, culture and lifestyle of the people of plains and mountains can be traced in the physical environment.

Subsequently, the French geographers, especially Vidal de la Blache, Brunhes, Martone, etc. advocated geography as the science of man and environment relationship. The concept of *pays* (micro-region), developed by Vidal de la Blache, was also based on the concept of relationship. He also coined the concept of *genre de vie* (lifestyle). Vidal de la Blache was convinced that *genre de vie* were themselves reflective of nature (physical environment), even as they transformed it. He always conceived human geography as natural, not a social science. This approach was criticized on the ground that 'human genius' and 'culture' of the people may overcome geographical limitations at defiance. Temporary effects contrary to nature (physical environment) may be within human possibilities, but in the long run nature prevails and reasserts her supremacy.

Ultimately, geography as the science of relationship appeared in the form of environmental determinism. Environmental determinism is the doctrine according to which the human activities are controlled by the physical environment. The environmentalists considered natural environment as the 'geographic factor' and their geography was known as 'pure geography'. In the opinion of environmental determinists, human geography is the study of influence of physical environment on man.

Barrows, in his presidential address (1922), recommended that relationships in geography should be studied 'from man's adjustment to environment, rather than the reverse'. Hettner (1907) also supported the concept of geography as the study of relationship. Thus, both the physical factors and the human factors (cultural environment) are to be studied in their relations to each other. Geography is, therefore, exclusively human geography, or as Barrow stated, geography is 'human ecology'. Geography is a natural science in the same way as plant ecology is a biological science. Sauer, in his book *Agricultural Origins and Dispersals* (1952), focused upon the patterns of human culture in relation to the natural environment. He also tried to explain how human interactions with physical environment have resulted into various cultural patterns in the different parts of the world.

While examining the lifestyle and history of the people in the different regions of the world, it may be said that there is a close relationship between the environment and the mode of life of the people. Undoubtedly, terrain, topography, temperature, rainfall, natural vegetation and soils have a direct bearing on the culture, economy and society of the people, yet the role of man as the transforming agent of his physical surroundings cannot be ignored. In fact, works of man reveal many facts for which environmental forces alone can give no satisfactory explanation. For example, similar locations may not lead to similar mode of life. The Eskimos of Tundra region differ markedly in their economic activities, and cultural practices from the Tungus, Yakuts and Yukaghirs, etc. The Khasis and Nepalis, living in Meghalaya (India) in the same physical environmental conditions, have the different cultural ethos. Same is the case with the Gujjars and Bakkarwals of Kashmir Valley and the Kashmiris in the state of Jammu and Kashmir. Likewise, the Hanjis (water dwellers) of Dal Lake and Jhelum river in Srinagar have different attitude and mode of life from the inhabitants of Srinagar city. Geography, as the discipline of relationship, though was quite prominent approach, it lost its position after the Second World War. The advocates of spatial science, locational analysis, behaviouralists, radicalists and humanists criticized this approach and declared it just deterministic and unscientific.

Geography as the Science of Distribution

It was during the pre-classical period (before Varenius and Kant) of the modern geography, when geography was considered as the science of distribution. According to the followers of this philosophical approach, geography is a discipline whose purpose is to study the distribution of different phenomena, separately and in relation to each other, over the earth surface.

The point of view that geography is the study of distribution was presented with the greatest emphasis by Marthe, who described geography as the study of 'the where of things'. More recently, it has been represented by De Geer who stressed on the concept of geography as the science of distribution.

If the study of distribution is fundamental to the character of geography, not merely incidental to other purposes, it must form a characteristic distinguishing geography from other fields. This necessary conclusion has led many to suppose that when a botanist determines the regional distribution of a particular plant, or a geologist determines the location of a volcano, or a sociologist maps the distribution of population in a country, instead of merely using a statistical table, the botanist, geologist or sociologist becomes thereby a geographer – or, at least, is working in geography. But, as Michotte concludes from this series of examples, each of these students is carrying on work necessary to the understanding of the particular kind of phenomena that he is studying from the point of view, not of geography, but of his own science. The use of inductive method, in any of the sciences dealing with phenomena located within the earth surface, in the effort to establish the relationship that governs the character and development of the phenomena of that particular science, will often require the determination of the phenomena before any principles can be determined.

Geography as the study of distribution is a derivation from Humboldt's description of geography as the study of 'that which exists together in area'. Humboldt, during the expedition of the Orinoco river, Amazon basin, Andes mountains, Colombia, Ecuador, Cuba and West Indies, established the fact that there are variations in the distribution of vegetation, soils, temperature and precipitation. Consequently, there are variations in the distribution of population and their economic activities. Ritter, a teleologist, also discussed the spatial distribution of physical phenomena in his monumental work – *Erdkunde*.

Hettner, in some of his writings, declared that geography deals with 'distribution by place'. He, while dealing with the distribution of animals and plants, emphasized the significance of *Wallace Line* which separates the plants and animals of Australia from those of Asia and South-East Asia. Hettner, however, explained that such a distribution of animals falls in the domain of zoological geography or simply animal geography. The same distinction in point of view, Hettner illustrated by several other examples of natural features. In each case, a view that the systematic science is focused on the phenomena, which are studied in their distribution, that of geography on the areas that differ from each other in their mineral, floral, or fauna contents.

Sauer, in his *Cultural Morphology* (Morphology of Landscape) and *Agricultural Origins and Dispersal*, focused upon the regional patterns and interactions of human culture – both material and non-material – in relation to natural environment. In another publication, Hettner asserted that if history may be considered as the science of *when*, geography would be logically the science of the *where*. The main objective and function of geography is to know where the phenomena are (the major thrust of systematic geography). Only the problem of determining and describing the *where* of things is not the distinctive function of geography; it is concerned to provide accurate, orderly and rational description and interpretation of the variable character of the earth's surface. The concept of geography as the discipline dealing with distribution has therefore not been accepted by the geographers as the main sphere of geography.

Geography as the Science of the Planet Earth

In the opinion of some of the geographers like Gerland, Ricthofen, Lehmann and Penck, geography is the science of the planet earth. There had been a puzzling question in the mind of geographers, whether the field of geography includes the earth as a whole or whether it is limited to that part of the earth which we know directly – the earth surface. While in some of the textbooks they take the earth planet as the sphere of geography, others opine that only that part of the earth is the sphere of geography with which we are more intimately familiar, i.e. the earth surface. In practice, geography is to study the world, as that term is most commonly understood.

Some of the geographers have arbitrarily restricted geography as a science which studies the outer shell of the earth. Richthofen, in his presidential address at Leipzig, in 1883, declared that the geographer's main concern is the outer shell of the earth. This concept was, however, declared illogical by Hettner.

Gerland viewed the earth as the subject of study in geography but he made a distinction between geography and geophysics. About the question what is the earth, Gerland discarded not only the teleological and anthropocentric views, but also, by implication, the geocentric viewpoint. Looking at the earth from the cosmic point of view, he saw it as but one of many similar units in the universe. From this point of view, the earth body would be included in an earth science, and thus the interior of the earth would appear vastly to outweigh the surface. Since we cannot make a direct observation on the interior of the earth, the geographers should concentrate on that part of the earth which can be observed directly, i.e. external surface, which is the result of the internal and external forces. The very fact that we have the science of geography, but no science of Mars-ography, reflects that geography is the study of the outer shell of the earth. Our earth is a unique planet. It is the one body in the universe on which surface temperatures fall within that very narrow range in which different materials exist simultaneously in solid, liquid and gaseous states. Consequently, earth is the body whose surface shows the greatest multiplicity of varying interactions of factors and differential results. Moreover, on it alone, life is found in millions of generic and individual variations. All these variations are, however, near the surface of the earth and, therefore, the unique interest of geographers is in the surface of the earth – the thin shell of outer earth in which are found the unique phenomena of the planet earth.

Lehmann took different point of view about the earth as the sphere of geography. He explained that outside of the learned circles the word 'world' particularly in the many combinations in which it is commonly used 'world tour', 'world map', 'world trade', 'worldwide', 'world war', etc. In all of these cases, there is no uncertainty as to what physical extent of space is included: they, however, do not include the centre of the earth, nor do they include the moon or other 'heavenly' bodies, but simply that outer shell of our planet as high in the atmosphere and as deep in the ground as man can experience.

If earth surface is regarded as the domain of geographic study, it is also the meeting place for all sciences that are concerned with earth (meteorology, climatology, geomorphology, geology, ecology, geophysics and environmental studies), history, economics and sociology. But each of these is interested in the earth surface from a different point of view, just as mineralogy and paleontology may study the same rocks from entirely distinct point of view.

It may be concluded from the above discussion that the domain of geography is not the entire planet but only the outer shell of the earth, where we may see the convergence of the atmosphere, hydrosphere, lithosphere, and biosphere. Geography, in fact, is concerned to provide accurate, orderly and rational description and interpretation of the variable character of the earth surface.

Geography: An Idiographic or Nomethetic Discipline

A number of dichotomies have earlier been discussed in Chapters 8 and 9, which show that geographical thought right from its inception has been confronted with the problems of dichotomies and dualism. According to some scholars, geography is an idiographic science, while others consider it as a nomethetic science. Kant, Hettner and Hartshorne considered geography as an idiographic science. In the words of Hartshorne, geography deals with the spatial differentiations of phenomena. He conceives geography as the study of peculiar phenomena of regions. Since geography deals with peculiar phenomena, generalization in the form of laws is useless, and any prediction in geography is of insignificant value.[23] For Kant, geography is description; for Hartshorne, it is a 'naive science'. If we accept this meaning of science, geography is 'naive description'.[24]

Nomothetic sciences deal with general laws. Systematic geography or general geography comes under this category. Humboldt, Sauer and many of their followers laid stress on systematic geography and the formulation of general laws. James, while discussing the idiographic and nomothetic approaches, emphatically declared that there is no such thing as a 'real region'. The region exists only as an intellectual concept which is useful for a particular purpose.[25]

Despite many upheavals, crisis, developments and revolutions in geography, it did not succeed in the formulation of nomothetic laws. It was only after the Second World War that geographers concentrated on theoretical issues and prepared diffusion models, location theory and gravity models as well as geometrical models to explain geographical patterns. This process has passed the take-off stage and it is hoped that new nomothetic (general) laws will be formulated which shall put geography on a sound footing and will bring it greater recognition in sister disciplines.

Geography: A Discipline of Synthesis

It has often been alleged that geography does not have its own area of study. It does not have its laws and thus encroaches upon the sphere of other disciplines. Some geographers argue that the subject matter of geography is largely shared with other disciplines. Some others are of the opinion that while the subject matter of geography is exclusive, geographers alone study places. There is, however, a consensus on the definition given by Ackerman.[26] Ackerman asserted that the fundamental approach in geography 'is the differentiation of the content of space on the earth's surface and the analysis of space within the same universe'.

In order to overcome the controversy of the subject matter of geography, Fenneman in 1919, prepared a model showing the circumference of geography. Fenneman's model has been shown in Figure 10.2.

It may be noted from Figure 10.2 that sciences overlap and that each one of the overlapping zones, which also represent specialized, systematic branch of geography, belongs equally to some other science. For example, the geologists deal with rocks, the botanists with plants, the meteorologists with climates, the demographers with population conditions, and the economists with economic conditions. Such studies, unfortunately, fail to consider the overall

Figure 10.2 The Circumference of Geography (Adopted from Fenneman)

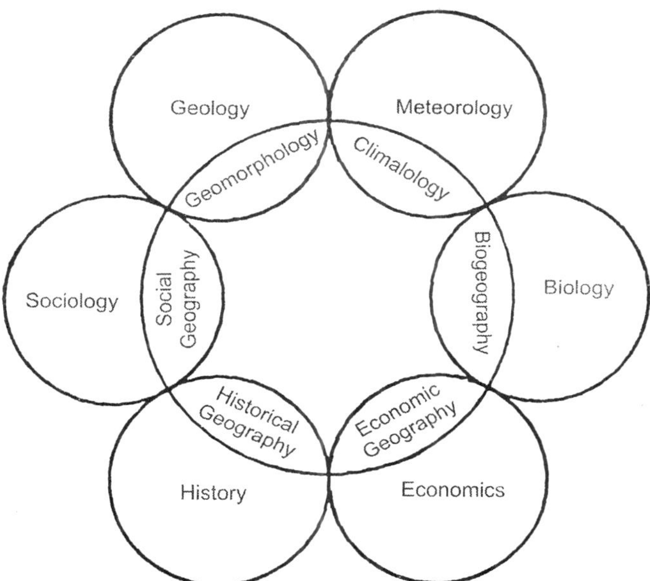

interaction between phenomena. It is geography which deals with an overall interaction of the biotic and abiotic phenomena in time and space with an anthropocentric approach. The sphere where atmosphere, lithosphere, hydrosphere, and biosphere converge is the region where geographers study the man-nature relationship. Thus, geography has its own area of study which may appear to be in the nature of an encroachment upon other disciplines. It is because of this peculiarity that geography is regarded as a *science of synthesis*. The geographers attempt a synthesis of the physical and human phenomena within an area or region. The models and laws formulated in general geography are tested by them in the regional studies.

The concept of geography as a science of regional synthesis is not very old. The British and American geographers did rigorous exercises to define the objectives, scope and nature of geography. Before the advent of 'quantitative revolution' geography used to be considered as 'short a theory and long a fact'.[27] Notwithstanding this definition, formulation of theory in geography is essential. In the words of Burton, theory 'provides the sieve through which myriads of facts are sorted, and without it the facts remain a meaningless jumble'.[28]

Explanations in Geography

In geography, where explanations have been attempted, these have tended to be *ad hoc* and unscientific in form. This point has been carefully examined by Harvey, who has offered the following explanations. Harvey recognized six forms of scientific explanations, covering methodological statements and generalizations from empirical studies. The explanations in geography put forward by Harvey to explain the general and empirical laws are as follows:[29]

(i) Cognitive Description

Under cognitive description are included collection, ordering, and classification of data. No theory may be explicitly involved in such procedures, but it is important to note that a theory of some kind is implied. Thus, classification involves some kind of a priori notions about structure and these notions really amount to a primitive theory. In the early stages of a discipline's development such theoretical assumptions may be amorphous and ill-defined.

(ii) Morphometric Analysis

Morphometric analysis is a type of cognitive description. It involves a space-time language rather than a property language. Morphometric analysis thus provides a framework within which the geographer examines shapes and forms in space. Morphometric analysis can lead to certain types of predictive and simulation models. In this analysis, stress is on measurement whereas studies of landscape morphology usually take the form of cognitive description. The locational theories and the central places are the results of this type of analysis. Geometrical predictions of this nature have had increasing significance in geography.

(iii) Cause and Effect Analysis

The cause and effect analysis develops from the assumption that previous causes can explain observed phenomena. We look for causal relationships which are, in their simplest form, of the type 'cause A leads to effect B'. This implies that cause B cannot lead to result A. Causal laws may be discovered by hypothetic deductive method, or more simply, by comparing data from different phenomena in a region. By comparing a map, showing the black-earth soil region of India (Maharashtra, Gujarat and Western Madhya Pradesh), with a map showing cotton concentration and its yield pattern, we may arrive at the result that there is close relationship between the black-earth soil and the quantity of cotton output. From this example, it may be inferred that black-earth (soil) influences the cotton yield; but that a high yield of cotton will make the soil or a region black cannot be inferred. The causal relationship is not a simple device; it demands multiple regression or factor analysis.[30]

(iv) Temporal Analysis

Temporal analysis may also be termed as historical analysis. This analysis is a type of cause and effect analysis, which is established over a long period of time. The assumption is that a particular set of circumstances may be explained by examining the origin and subsequent development of phenomena by the operation of process laws.[31] In the words of Darby, 'the foundations of geographic study lie in geomorphology and historical geography' – both of which were dominated by temporal modes of explanation.[32]

Temporal analysis helps in understanding the spatial distribution of many of the phenomena, but it cannot be taken as the only approach of geographic explanations. History can be seen as a causal series which started at the vaguely defined 'dawn of history' and ends today. In practice it will never be possible to understand such a comprehensive causal series; the analysis, therefore, must be restricted to some determined period of time.[33]

(v) Functional and Ecological Analysis

The cause and effect analysis was rejected by the positivists to avoid the metaphysical and normative trappings. To counter the cause and effect relationship functional analysis was developed. Functional analysis attempts to analyse phenomena in terms of role they play within a particular organization. Towns may be analysed in terms of function they perform within an economy (thus functional classification of towns is developed), rivers may be analysed in terms of their role in denudation and so on.[34] Ecological and functional thinking have been important in geography. At the present time there are numerous geographers who regard ecological concepts as providing an important basis for geographic explanation.[35]

(vi) Applications of Geography

Harvey emphasized on the application of geographical laws for the solution of the problems of resources, environment and society.

SYSTEM ANALYSIS

System has been defined differently by different scientists. In the words of James, a system may be defined as 'a whole (a person, a state, a culture, a business) which functions as whole because of the interdependence of its parts'. If we accept this definition, then it can fairly be said that geographers have been using forms of system concepts since the dawn of the subject. However, till the outbreak of the Second World War no technique had been developed to enable geographers to analyse complex systems.

Geography deals with complex relationships of living and non-living organisms in an ecosystem. System analysis provides a framework for describing the whole complex and structure of activity. It is, therefore, peculiarly suited to geographic analysis since geography deals with complex multivariate situations.[36] It was because of this advantage that Berry and Chorley suggested system analysis and general system theory as the basic tools for geographic understanding. In the opinion of Chorley (1962), there is great significance of system analysis in geographical studies. The main advantages of system analysis are:

1. there is need to study systems rather than isolated phenomena;

2. there is need to identify the basic principles governing systems;
3. there is value in arguing from analogies with subject matter; and
4. there is need for general principles to cover various systems.

General System Theory

The concept of general system theory was developed by biologists in the 1920s. It was Ludwig von Bertalanffy who declared that unless we studied an individual organism as a system of multifarious associated parts we would not really understand the laws which govern the life of that organism.[37] After some time he realized that this idea could be applied to other non-biological systems, and that these systems had many common characteristics over a range of sciences. It was possible to develop a general system theory which gave the same analytical framework and procedure to all sciences.

A general system is a higher order generalization of a multiplicity of systems which individual sciences have recognized. This is a way of unifying the sciences.[38] This led to interdisciplinary approach in research. In other words, the general system theory is a theory of general models.

According to Mesarevic's definition, the general system theory is concerned not merely with isomorphism and analogy in system analysis, but with setting up some general theory for which characteristics of various systems can be deduced. It is thus concerned with the deductive unification of system analytic concept.

The general system theory provides a framework for relating individual systems and types of systems within a unified hierarchical structure. Such a structure is useful in that it allows us to understand better the relationships that exist between various types of systems; to state categorically the conditions under which one system approximates another, and to identify types of systems that may be useful to us even though we have not yet identified real system to match them.

The general system theory can be understood in the light of a new concept of mathematics and physics. This concept is known as 'cybernetics' (from the Greek *kybernete* – helsman). Cybernetics may be defined as the study of regulating and self-regulating mechanisms in nature and technology. A regulatory system follows a programme, a prescribed course of action which produces a predetermined operation. In nature, there is a very large number of self-regulating mechanisms, such as the automatic regulation of body temperature. These self-regulating mechanisms follow certain common laws and these can be described mathematically in the same way.[39] Whilst regulation is very precise in nature, in human societies it is defective.

Cybernetics places emphasis on the interaction between components rather than making sharp distinctions between cause and effect. Between two components, causal mechanism may work both ways. An impulse which starts in one part of the system will work its way back to its origin after being transformed through a range of partial processes in other parts of the system.

This cybernetic theory enables us to understand the operation of the general system theory.[40]

The abstract character of a system is emphasized when we realize that a system, if it is to be analysed, must be 'closed'. An 'open system' interacts and interconnects with the surrounding systems, and therefore, becomes difficult to analyse. All real systems (such as landscapes) are open systems. When we analyse a system we can only consider a finite number of elements within the system and the reciprocal relations between them The elements and connections which we are not able to consider in such an analysis must be disregarded completely. We have to assume that they do not affect the system. In the analysis of a region, we can of course take into account individual influences and single elements which are not geographically located within the predetermined area or region. The abstract system remains closed all the same because we enclose these elements and relationships in our conceptual model. The system is not synonymous with the model we have made for it, represented by the elements and connections we have chosen to enclose or consider.

In other words, we can only study a system after we have determined its boundaries. This presents no mathematical problem since the boundaries draw themselves insofar as some lying outside it, although it is not all that easy to choose those elements, in practical geographical research. As an example, Harvey[41] describes a firm which functions within an economy on the basis of a particular set of economic circumstances. When we analyse the internal relations and elements within the firm as a closed system, we must regard these circumstances as unchangeable. To extend the boundaries of the system so as to include the changing social and political relationship in the society of which the firm is a part may well alter the result of the analysis. So, even in this simple case, the drawing of boundaries creates problems.

By identifying the set of elements which we believe best describe the real system we can model a real situation. For example, in a large industrial company engaged in several branches of activity, the head office and each of the branch offices form its constituent elements. Mathematically expressed, the system consists:

$A = (a1, a2, a3 \ldots an)$

To this expression should be added an element a° which represents the environment of the system within which the firm operates. We can then infer a new set of elements:

$B = (a^\circ, a1, a2 \ldots an)$

This includes all the elements in the system plus an extra element which represents the environment. We can then investigate the connections between these elements. Analysing the company we can see whether there are any connections between the branches, and, if so, between which branches. We can observe whether the contacts go both ways and what the contact model implies.

Thus, a system consists of:

1. A set of elements identified with some variable attributes of objects.

2. A set of relationships between these attributes of objects and the environment.[42]

Merits of Abstract Construal of a System

The abstract construal of a system has a number of important advantages, which are given below:

1. Any geographical region (landscape) has a number of phenomena. System analysis attempts to reduce this complexity to a simpler form, in which it may be more easily comprehended and which models can be constructed.
2. It allows, for example, the development of an abstract theory system which is not tied down to any one particular system or set of systems.
3. This theory provides us with good deal of information about the possible structures, behaviours, states, and soon, that might conceivably occur.
4. It provides us with the necessary technical apparatus for dealing with interactions within complex structures.
5. System theory is associated with an abstract mathematical language, which, rather like geometry and probability theory, can be used to discuss empirical problems.

Structure of a System

A definition of 'system' has been given in the foregoing paras. Given the definition of a system it is possible to elaborate its 'structure'.[43] A system is composed essentially of three components:

1. a set of elements;
2. a set of links; and
3. a set of links between the system and its environment.

Elements of a System

Elements are the basic aspects of every system, structure, function, and development. From the mathematical point of view, an element is a primitive term that has no definition, like the concept of point in geometry. Nevertheless, the structure of a system is the sum of the elements and the connections between them. Function concerns the flows (exchange relationships) which occupy the connections. Development presents changes in both structure and function which may take place over time.[44]

The definition of an element depends on the scale at which we conceive of the system. For example, the international monetary system may be conceptualized as containing countries as elements; an economy may be thought of as being made up of firms and organizations; organizations themselves may be thought of as system made up of departments; a department may be viewed as a system made of individual people; each person may be regarded as a biological system; and so on.[45] Similarly, a car may be an element in the traffic system, but may also be regarded as constituting a system. It is clear from these examples that the definition of an element depends on the scale at which we conceive of the system.

The concept of element as a component unit of a system has been plotted by Blalock and Blalock which has been shown in Figure 10.3. This figure shows two different views of interaction. The upper diagram shows System A and System B interacting as units, with smaller system interactions going on within each system. The lower diagram shows Systems A and B interacting at lower levels.[46]

Figure 10.3 Systems and Subsystems. The upper diagram shows System A and System B interacting as units, with smaller system interactions going on within each system. The lower diagram shows systems A and B interacting at lower levels (From Blalock and Blalock, 1959)

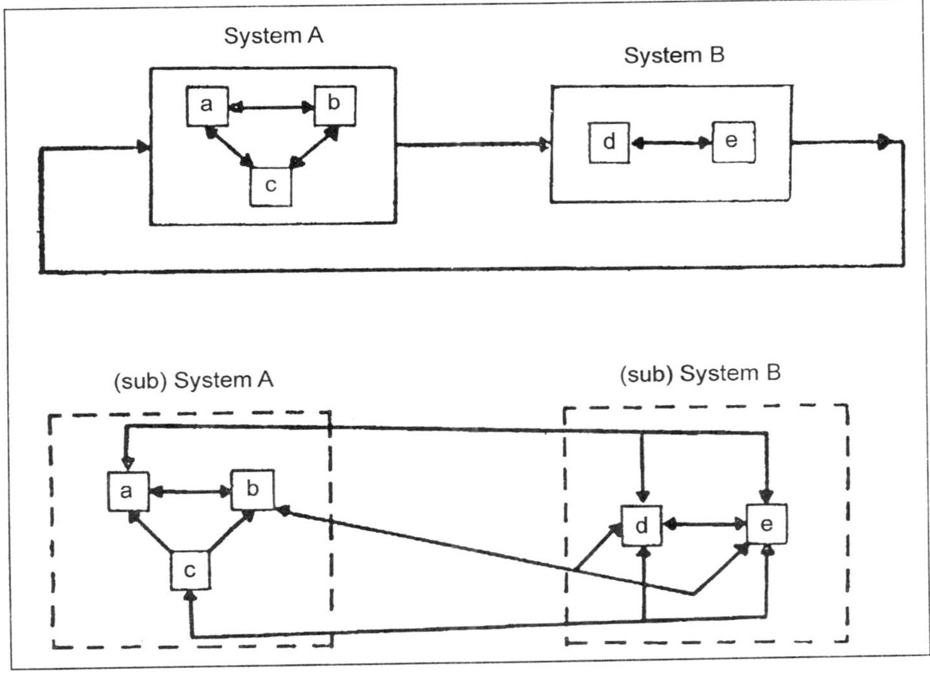

After it has been decided which scale to use, another problem in system-building is how to identify the elements. Identification is particularly difficult when we are dealing with phenomena which have continuous distribution, e.g. when precipitation forms an element in system. Identification is the easiest with elements which are clearly separated, such as farms. But, from the point of view of mathematical systems theory, an element is a variable. It follows, therefore, that in seeking for a translation of the mathematical element in geographical context we must construe the element as an attribute of some defined individual rather than as the individual itself.[47]

Links or Relationships

The second component of a system is links (relationships). The links in a system which connect the different elements in it have been shown in Figure 10.4. These are as follows:

1. Series relation.
2. Parallel relation.
3. Feedback relation.
4. Simple compound relation.
5. Complex compound relation.

Three basic forms of relationships can be defined as under:

1. *Series relation*: This is the simplest and is characteristic of elements connected by an irreversible link. Thus, *ai* – *aj* forms a series relation and it may be observed that this is the characteristic cause and effect relation with which traditional science has dealt. This relationship can be explained by taking an example from India. The productivity of rice in Punjab depends on irrigation available or cultivation of saffron in the valley of Kashmir is due to the Karewa soil.
2. *Parallel relation*: This relationship occurs when two or more elements affect a third element, or inversely when one element affects two or more others. It may be noted from Figure 10.4 that *ai* and *aj* are affected by some other element *ak*. For example, the precipitation and temperature variables influence vegetation and vegetation, in turn, influences the amount of rainfall received and the general temperature conditions.
3. *Feedback relation*: A feedback relation is the kind of link that has been newly introduced into analytic structures. It describes a situation in which one element influences itself.[48] For example, the leguminous crops sown in a field enrich nitrogen in the soil and thus the crops get themselves affected (Figure 10.4.3). The feedback relationship may be direct, positive, negative or no feedback. An example of the direct feedback is: A influences B which in turn influences A, or it may be indirect, with the impulse from A returning to it via a chain of other variables. With negative feedback, the system is maintained in a steady state by self-regulating processes termed as homostatic or morphostatic. A classic example is provided by the process of competition in space, which leads to progressive reduction in excess profits until the spatial is in equilibrium. But, with positive feedback, the system is characterized as morphogenetic, changing its characteristics as the effect of B on C leads to further changes in B via D.

It is possible to combine these relationships in a number of ways (Figure 10.4.4) so that two elements may be connected in different ways simultaneously. The links thus form a kind of 'wiring system' connecting the elements in various ways (Figure 10.4.4–5).

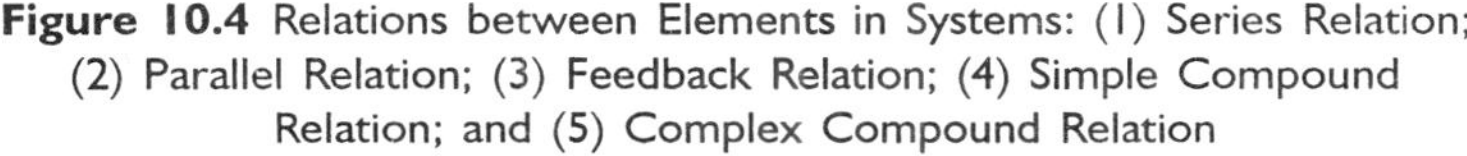

Figure 10.4 Relations between Elements in Systems: (1) Series Relation; (2) Parallel Relation; (3) Feedback Relation; (4) Simple Compound Relation; and (5) Complex Compound Relation

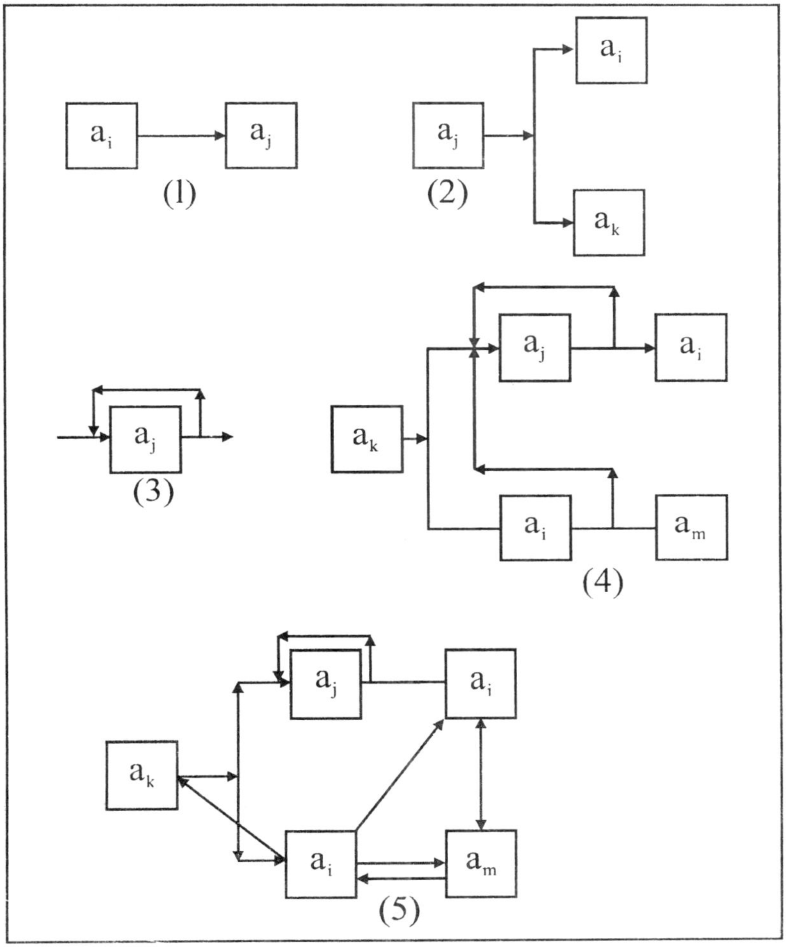

Behaviour of a System

Behaviour of a system means interrelationships of the elements, their reciprocal effect on each other. Behaviour has to do therefore, with flows, stimuli, and responses, inputs and outputs and the like. We can examine both the internal behaviour of a system and its transactions with the environment. A study of the former amounts to a study of functional laws that connect behaviour in various parts of the system. Consider a system that has one or more of its elements related to aspect of the environment. Suppose the environment undergoes a change. Then, at least one element in the system is affected. The effect of this affected elements is transmitted throughout the system until all connected elements in the system are affected. This constitutes a simple stimulus response, or input-output system without feedback to the environment:

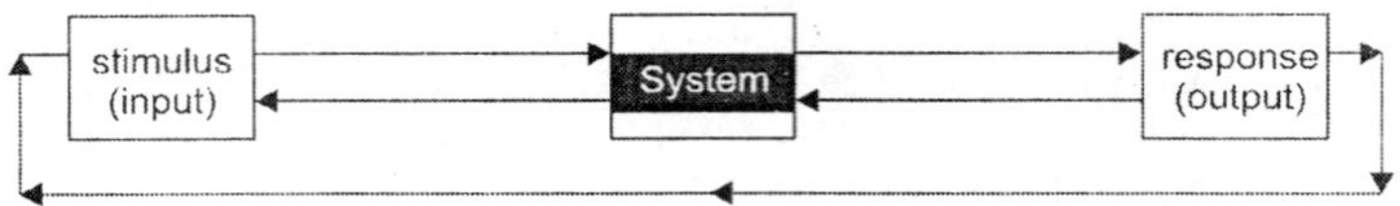

The behaviour is described by the equations (deterministic or possibilistic) that connect the input with the output.[49]

Geographical System

A system where one or more of the functionally important variables are spatial may be described as a geographical system. Geographers are primarily interested in studying systems whose most important functional variables are spatial circumstances, such as location, distance, extent, sprawl, density per areal unit, etc.[50] In the last few decades, system approach has drawn the attention of geographers. Chorley attempted to formulate thinking in geomorphology in terms of open system; Leopold and Langbein used entropy and steady state in the study of fluvial systems; and Berry attempted to provide a basis for the study of 'cities as systems within systems of cities' by the use of two concepts of organization and information in spatial form.[51] Recently, Wolderberg and Berry have used systems concept to analyse central-place and river patterns, while Curry has attempted to analyse settlement-locations in systems framework. Those geographers who focus attention upon spatial organization invariably invoke systems as Haggett's account of locational analysis in human geography demonstrates.[52]

In geography, static or adaptive systems can be easily constructed. It is difficult to make a geographical system dynamic, for that we must combine time and space in the same model. Space may be expressed in two dimensions by cartographical abstraction. We may be able to present a satisfactory explanation for such a system but it is very difficult to handle and analyse it. Lund has analysed these problems in his time-space model.[53]

Some of these problems can be solved by developing geographical models which may be classified as 'controlled systems' (discussed above). Controlled systems are particularly useful in planning situations when the objective is known and the input in the economic geographic system has been defined. In most of the cases, we can control some of the inputs, but others are either impossible or too expensive to manipulate. For example, if we wish to maximize agricultural production, we may be in a position to control the input of artificial fertilizers, but we cannot control the climate. Partially controlled systems are therefore of great interest. Our increased knowledge of environmental conditions leads us to appreciate the extent of the need for the development of planning and control systems. Many of the scientists engaged in research into possible future conditions fear that the positive feedback mechanism in the form of technological development and control which have led to an exponential increase in population, industrial production, etc. will, in the long run, result in a dramatic crisis of pollution, hunger and shortage of resources. One of the causes of such a crisis would be the long-term suppression of natural negative feedback mechanism.[54]

System analysis may provide a useful systematization of our models, theories of structured ideas, but it is not necessary to refer to system analysis and its mathematical implications when we are doing practical research. For instance, a world map of iron ore production and trade may be described in systematic terms: the elements are the producing and consuming centres, the links or relations are the trading lines, the amount of iron transported along different lines depicts the function, and maps showing these situations at specific time intervals would describe the development of the system.[55] Moreover, the system approach was technically much more demanding, and perhaps for that reason attracted fewer active researchers.

Both system analysis and general system theory have been criticized on the ground that they are instrinsically associated with positivism, i.e. these do not take into consideration the normative values (aesthetic values, beliefs, attitudes, desires, hopes and fears), and thus do not give a real picture of a geographical personality.

The development of geographical research has been discussed in the foregoing paras. It has passed through three different phases of development. The development of a science covers three broad stages: (i) descriptive, (ii) analytical, and (iii) predictive. Description is the first step and the simplest; it is concerned with description and mapping of phenomena. Geography from antiquity to the middle of the 18th century was in this phase. The analytical stage moves a step further by looking for explanation and seeking the laws which lie behind what has been observed. The period of Alexander von Humboldt falls in this phase. It was during this period that analysis of the spatial distribution of phenomena started. The third stage in the development of a science is the predictive stage. By the time predictive stage was reached, the laws had been studied so thoroughly that we could use models to predict occurrences. This stage was partly reached with the advent of geomorphology and climatology in the closing decades of the 19th century. But, the real upheaval in the field of human geography is a post-Second World War phenomenon. Many locational theories have been formulated which are predictive in nature, and thus we can say that geography has entered the third stage of its development. Geographers are trying to develop models for controlled systems which may be used to guide development in the future. It is clear from the above discussion that geographers are now moving into the predictive stage.

Geography: A New Synthesis

A new school of geographers is emerging who are stressing that geography must be a synthetic science. The synthetic approach in geography can provide practical forecasts of the consequences of man's interference with natural forces.[56]

Historically, geography has been divided into physical versus human, systematic versus regional, deterministic versus possibilistic and normative versus positive categories. All these dichotomies have already been briefly discussed in Chapter 9. Haggett, however, felt that it was more valuable to divide the subject in relation to the way it analysed its problems.[57] Haggett, on the basis of analysis technique, divided geography into (i) spatial analysis, (ii) ecological analysis, and (iii) complex regional analysis.

The internal structure of geography, designed on the above criteria, is given in Figure 10.5. It will be interesting to make a brief comment on these three styles of analysis.

1. *Spatial Analysis*: It concerns itself with varieties in the localization and distribution of a significant phenomenon or group of phenomena; the analysis of spatial distribution of rainfall, or the average yield of wheat or rice in a given region. If a geographer tries to ascertain the factors which control the distribution of rainfall and yield levels, in order to make the distribution more effective or just, he is doing a spatial analysis.
2. *Ecological Analysis*: It concerns itself with the study of connections between human and environmental variables. This type of analysis is done in a closed or partly closed ecosystem, i.e. it is an intra-regional and not an inter-regional analysis of the spatial distribution of phenomena.
3. *Regional Complex Analysis:* It combines the results of spatial and ecological analysis. Appropriate regional units are identified by areal differentiations. Connecting lines and flows between individual regions may then be observed. The peculiarity of complex regional analysis is that the inter-regional analysis of the distribution of phenomena, their positive and feedback relationships are examined.[57]

In his scheme, Haggett tried to arrange his themes under the primary headings of ecological and complex regional analysis (Figure 10.5). It may be seen from this figure that all the three types of analysis, i.e. spatial, ecological, and complex regional have theoretical and applied aspects. The theoretical aspect of spatial analysis deals with spatial interaction theory, diffusion theory, and others; while its applied aspect covers watershed development, urban places, and others.

Ecological analysis also has theoretical and applied aspects. Under the theoretical aspect, environmental structure, ecosystems, etc. are analysed; while its applied side analyses natural resource geography, hazard appraisal and others (Figure 10.5).

So far as regional complex analysis is concerned, its theoretical aspect deals with regional growth theory, inter-regional flow theory, etc.; while the applied side looks into regional forecasting, regional planning and other relevant matters (Figure 10.5). This analysis is involved with functional regions. A functional region has been defined with reference to the contact of relationship between a centre and its tributary surrounding regions (see Chapter 9). The boundary of this region is established at the point where the influence of the centre is not stronger than that of another centre. Functional regions are linked through complex hierarchical models. Such an analysis becomes possible through the use of sophisticated quantitative techniques.

Figure 10.5 The Internal Structure of Geography (From Haggett, 1779)

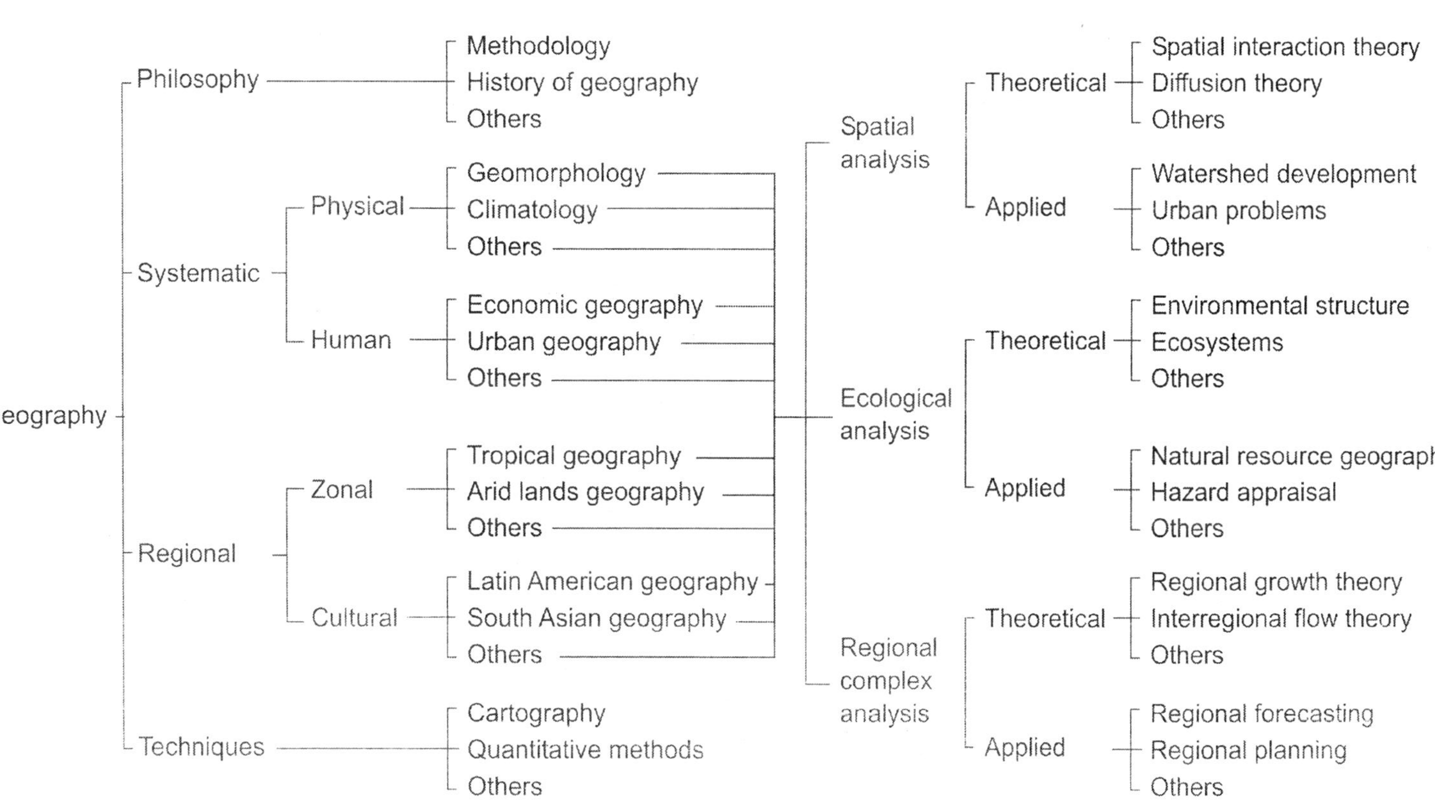

In brief, modern geography focuses its attention on spatial analysis, a systematic geography constructed with newer, expanded models; ecological analysis, a regional study based on homogeneous regions; and regional complex analysis, a regional geography based on functional regions. Thus, geography is a science of synthesis (Figure 10.5).

Notwithstanding the fact that geography is a science of synthesis, it has not been able to make much advance in regional synthesis. The main reason for this shortcoming is that it deals with physical and human components, biotic and abiotic factors which are governed by different physical and social laws. There are some basic constraints which come in the way of making geography a discipline of synthesis. Some of the recent approaches which are trying to deal with this question have been described in Chapter 12. The various schools of thought which are emerging and trying to answer the man-nature relationship question may perhaps provide a temporary solution of the problem, but we should continue to explore a more dependable and more precise objective approach to make geography a discipline of regional synthesis. Undoubtedly, in the near future, a geographer may not be the master of his fate; his efforts will, however, lead him into the unknown and unforeseen where he learns new things.[58]

Some of the important approaches being adopted by the geographers have been discussed above. In recent years, geographers, confronted with many philosophical issues in the discipline, are trying to find an answer to the geographical problems with the positivistic, idealism, realism and dialectical materialism approaches which have been examined in the next chapter.

REGIONAL CONCEPT

Region is a dynamic concept which has been defined differently by different geographers. In the concluding part of the 19th century, the French geographers like Vidal de la Blache called the areas of similar physical and cultural characteristics as *pays.* A more comprehensive and widely acceptable definition of region can be given as 'an area having the homogeneity of the physical and cultural phenomena'. It has also been defined as 'an area that is differentiated from other areas according to the specified criteria'.

Herbertson (1905) was the first to divide the earth into major natural regions on the basis of climatic parameters and thus having some links with determinism. On the smaller scale, geographers attempted to identify individual areas with particular characteristics.

The fundamental idea behind the small geographic regions was to show some distinct individuality, if not necessarily entire homogeneity through a study of all its geographical features, i.e. structure, climate, soils, vegetation, agriculture, mineral and industrial resources, settlement and distribution of population. Some of the work, exemplified by Herbertson (1905), was the precursor of the concept of ecosystem.

The basic objective of the discipline of geography is to provide accurate, orderly, and rational description and interpretation of the variable character of the earth surface (Hartshorne in *Perspective on Nature of Geography*, 1959). In fact, the geographer's main concern is how things are distributed over the surface of the earth, how physical and cultural features of areas are alike or different from place to place, how the varying content of different places came about, and what all these differences and similarities mean for people.

Thus, region is a device of area generalization. It has also been defined as 'a differentiated segment of earth surface' (Whittlesey, 1929). The generalization of the features of the earth surface is to make the infinitely varying world around us understandable through spatial (regional) summaries. It is, however, difficult to divide the earth surface into neat regions of uniformity.

Although as many possible regions exist as there are physical (topography, relief, climate, soils, natural vegetation, minerals, etc.), cultural (religion, language, population, agriculture, industries) or organizational (socio-economic institutions) phenomena.

The number of regions may be enormous and vary greatly, yet they may all share certain common characteristics which are given below:

Attributes of Region

1. *Regions have location*: All region - physical or cultural - are often expressed in the regional name such as the Middle East, the South-East Asia, the North-West Europe, the Far East, etc.
2. *Regions have spatial extent*: The homogeneous physical and cultural attributes of the earth surface have spatial (areal) extent. For example, the Thar Desert, the Sahara Desert, the Latin America and Anglo-America cover certain areas of the earth surface. Thus, regions are not in abeyance; they have a personality on the ground.
3. *Regions have boundaries*: Each region - physical or cultural - has a boundary. The boundary of a region is drawn at the outer edge where the phenomenon (feature) no longer occurs or dominates. For example, where the Himalayas and the Siwaliks end, the Indo-Gangetic plains begin, and where the Gangetic plains end, the Deccan plateau begins. Similar may be the case of the language regions in India. There is a line of demarcation between the Telugu, Tamil and Malayalam regions in India. Similarly, we find the tribal and non-tribal regions in the different states of India. Likewise, at some defined point, urban is replaced by rural, or the rain forest ceases and the savanna emerges. Regions boundaries, are however, not like the walled medieval city.
4. *Regions may be either formal or functional*: Formal regions are areas of uniformity throughout in one or limited combinations of physical or cultural features. The equatorial region, the monsoon region, the Sahel region (Africa), the Tundra region, the mountainous region are the examples of formal physical regions. Similarly, we may observe the

homogeneity of language, religion, ethnicity and lifestyle in certain areas. Such regions are known as the formal cultural regions.

Whatever the basis of its definition, the formal region is the largest area over which a valid generalization of attribute uniformity may be made. Whatever is stated about one part of the region holds true for its other parts also. Up to 1960, most of the regions demarcated by geographers used to be formal regions.

Functional region, in contrast, is a spatial system defined by the interactions and connections that give it a dynamic, organizational basis. Its boundaries remain constant only as long as the interchanges establishing it remain unaltered. 'City region' may be cited as a good example of functional region. The city region is 'an area of interrelated activities, kindered interests and common organizations, brought into being through the medium of the routes which bind it to the urban centres'. We can delineate the commuting regions of Delhi, Bombay, Calcutta or any one of the metropolitan and mega cities. Similarly, functional regions of the national capital of India may be demarcated by taking the supply of milk, fruits, vegetables and newspapers. The functional region is, however, a dynamic concept which changes in space and time.

5. *Regions are hierarchically arranged*: Although regions vary in scale, type and degree of generalization, none stands alone as the ultimate key to areal understanding. Each defines only a part of the spatial (regional) reality.

   On a formal regional scale of size progression the Ganga-Yamuna Doab may be seen as part of the Upper Gangetic Plain, which in turn is a portion of the Sutlej-Ganga Plain. Similarly, the Central Business District (CBD) of Delhi (Connaught Place) is one land use complex in the functional regional hierarchy that describes the spatial influences of the city of Delhi and the National Capital Region of which it is the core. Each recognized regional entity in such progression may stand alone and at the same time exists as a part of a larger, equally valid, territorial unit.

   The above generalizations about regions and the regional concept show that regions are human intellectual creations designed to serve a purpose. Regions focus our attention on spatial uniformities. They bring clarity to the seeming confusion of observable physical and cultural features of the world we inhabit. Regions provide the framework for the purposeful organization of spatial (regional) data and information.

6. *Regions have transitional boundaries*: Generally, regions do not have sharp boundaries. In most of the cases their boundaries are transitional. It means there is some overlapping of one phenomenon over the other.

Classification of Regions

The regions may be classified as under:

1. Regions based on physical characteristics.
2. Regions based on cultural characteristics.
3. Regions based on an amalgamation of the physical and cultural variables.

(i) Physical Regions

The simplest of all regions to define, and easiest to recognize is the formal region based on a single variable or single characteristic. The island is land, not water, and its unmistakable boundary is naturally given where the one element (land) passes to the other (water). The dense forest may break dramatically upon the open grassland. The nature of change is singular and apparent.

The physical geographers, who explain the natural environment, generally deal with single factor formal regions. Many of the earth features (physical features) do not exist in simple, clearly defined units. They must be arbitrarily and statistically demarcated or regionalized by the application of boundary definition. For example, the rainfall regions, the variability of rainfall regions, the plants association regions, soil regions (black-earth region of India), etc. must be decided on as regional limits and all such limits are subject to change through time or by purpose of the regional geographer. These regions and their boundaries change with the passage of time.

(a) *Landform regions:* The landform regions are classified and demarcated on the basis of structure, relief, configuration, genesis and age. These regions are independent of human influence and unaffected by time on the human scale. Landforms constitute basic, naturally defined regions of physical geographic concern. The Himalayan system, the Vindhyan system, the Aravalli system, the Alpine system, the Kashmir valley, the Brahmaputra valley, plateaus and mountains, humid landform areas, dry landform areas, and glacial areas are some of the examples of the landforms regions.

The landforms regions are the result of the internal and external physical forces. These have a close influence on climate, natural vegetation, and soils. The basic human activities are also largely influenced by the landforms regions.

(b) *Climatic regions:* A specific area in various combinations of climatic elements (temperature, rainfall, etc.) may be recognized as a climatic region. Numerous attempts have been made to identify and classify climatic regions: (i) based on latitudinal temperature zones, we have the torrid, temperate, and frigid zones; and (ii) based on temperature, rainfall and climatic effects, we have Koppen's, Thornthwait's and Trewartha's climatic regions.

Although weather deals with the state of atmosphere over short duration, regions may also be recognized on the basis of weather conditions. For example, the air mass region or air masses.

(c) *Air masses:* An air mass is a vast body of air whose physical properties (temperature, humidity and lapse rate) are more or less uniform in the horizontal plane. On the basis of these characteristics we have: (i) Arctic, (ii) polar-continental, (iii) polar-maritime, (iv) tropical-continental, (v) tropical-maritime, (vi) equatorial air masses, and (vii) the monsoon. The climatic regions are dynamic in nature. Climates like vegetation and soils change through time by natural process or by the action of humans. Boundaries shift as witnessed by the recent migration southward of the Sahara.

(d) *Ecosystems as regions:* Ecosystem is an ecological concept defining the relationship between a set of living and non-living objects. One of its clearest definitions was given by Fosberg (1963) who states:

> Ecosystem is a functioning, interacting system composed of one or more living organisms and their effective environment, in a biological, chemical and physical sense. It is a concept applicable at any scale ranging from the planet earth as an ecosystem down to the smallest patch of moss and lichen on a rock surface.

The main advantage of the ecosystem as region is that it brings together in a single framework environment, humans and biological realm, permitting an analysis of the relationship between biotic and abiotic components of area. The ecosystem concept or ecosystem region provides a point of view for investigating the complex consequences of human impact upon the natural environment.

(ii) Cultural Regions

A cultural region refers to an area over which the cultural traits of human group may be identified. The culture and cultural environment of human groups (ethnic groups) varies from place to place. This variation in cultural traits results into variation in human occupation and his organization of space. Some of the important cultural regions are: (a) population regions, (b) linguistic regions, (c) religious regions, (d) political regions, (e) economic regions, (f) natural resource regions, (g) urban regions (megalopolis), (h) agricultural regions, (i) industrial regions, and (j) mental regions (mental map).

(a) *Population regions:* Population and its demographic attributes constitute an important aspect of cultural landscape. To delineate an area into the high density and low density areas is known as population regions. The age and sex composition, birth, death and growth rate patterns, the literacy, occupations and patterns of migration may also be delineated. All these are known as population regions.

The delimitation of population regions was emphasized by the American geographer, Glenn Trewartha. He also suggested the tools and techniques of delimiting the population regions.

If we divide India into population regions, it may be stated that the alluvial plains are densely populated, while the mountains, deserts and plateaus of the country are less densely populated. Similarly, the world may be divided into the areas of high and low concentrations of population. For example, all the plains of the world, especially those of the lower latitudes, are densely populated, while the deserts, equatorial forests, tundra and mountainous areas are less densely populated. The regional distribution of population in Latin America has a strong nucleated character as most of the population is confined in isolated clusters. Moreover, the individual clusters show considerable variation in density. The clustered nature of Latin American countries may be attributed to the search of gold and the zeal of missionaries that imbued the Spanish colonialists.

(b) *Linguistic regions:* All over the world, different social groups speak different languages. The delineation of different language areas on a map is known as language region. Taking language as the criterion, the world may be divided into: (i) Indo-European, (ii) Indo-Iranian, (iii) Sino-Tibetan, (iv) Afro-Asiatic (Arabic), (v) Austro-Asiatic, (vi) Amerindian, and (vii) Negro language regions.

A typical example of linguistic regions may be cited from India. The Indian states have been demarcated on the basis of languages. For example, the states of Gujarat, Maharashtra, Andhra Pradesh, Karnataka, Tamil Nadu, Kerala, Assam, West Bengal, Orissa and Punjab have their own regional languages.

We may delineate macro, meso and micro level language regions to examine and interpret the cultural personality of a country, nation or area.

(c) *Religious regions:* The world may be divided on the basis of religions, e.g. religions of Christianity, Islam, Judaism, Hinduism, Buddhism, Sikhism and Jainism. Each religion has its own basic tenets.

(d) *Political regions:* The most rigorously defined formal cultural region is the national state. Its boundaries are carefully surveyed and in many cases are marked by fences and guard posts. There is no question of any arbitrarily divided transition zone. This rigidity of a country's boundaries, its unmistakable placement in space and the trapings – flag, anthem, army, government – that are uniquely its own give to the state an appearance of permanence and immutability not common in other, more fluid cultural regions. But its stability is often more imagined than real. Political boundaries are not necessarily permanent. The artificial political boundary in the form of Berlin Wall was demolished in 1990. They are subject to change, sometimes violent changes, as a result of internal and external pressure. The Indian subcontinent illustrates the point.

The history of the subcontinent since about 400 BC has been one of the alternating creation and dissolution of empires, of the extension of central control based upon the Ganga Basin, and of resistance to that centralization by the marginal territories of the peninsula. In 1947, the British rulers partitioned the subcontinent of India on the basis of religion. The independent state of India was created out of the largely Hindu areas, while the Muslim majority area was declared as Pakistan. The East and West Pakistan wings of that country could not remain united as there was great diversity in the ethnicity, language, customs, traditions, values, and economy. Consequently, the East Pakistan became Bangladesh in 1971. The violent history of the subcontinent shows that nationalism may be sought on the religious basis, but its maintenance may be difficult because of the physical and socio-economic and cultural regions.

(e) *Economic regions:* Economic regionalization is among the most frequent, familiar and useful employment of the regional method. The economic regions in geography identify economic activities, and resources over space. The economic region serves as a useful tool for planning and a framework for the manipulation of the people, resources and economic structure of a formal region. Problems like poverty, hunger, outmigration, cultural deprivation, underdevelopment and malnutrition may be systematically examined with the help of economic regions. Economic regions are generally delineated with the help of several socio-economic indicators.

(f) *Natural resource regions:* The unevenly distributed resources upon which the people depend for existence are logical topics of interest in regional concept. Resource regions are mapped, and their raw material qualities and quantities are discussed. The availability of resources, their role in the development industries and tertiary sector have great planning importance. We can delineate world into oil, natural gas, coal and iron ore regions.

(g) *Urban regions (megalopolis):* Urban centres may be formal or functional regions. Cities and urban centres are the areas of production, exchange, administration, distribution and consumption. They have hierarchical structure. Internally, they show complex patterns of land use and functions.

The north-eastern seaboard of the United States is the leading megalopolis in the world. It is a continuous stretch of urban and suburban areas from New Hampshire to Northern Virginia, and from the Atlantic coast to the foothills of Applachian mountain. The people of this area have a unique way of life. Here, there has been developed a kind of supremacy in politics, in economics, and in cultural activities. This is an urban system (region) of deeply interwoven urban and suburban areas with a total population of over 80 million.

(h) *Agricultural regions:* These are: wheat, rice and jute regions of India; and cotton belt, corn belt and wheat belt of USA.

(i) *Industrial regions:* Industrial regions are demarcated by taking into consideration the industrial parameters. Hoogly basin, Bombay-Ahmedabad region, Chennai-Coimbatore region, Delhi industrial region, Donetz Basin (Ukraine), Central Industrial Region (Moscow), Rhine region, Saar Basin (Germany), Tokyo region, Kobe and Yakohama region (Japan), and Birmingham and London industrial regions of UK are some of the examples of industrial regions.

(j) *Mental regions (mental map):* Mental region is a distorted ego-centric image of a place. The images which constitute mental maps (the political, social, cultural, and economic values) held by men blend into an overall image about the space around him. This image or mental map differs from person to person. For example, primitive societies, particularly, have distinctive views about place and earth. The mental map of Nagas differs from that of Khasis. The Nagas in remote villages still measure the linear distances in their conventional way of smoking pipe, as they are not familiar with the mile or kilometre scale. The Khasis demarcate their territory on the basis of matrilineal society. The perception of space differs from rural to urban areas and from developed to developing countries.

To summarize, region is a mental construct, whose sole purpose is the purposeful organization of spatial data. The regions help in understanding the man-nature interaction and provide a good tool for the planning and socio-economic development of different segments of a country/world.

Regionalism

Regionalism, also known as sub-nationalism, is a concept of political geography. It is a movement which seeks to politicize the territorial predicaments of its regions with the aim of protecting or furthering its regional interest. Generally, regionalism is based on ethnicity, caste, creed, language, colour or culture. Regionalism also aims at achieving autonomy and local power (political and economic both). It is a political rhetoric and self-assertiveness based on a deep-seated mistrust of a more centralized government. All sorts of regionalism have in common a counter culture. Despite industrialization, modernization and globalization, it is an international phenomenon and may be observed in almost all the developed and developing countries of the world.

Regionalism may help in overcoming and solving some of the socio-political and economic problems of its supporters, but it is a big barrier in the process of national integration. In India also, there seems to be an emerging trend of regionalism. The people of Assam, Manipur, Mizoram, Nagaland, Maharashtra, Punjab, Kashmir, Tamil Nadu, etc. occasionally give the impression of a strong feeling of regionalism. This type of sub-nationalism has posed some serious problems to India, like that of *Punjabiat* and *Kashmiriat*. The Ahoms (Assamese) and Maharashtrians also occasionally give a strong impression of regionalism or sub-nationalism. Whatever the objective of the regionalists may be, it creates many problems for the government. In recent years, we have seen the disintegration of the mighty Soviet Union and the creation of several independent republics in the erstwhile Soviet Union. For geographers, regionalism may be a good area for research, especially to identify the basis of strong regional biasness in a particular area and to predict the future pattern of behaviour of the people of that region.

Notes

1. Holt-Jensen, A., 1981, *Geography: Its History and Concepts*, London.
2. Minshull, R.,1970, *Changing Nature of Geography*, London.
3. Berry, B.J.L., 1961, 'A Method of Deriving, Multifactor Uniform Region', *Przeglad Geograficzny*, Eng. Trans. 20, pp. 45–52.
4. Ibid.
5. Chorley and Haggett, 1969, *Frontiers in Geographical Teachings*, London.
6. Ibid.
7. Harvey, D., 1973, *Social Justice and the City*, London.
8. Stamp, L.D., 1966, 'Ten Years' in *Trans. Br. Geogr.*, Vol. 40, pp. 11–20.

9. Minshull, R., op. cit., p. 65.
10. Houston, J.M., 1963, *A Social Geography of Europe*, p. 18.
11. Einstein, A. and Infeld, L., *The Evolution of Physics*, Cambridge.
12. Minshull, R., op. cit., p. 120.
13. Ibid.
14. Kuhn, T.S., 1962, 1970, *The Structure of Scientific Revolutions*, Chicago.
15. Haggett, P.,1979, *Geography a Modern Synthesis*, New York, p. 21.
16. Holt-Jensen, A., op. cit., p. 38.
17. Ibid., p. 40.
18. Ibid., p. 41.
19. Taylor, P.J., 1976, 'An Interpretation of the Quantification: Debate in British Geography', *Transaction Institute of British Geographers*, New Sr. 1, pp. 129–142.
20. Braithwaite, R.B., 1953, *Scientific Explanation*, London, p. 12.
21. Harvey, D., 1969, *Explanations in Geography*, London, p. 31.
22. Chabot, G., 1950, *Les Conceptions Francaisis de la Science Geographique, Nersky Geogr, Tidsskr*, Vol. 12, pp. 309–321.
23. Holt-Jensen, A., op. cit., p. 53.
24. Schaefer, F., 1953, Exceptionalism in Geography, *Ann. Assoc. Amer. Geogr.*, Vol. 43, pp. 226–249.
25. James, P.E. and Martin, G.J., 1972, *All Possible World*, p. 121.
26. Ackerman, E.A., 1958, *Geography a Fundamental Research Discipline*, University of Chicago, Department of Geography, Research Paper No.53, p. 8.
27. Ballabon, M.G., 1957, 'Putting the 'Economic' in Economic Geography', *Economic Geography*, Vol. 33, pp. 217–223.
28. Burton, I., 1963, 'The Quantitative Revolution and Theoretical Geography', *Canadian Geography*, Vol. 7, pp. 151–162.
29. Harvey, D., op. cit., p. 78.
30. Haggett, P., op. cit., pp. 297–303.
31. Harvey, D., op. cit., p. 80.
32. Darby, H.C., 1953, 'On the Relations of Geography and History', *Transaction*, Institute of British Geographer, Vol. 19, pp. 1–11.
33. Holt-Jensen, A., op. cit., p. 107.
34. Harvey, D., op. cit., p. 81.
35. Stoddart, D.R., 1965, 'Organism and Ecosystem as Geographical Models', in Chorley, R.J., and Haggett, *Frontiers in Geographical Teaching*, London.
36. Harvey, D., op. cit., p. 81.
37. Holt-Jensen, op. cit., p. 127.
38. Harvey, D., op. cit., p. 81.
39. Holt-Jensen, A., op. cit., p. 126.

40. Von Bertalanffy, L., 1968, *General System Theory Foundation, Development, Applications*, New York.
41. Harvey, D., op. cit., p. 457.
42. Ibid.
43. Ibid.
44. Holt-Jensen, A., op. cit., p. 130.
45. Klish, J. and Valach, M., 1967, *Cybernetic Modelling*, London, p. 35.
46. Blalock, H.M. and Blalock, A.,1959, 'Towards Clarification of System Analysis in the Social Sciences', *Philosophy-Sciences.,* Vol. 26, pp. 84–92.
47. Kuhn, op. cit., p. 50.
48. Wiener, N.,1961, *Cybernetics*, Cambridge.
49. Klir and Valach, 1967, op. cit., pp. 31–32.
50. Harvey, D., op. cit., p. 468.
51. Ibid.
52. Haggett, P., 1956, *Locational Analysis in Human Geography*, Arnold.
53. Carlstein, T., 1978, *Swedish Geography at Lund*, 58, pp. 130–135.
54. Meadows, D.H., Meadows, D.L., 1972, *Limits of Growth*, London.
55. Norgdard, A., 1972, *Kerologisko Metoder*, Oslo.
56. Anuchin, V.A., 1973, 'Theory of Geography', in *Directions in Geography*, ed. R.J. Chorley, London.
57. Haggett, op. cit., p. 22.
58. Bird, J., 1979, 'Methodology and Philosophy', Progress Paper, in *Progress in Human Geography*, 3, pp. 117–120.

11

Models in Geography

In the post-Second World War period, the definition of geography, geographic thought, and geographic methodology have undergone great transformation. In order to put the subject on a sound footing and to command respect in sister disciplines, geographers have increasingly concentrated in the last few decades on the theme of geographic generalization, formulation of models, theories and general laws. This geographic generalization is also known as 'model-building'.

The term 'model' has been defined differently by different geographers. In the opinion of Skilling (1964), a model is 'either a theory, a law, a hypothesis, or a structured idea. Most important, from the geographical point of view, it can also include reasoning about the real world (physical and cultural landscape) by means of relation in space or time. It can be a role, a relation or an equation'.

In the opinion of Ackoff, 'a model may be regarded as the formal presentation of a theory or law using the tools of logic, set theory and mathematics'. According to Haines-Young and Petch, 'any device or mechanism which generates a prediction is a model'. Accordingly, modelling, like experimentation and observation, is simply an activity which enables theories to be tested and examined critically.

Most of the geographers of the post-Second World War period have widely conceived models as idealized or simplified representation of reality (geographic landscape and man-nature relationship).

Significance of Model

Geography is a discipline which deals with the interpretation of man-nature (resource) relationship. The earth – the real document of geographical studies – is however, quite complex and cannot be comprehended easily. The earth's surface has great physical and cultural diversity. In geography, we examine location, landforms, climate, soils, natural vegetation and minerals' spatial distribution and their utilization by mankind which lead to the development of cultural landscape. Moreover, geography is a dynamic subject as the geographical phenomena change in space and time. The subject matter of geography, i.e. the complex relationship of man and environment can be examined and studied scientifically by means of hypotheses, models and theories. The basic aim of all models is to simplify a complex situation and thus render it more amenable to investigations. In fact, models are tools which allow theories to be tested. A more restricted view of models is that they are predictive devices.

Need of Modelling in Geography

Geographers are interested in making laws and theories in their discipline like those in physical, biological and social sciences. Model is a device for understanding the vast interacting system comprising all humanity and its natural environment on the surface of the earth. This is of course not attainable except in a highly generalized manner. Modelling in geography is, therefore, done due to the following reasons:

1. A model-based approach is often the only possible means for arriving at any kind of quantification or formal measurement of unobserved or unobservable phenomena. Models help in estimations, forecasts, simulations, interpolation and generation of data. The future growth and density of population, use of land, intensity of cropping, migration pattern of population, industrialization, urbanization and growth of slums may be predicted with the help of such models. These are very useful in the forecast of weather, change of climate, change in sea level, environmental pollution, soil erosion, forests depletion and evolution of landforms.
2. A model helps in describing, analysing and simplifying a geographical system. Locational theories of industries, zoning of agricultural land use, patterns of migration and stages of development of landforms can be easily understood and predicted with the help of models.
3. Geographical data are enormous and with every passing day these data are becoming more and more difficult to understand. Modelling is undertaken for structuring, exploring, organizing and analysing the obtained enormous data through discriminating pattern and correlation.
4. Alternative models can be used as 'laboratories' for surrogate observation of systems of interest which cannot be observed directly, and for experimenting and estimating the effects and consequences of possible changes in particular components as also for generating future scenario of evolution and end states of system of interest.
5. Models help in improving the understanding of causal mechanism, relationships between micro and macro properties of a system and the environment.
6. Models provide framework within which theoretical statements can be formally represented and their empirical validity then put under scrutiny.
7. Modelling provides linguistic economy to geographers and social scientists who understand their language.
8. Models help in the building of theories, general and special laws.

Features of a Model

The main features of a model are as under:

1. The geographical reality of the earth's surface and man-environment relationship are quite complex. Models are the selective pictures of the world or part of it. In other words, a model does not include all the physical and cultural attributes of a macro or micro region. In fact, model is a highly selective attitude to information.
2. Models give more prominence to some features and obscure and distort some others.
3. Models contain suggestions for generalization. As stated above, predictions can be made about the real world with the help of models.
4. Models are analogies as they are different from the real world. In other words, models are different from reality.
5. Models tempt us to formulate hypothesis and help us in generalizing and theory-building.
6. Models show some features of the real world in a more familiar, simplified, observable, accessible, easily formulated or controllable form, from which conclusions can be drawn.
7. Models provide a framework wherein information may be defined, collected and arranged.
8. Models help in squeezing out the maximum amount of information from the available data.
9. Models help to explain how a particular phenomenon comes into existence.
10. Models also help us to compare some phenomena with the more familiar ones.
11. Models cause a group of phenomena to be visualized and comprehended which otherwise could not be comprehended because of its magnitude or complexity.
12. Models form stepping stones to the building of theories and laws.

Types of Models

As described earlier, the term 'model' has been used in a great variety of contexts. Owing to the great variety, it is difficult to define even the broad types of models without ambiguity. One division is between the descriptive and the normative. The *descriptive* model is concerned with some stylistic description of reality whereas the *normative* model deals with what might be expected to occur under certain stated or assumed conditions. Descriptive models may be concerned with the organization of empirical information, and termed as data, classificatory (taxonomic), or experimental design models. Contrary to this, normative models involve the use of a more familiar situation as a model for a less familiar one, either in a time (historical) or a spatial (geographical) sense and have a strongly predictive connotation.

On the basis of stuff (data) from which they are made, models may also be classified into *analogue*, *physical* or *experimental* models. The physical or experimental model may be iconic (idol shaped) in which the relevant properties of the real world are presented with the same properties with only a change in scale. For example, maps, globes and geological models are physical or experimental models. Models may be an analogue (simulation) having real world properties represented by different properties. Analogue or simulation models are concerned with symbolic assertion of a verbal or mathematical kind in logical terms.

General Classification of Models

As stated at the outset, complexity of geographical landscapes and geographical situations is such that models are of particular importance in studying geography. A large number of models have been designed, adopted and applied by geographers. A more simple classification of models illustrated with examples has been given as follows:

Scale Models

Scale models, also called hardware models, are perhaps the easiest type to appreciate as they are direct reproductions, usually on a smaller scale of reality. Scale models may be either static, like the model of a land surface of a geological model, or dynamic, like a wave tank or river flume. Dynamic models are perhaps more interesting and useful in geographical work. The great advantage that a dynamic model has over reality is that the operative processes can be controlled. This allows each variable to be studied separately. In a wave tank, the effect of material size, wave length and wave steepness on a beach slope can be measured quite accurately if two variables are held constant while the third is varied. If the resultant beach slope angle is plotted against each variable in turn the points obtained in each case may either fall in a nearly straight line indicating a significant relationship, or in a diffused scatter suggesting little or no relationship. Close relationships revealed by the model may not be apparent on a natural beach where the wave variables cannot be controlled.

There are, however, difficulties in applying the results of model studies of this type to a natural situation. One of these is the problem of scale. If wave size and material size are scaled up in the same proportion, then the sand of the model would become large *cobbles* in nature – and these two materials do not react similarly to waves. Again if sand in nature is scaled down to model size, it would be silt or clay which also responds differently from sand under wave action. Despite such difficulties scale models have yielded very useful results in many fields of enquiry. The fact that engineers make a scale model before embarking upon any major project such as river improvement, dam construction, canal excavation, landslides, tidal surges, flood forecast, or harbour works scheme, demonstrates the value of this type of model.

Scale models are often used by physical geographers and especially by geomorphologists. In fact, geomorphologists have carried out fundamental research with scale models in order to investigate processes that are difficult to observe under natural conditions, such as river action, glacial movement, wind erosion, marine processes and erosion by underground water.

Maps

Maps are the models that are most familiar to geographers. They are a special type of scale model which become increasingly abstract as the scale becomes smaller. At one end of the spectrum is the stereo-pair, vertical air-photograph which provides virtually a true scale model of the real world. It is, however, static and represents only the area shown at one instance of time. A simple vertical air photograph loses the impression of height but still shows all the visible elements of the landscape virtually true to scale. A large-scale map loses much of the detail of the landscape although it can show buildings, roads and other features of this size accurately. As the scale is reduced the information becomes more symbolic and can no longer be shown true to scale; even more detail must be omitted. The map can, however, give an indication of the relief by means of contours, hill shading and hachures; this is missing from the simple vertical air photograph. Another advantage which maps also have over reality is that they show a very large area simultaneously, so that mutual space relationships can be much more easily appreciated and compared than on the ground. Many maps use symbols to show specific features or distributions such as population density; these are even more abstract and further removed from reality that they represent. A new insight into a familiar area can be given by drawing a diagrammatic map where the scale is not correct for an area, but is adjusted to show population or some other variable to scale. Modifications in area, distance and direction are also needed in maps covering the world or large parts of it. A curved surface cannot be correctly reproduced on a plane or flat piece of paper. In fact, it is impossible to show a three dimensional earth on a two dimensional plane or sheet of paper. The earth may be truly represented on a globe, but globes have very little utility in geographical studies.

Simulation and Stochastic Models

Simulation means imitating the behaviour of some situation or process by means of a suitably analogous situation or apparatus, especially for the purpose of study or personal training. Stochastic means: randomly determined or that which follows some random probability distribution or pattern, so that its behaviour may be analysed statistically but not predicted precisely.

Simulation and stochastic models have been developed to deal with dynamic situations rather than with a static state shown on a map. This type of model simulates particular processes by means of random choices, hence the term 'stochastic', one which is connected with chance, occurrences. It can be illustrated by its application to drainage development. Starting with a pattern of grid squares it is assumed that a stream source exists at the centre of certain randomly chosen squares. Random numbers are again used to determine in which of the four possible directions, each stream will flow and a line is drawn to represent its course as far as the centre of the adjacent square. By repeating the process (with certain reservations that approximate to reality) there emerges a complete drainage network which shows many similarities to natural drainage patterns. Thus a conclusion can be reached that the natural drainage pattern has some element of chance about its make up.

Simulation models can also be of use as a means of analysing a large number of variables, which is a recurring problem in geography. For instance, the development of coastal spit can be shown to depend on a number of distinct processes or wave types. These different processes can be built into a model in such a way that each of them is allocated a specific range of random numbers. Each random number that comes up results in the operation of the appropriate process. In this way, the spit can be built up by the action of different processes in a random order, but in specific proportions. If the simulated spit resembles the real one, then one can conclude that the processes probably operate in the proportion specific in the model. Once a realistic model has been found it can then be used to predict future development of the spit provided the processes continue to operate in similar proportions.

Stochastic simulation models have also been successfully used in the field of human geography to study the spatial diffusion of a variety of phenomena, including the spread of population diseases such as malaria, smallpox, dengue fever, chickenpox, and AIDS or innovations such as the use of a particular piece of machinery, tractors, chemical fertilizers, pesticides and weedicides. The simulation is made realistic by imposing barriers that can be crossed with a varying degree of difficulty. Random numbers are used to determine the direction of spread and the effect of the barriers can then be assessed. The term 'Monte Carlo' is used to describe some stochastic models, in which chance alone determines the outcome of each move within the conditions of the model. The Monte Carlo model may be compared with the Markov Chain model in which each move is partially determined by the previous move. The Markov Chain is exemplified in the random walk drainage development model described above. Both types have been applied in many fields of geographical research.

Mathematical Models

Mathematical models are considered to be more reliable but difficult to construct. They obscure many of the human values, normative questions and attitudes. Yet, they have symbolic assertions of a verbal or mathematical kind in logical terms. For example, suppose I offer the following arguments:

1. A is larger than B, and
2. B is larger than C.

   Now by virtue of (1) and (2) together, I offer the following theorem or conclusion:
3. Therefore, A is larger than C.

The logical validity of this conclusion will not change with the change in time. Logically, it had to be true in 3000 BC, 2000 BC, 1000 AD, and it will be true in 2025 AD, 3000 AD, 4000 AD. Thus, the validity of the conclusion does not depend on specific historical period. It is ahistorical.

In the same way, the logical validity of a theory is also spatial. If a theorem is logically valid, it must be locally valid in the United States, Germany, Russia, France as well as in India, Pakistan, China and Japan.

Mathematical models can be further classified according to the degree of probability associated with their prediction into deterministic and stochastic.

Mathematical models represent the equation of specific processes by means of mathematical equations which relate the operative process to the resultant situation. It is necessary, however, to have a sound knowledge of the physical processes concerned, and consequently, this type of model-building has been mainly the work of physicists. For example, a dynamic mathematical model of glacier flow has been constructed by J.F. Nye. He simplifies the basic assumptions as far as possible to make the equations sufficiently simple to solve. Thus, the glacier bed is assumed to have a rectangular cross profiled (U-shaped valley) of uniform size and specific roughness. The ice is assumed to be perfectly plastic in its response to stresses. Then, given certain stresses, the response of the ice can be calculated by means of differential equations. These can predict specific flow patterns and ice profiles for given values of the assumed conditions. The geomorphologist can play his part by measuring flow patterns and glacier dimensions in the field. The closeness with which these approximate to the calculated values is a measure of the success of the mathematical model. If the observed flow pattern agrees closely with the predicted one, then the model can be used with some confidence to provide values for flow in parts of the glacier that cannot readily be measured in the field, but which are very important in studying the effect of glaciers on the landscape. The speed of basal flow is important in this context. Mathematical models have also advanced our knowledge of how rivers move their load and adjust their beds, and how waves operate on the coast. These models are usually in the form of differential equations largely based on known physical relationship, and it is essential to test their numerical results against observations made under natural conditions or in a scale hardware model. The models are only as successful as the assumptions and simplifications on which they are based. They provide a very simplified situation, but one that can be expressed in precise numerical terms and hence is capable of suitable mathematical manipulation. For this reason such models are more suited to problems in physical geography.

There has, however, been somewhat different development of mathematical model in human geography. These are more in the nature of empirical relationships that can be expressed in mathematical terms. An example is the rank size relationship. This relationship shows that within any class of occurrences there are usually a few large items and many small ones with a fairly regular distribution between them. It has been applied to towns in many parts of the world. There are a few large towns but many more small ones, and between the two, a moderate number of medium ones; the relationship is approximately linear on a double logarithmic scale.

Mathematical models have also been developed in economic geography, which is more susceptible to quantitative formulation than other branches of human geography. Such models are often not dynamic in the same way as are the differential equations in physical geography, although some deal with flow of goods, etc. from one region to another.

Another mathematical model is linear programming, which is relevant to many situations in economic geography. It is a method of finding the optimum solution to a problem in which several conditions must be fulfilled. A factory will have certain requirements of labour, capital, raw materials, transport and access to markets and each of it determines conditions that can be expressed as mathematical equations and represented graphically on straight lines. When all the equations have been plotted they reveal the point of optimum value in terms of location. The procedure provides a definite solution based on the values assigned to the equations. If the values are accurate, then the optimum solution will be obtained.

Analogue Models

Analogue models differ from those type of models which have already been described. In the analogue models, instead of using limitations of the original or symbols to represent it, the feature being studied is compared with some completely different feature by means of an analogy. An analogue model uses a better known situation or process to study a less well-known one. Its value depends on the researcher's ability to recognize the element common to two situations. These elements constitute the positive analogy; the dissimilar or negative analogy and the irrelevant or neutral analogy are ignored.

Reasoning from analogy has long been a part of geographical study. James Hutton (1726–1797), in his major work published in 1795, recognized the similarity between the circulation of blood in the body and the circulation of matter in the growth and decay of landscapes. A similar circulation can also be seen in the hydrological cycle. The Davis' concept of 'normal cycle of erosion' and Ratzel's concept of 'state as a living organism' are important examples in which landforms and state have been compared to a living organism. Both these concepts are thus analogies.

The analogy used to further geographical knowledge must be better understood than the feature being investigated. The behaviour of metals under stress has been intensively studied, and this has allowed useful analogies to be drawn between metals and ice. Methods of dealing with one problem can often be transferred by analogy to a completely different situation. The study of kinetic waves has been applied to the movement of vehicles on crowded roads, to the movement of stones and flood waves in rivers, and to the formation of surges at a glacier snout. These very dissimilar problems have a common fact that they are one-dimensional flow phenomena and from this point of view they can be treated with the same technique.

Analogies have also proved fruitful in the study of problems in human geography; for example, those that draw on certain well-established relationships in physics. The gravity model is a good example of this type. It is based on the physical observation that the attractive force between two bodies is proportional to the product of their masses divided by the square of the distance between them. The value for the distance in the model is often squared to approximate more closely to the force of gravity as observed in physics. The attractive force may be considered in terms of transactions between two places. The number of transactions is likely to increase as the size of the places, often measured in terms of population number, increases and as the distance between them decreases. This model presupposes that there is no other force involved to limit the transaction, such as an international or language barrier. Various other physical relationships used as analogue models include the patterns of a magnetic field and the second law of thermodynamics.

Theoretical Models

Theoretical models can be divided into two categories. The conceptual models provide a theoretical view of a particular problem allowing deductions from the theory to be matched against the real situation. This can be exemplified by the theoretical consideration of the effect of a rising and falling sea level upon the coastal zone if certain specific conditions are fulfilled. It is assumed that wave erosion is the only process operating, that waves can only erode rock to a certain depth of the order of about 13 metres (40 feet) and that the waves erode a wave-cut platform to a certain gradient below which they cannot operate effectively. It is also assumed that the initial coastal slope is steeper than this gradient. A consideration of the prolonged action of waves eroding under these conditions, with a rising and falling sea level, leads to the conclusion that only with a slowly rising sea level, can a wave-cut platform of great width be produced. The theoretical forms of the coastal zone under the various conditions specified can be established and then compared with actual coastal zones. Much more elaborate theoretical models of this conceptual type have been developed in the study of the evolution of slope profiles. These are based on the known or assumed effect of different slope processes.

A long series of stages of modification can be derived from this type of theoretical model, and these can again be matched with actual slopes.

The second type of theoretical model is associated with the word 'theory', when this is used to denote the overall framework of a whole discipline. The framework must not be too rigid or it will cramp the growing edges of the subject, where the most exciting work is going on. The ideal is a flexible framework that can contain a wide variety of geographical endeavour and yet give it coherence and purpose. Models are particularly valuable in this context as they are often common to all branches of the subject and so help to give it unity.

An analogy may help to illustrate the way in which the vast and growing amount of geographical data may be organized within a theoretical framework. Geography may be compared with a five-storey building, each storey being supported by the one below and supporting the one above (Figure 11.1):

Figure 11.1 Theoretical Framework of Geographical Model

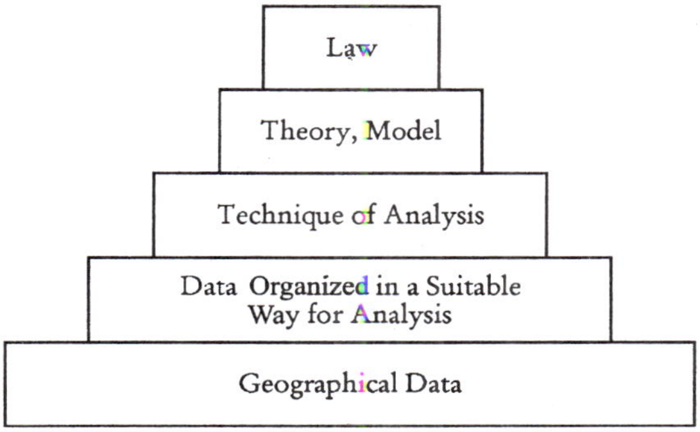

1. The lowest storey is the one which accommodates the data, the raw material of geographical study.
2. The data lead up to the level of model where they are organized in a suitable way for analysis.
3. The techniques of analysis, lying on the next storey, depend on the model adopted for the study.
4. Analysis leads up to the next floor, concerned with the development of theories.
5. The theories in turn lead up to formulation of tendencies and laws. These are located at the top as they are the ultimate aim of geographical methodology.

Critical Views

For understanding and explaining complex geographical phenomena, models are of great importance. Modelling has, however, been criticized on many counts. Critical views on modelling vary from those who accept modelling but criticize the way in which modelling is done to those who reject modelling as a worthwhile activity in geography.

Those who agree with modelling in geography but do not agree with the way models are being prepared and hold the view that most of the models are prepared badly. The basic aim of the modeller is to represent complexity by something simpler. In the exercise of modelling, the modeller may simplify the complexities of geographical realities too much or too little. Oversimplification may mislead students and generate misunderstanding which may ultimately lead to bad prediction. Undersimplification is of little use in teaching as it does not explain the reality and gives insufficient basis for prediction.

The second objection to modelling is that the modellers may concentrate on the wrong things. Sometimes models may neglect to fulfil the basic criterion of simplifying. They go for the principal component analysis, stepwise regression and Q-analysis. These techniques often produce models more complicated than the original data. Moreover, models may incorporate some of the salient points and omit others.

There are scholars who do not question the appropriateness of modelling as a generally applicable strategy in geography. There is a group of geographers who consider modelling as a worthwhile activity but hold the view that geographers should not be forced to apply modelling techniques to everything. According to them, modelling is not appropriate in some branches of geography, especially in human geography, regional geography, cultural geography and historical geography. In various branches of regional, cultural and historical geography, modelling strategies have distorted the subject by putting overemphasis on some topics and underemphasis on others. By this strategy, generalizations have been made on the basis of few cases and many a time at the expense of specific cases.

Those who outrightly reject modelling in geography say that geography is not a pure physical science, it has a very strong component of human beings and models may not properly adjust and interpret the normative questions like beliefs, values, emotions, attitudes, desires, aspirations, hopes, and fears, and therefore, models cannot be regarded as dependable tools for explaining correctly the geographical reality.

Criticism of modelling may also be based on objections to the generalization that modelling usually involves. It may be considered futile to construct general models to apply to geographical events, especially where idiosyncratic (regional) human actions and free will are concerned. Or, it may be that the geographer's purpose is to predict or understand specific events and situations, his or her interests may be in the unique (specific, regional) case for which a general model is thought irrelevant.

Many of the models in geography have also been criticized on the grounds of application of sophisticated mathematical and statistical tools and techniques. Despite the quantitative revolution, few geographers feel comfortable with mathematical symbolism and ideas, and are thus largely unconscious of the generality, clarity and elegance that mathematical modellers appreciate in a good model. Geographers apart, even students, policy makers, clients and the public at large, may find mathematical models difficult to understand.

Another criticism is that no model is adequate by itself; any model must be continually subject to reassessment, modification and replacement. In Feyerabend's words (1975):

> Knowledge ... is an ever-increasing ocean of mutually incompatible (and perhaps incommensurable) alternatives, each single theory, each fairy tale, each myth that is part of the collection forcing the other into greater articulation and all of them contributing, via this process of competition to the development of consciousness. Nothing is ever settled, no view can ever be omitted from a comprehensive account.

In fact, accountable growth of knowledge is not a well-regulated activity where each generation automatically builds upon the results achieved by earlier workers. It is a process of varying tension in which tranquil periods characterized by steady accretion of knowledge are separated by crises which can lead to an upheaval within subjects, disciplines and break in continuity.

Model-building also demands considerable reliable data. Such reliable data are rarely attainable in the developing and underdeveloped countries. As a matter of fact, any set of data collected in the developing countries have many pitfalls and shortcomings. Any model, theory, or law developed on the basis of weak and unreliable data is bound to give only a distorted and faulty picture of the geographical reality. It has also been found that generalizations done with the help of models and structured ideas are bringing exaggerated results leading to wrong predictions.

Most of the models have been developed in the advanced countries of Europe and America, and theories and models were constructed in these countries on the basis of data collected there. There is certainly a danger that the models developed in Europe and America may be elevated to general truth, and given the status of universal models. In reality we do not have universal human, cultural, industrial, agricultural and urban geography. There are different socio-cultural and agro-industrial processes, working in different parts of the world, which result into different cultural landscapes. Owing to these constraints, generalizations made on the basis of models may be misleading and faulty. Moreover, the data used by the western experts are related to a period of about one hundred years. If these models, developed on the basis of data of developed countries, are applied in the developing countries, the results and predictions may be disastrous.

References

Ackoff, R.L., Gupta, S.K. and Minas, J.S. 1962, *Scientific Method: Optimizing Research Decisions*, New York.

Apostal, L., 1961, 'Towards the Formal Study of Models in the Non-formal Sciences' in Freudenthal, H. (ed.), *The Concept and the Role of the Model in Mathematics, Natural and Social Sciences*, Dordtrecht, Holland, pp. 1–37.

Berry, B.J.L., 1964, 'Approach to Regional Analysis: A Synthesis', *Annals of the Association of American Geography*, 54, pp. 2–11.

Braithwaite, R.B., 1953, *Scientific Explanation*, Cambridge.

Bunge, W., 1964, 'Geographical Dialectics', *Professional Geographer*, 16 (4), pp. 28–29.

Caws, P., 1965, *The Philosophy of Science*, Princeton.

Chorley, R.J. and Haggett, P., 1969, *Integrated Models in Geography*, Metheun, London.

Cole, J.P. and King, C.A.M., 1968, *Quantitative Geography*, John Wiley.

Hartshorne, R., 1959, *Perspectives in the Nature of Geography*, London.

Wrighley, E.A., 1965, 'Changes in the Philosophy of Geography', in Chorley, R.J. and Haggett, P., (eds), *Frontiers in Geographical Teaching*, London, pp. 3–20.

Part III

Contemporary Movements in Geography

12

Modern Themes in Geographical Thought

In any discipline, philosophies are essential because they create a set of subjectively imposed constraint under which an individual does research. Closely associated with philosophy is methodology. A methodology is the logic of an explanation, ensuring that our arguments are rigorous, that our inferences are reasonable, that our method is internally coherent.[1] Thus, the methodologist is concerned with the logical nature of an explanation, and the philosopher is concerned with thinking of 'beliefs' with respect to the nature of geography.[2] The outcome of the philosophy and methodology is a theory or paradigm. Today, geographers are busy in formulating theories and paradigms with scientific and humanistic approaches. The present chapter deals with these matters.

Geographical thought at any point of time is a manifestation of the interaction between the prevailing philosophical viewpoints and the major methodological approaches. Because of the extreme diversity of viewpoints with regard to both philosophy and methodology, there has been a constant extension, and even a shift, in the focus of the discipline.[3] The 1970s may be termed as a turning point in the writing of modern geography. It was during this period that a number of important works advocating different philosophical departures were published. Positivism, idealism, pragmatism, and functionalism were redefined; while idealism, existentialism, Marxism, radicalism, humanism, and behaviouralism were inducted in geographical researches as humanistic approaches.

All individual philosophies are extremely complex. A brief chapter, such as the present one, cannot possibly deal successfully with all aspects of these philosophies and methodologies which differ in nature from purely objective to idealistic. However, an attempt has been made in this chapter to explain the fundamental ideas of geographical themes and movements prevalent in geography at present.

POSITIVISM

Positivism is a philosophical movement, characterized by an emphasis on science and scientific method as the only source of knowledge (a sharp distinction between the realm of fact (data) and value (cultural), and a strong hostility towards religion and traditional philosophy, especially metaphysics. August Comte (1798–1857) declared metaphysics as useless branch of enquiry. He demanded a 'sociocracy' ruled by scientists, for the unity, conformity and progress of all humanity.

Positivism is also called empiricism. It is a philosophical viewpoint that limits knowledge to facts that can be observed and to the relationships between these facts. The proponents of positivism advocate that science can only concern itself with empirical questions. Empirical questions are questions about how things are in reality. In this context, reality is defined as the world which can be sensed. In empirical inquiry, it is assumed that facts 'speak for themselves'.

It means science is concerned with objects in the world. The subject or subjects for which there is a world or worlds are excluded from the field of interest. Thus, what is not derived from the evidences of the senses is not knowledge. Reliable knowledge can only come from basic observations of actual conditions. To be scientific is to be objective, truthful and neutral. The positivists also gave emphasis on the unity of science. Scientific status is guaranteed by a common experience of reality, a common scientific language and method ensures that observations can be repeated. Since science has a unified method, there can only be one comprehensive science.[4] In other words, the entire system of science grows under the principles of physics, chemistry, biology, psychology and social sciences which can be linked together logically.

Thus, positivism is a philosophy which is anti-idealism (a view that reality is mental or mind dependent). Positivists further stress that since we cannot investigate and test moral norms (e.g. values, beliefs, attitudes, prejudices, customs, traditions, taste, aesthetic values, etc.) we should keep away from normative questions. In other words, our tastes, traditions, likings, attitudes and aesthetic satisfactions cannot be justified scientifically. The essence of the positivist philosophy is that ideally speaking science is value-free, neutral, impartial, and objective. The followers of positivism regarded metaphysical (which lies outside our sense perceptions or is independent of them) questions also as unscientific. In the most general terms, positivism determined the scientific status of its statements through:

1. *Empiricism*: *Empeire* is the Greek word for experience. Their grounding is in a direct, immediate and empirically accessible experience of the world which gives observation statements a peculiar privilege over theoretical ones, and which guaranteed their generality through. The modern view of scientific method is that both experience and reason play an important role in science. Reason or imagination provides speculative hypothesis; experience helps to weed out those which are false.

2. *Unified Scientific Method*: A unitary scientific method, accepted and routinely drawn upon by the entire scientific community.
3. *Formulation of Scientific Laws*: The formal construction of theories capable of empirical verification; their successful proof would serve to identify universal laws.
4. *Exclusion of Normative Questions*: A strictly technical function, in that they revealed the effectivity or even the necessity (but not the desirability) of specific conjunctions of events; thus value-judgements and ethical utterances (beliefs, values, customs, attitudes, prejudices, aesthetic values, etc.) were ruled out of the scientific court because they could not be empirically tested, and the statements which remained could be brought together.
5. *Unification of Scientific Laws*: The progressive unification of scientific laws into a single and incontrovertible system.

The cumulative effect of these five claims was to move from the immediate through the unitary to the universal: to close the system around a particular version of the present and to refuse admission to alternative ways of being in and acting on the world.

Historically, the concept of positivism emerged after the French Revolution and was established by Auguste Comte during the 1830s in France. The revolution created disorder in the French society. Positivism began as a polemical weapon against the 'negative philosophy' prevalent before the revolution. The negative philosophy was a romantic and speculative tradition which was concerned more with emotional rather than with practical questions, and which sought to change society by considering utopian alternatives to existing questions. The positivists regarded such speculation as 'negative' since it was neither constructive nor practical; it also showed that philosophy was an 'immature' science. Philosophers, like other scientists, should not concern themselves with such speculative methods but should study what they could to grasp material objects and given circumstances. This approach was to be recommended as positive approach.[5] The positivist movement broke a range of taboos and religious beliefs against empirical investigations.

As stated above, the followers of positivism believed that alongside the natural sciences there should also be a science of social relationships (sociology) to be developed on the same principles. As natural sciences discovered the laws of nature, so scientific investigations of communities would discover the laws of society. Comte averred that social development took place in three stages: (i) theological, when man explains everything as God's will; (ii) metaphysical; and (iii) positive, when causal connections are discovered between empirically observed phenomena.

One of the main characteristics of positivists is that they are anti-authoritarian. Positivism suggested that we could not accept authority simply because it was authority, but only give credence to things for which there was scientific evidence. This empirical research led positivists into confrontation with dictatorial regimes.[6] In 1930, a group of scientists, known as 'logical positivists', was founded in Vienna – also known as 'Vienna Circle'. They were against everything which could not be verified empirically and investigated by controlled methods. They saw Nazism as a mixture of irrational prejudices and ideological dogma.

Positivist work in human geography has been criticized by realists, and Marxists, because it seeks 'laws' of the superstructure which are unrelated to the processes in the infrastructure, and which in any case cannot exist because of the change that is inherent in the infrastructure.

The assertion of the positivists that value-free, objective research is possible, has been vehemently criticized by the proponents of humanistic approaches, especially by the idealist structuralists, existentialists, Marxists, behaviouralists, and phenomonologists. The positivistic laws, mathematization and value-free analysis are difficult to achieve. According to the positivists, there are technical solutions of all problems and value-free research is possible. In practice, it has been observed that subjective elements enter at many stages of the research process, especially at the stage when researchers choose their research topic from the many available ones. We can, for example, guess that a researcher, starting from his own well-established and strong opinion as to what the distribution of world's food supply should be, will choose to investigate the empirical question as to how the food supply is actually distributed. Even if the research worker does not deliberately consider what the distribution should be, it would be difficult for him wholly to exclude his own view at the stage of problem formulation and the interpretation of results. Once the findings are available, the description of existing distribution will influence the view of many decision-makers, as to what the distribution should be.[7] In this way, it can be said that scientific activity in itself shapes reality and thus it is no longer a passive observer.

The positivists' assertion about the unity of science has also been criticized. So far, social scientists have not been able to develop the ideal of unified science. Each discipline (sociology, psychology, economics, political science, geography) has its own approach for the analysis of the world. They express the reality according to their cognitions and methodologies.

A serious criticism of positivism lies in the fact that natural and social sciences are not and cannot be of the identical nature from the experimental point of view. The same methods, however, cannot be applied in social sciences. In social sciences scientists deal with man who cannot be taken as a 'thing' because he has brain and possesses thinking process. In fact, we cannot consider human behaviour the same way as animal behaviour, because men have intentions, imaginations, beliefs which cannot be translated into 'thing' language of the natural sciences. Thus, the element of subjectivity is a must in the study of normative things with a view to make social laws.

PRAGMATISM

Pragmatism is a philosophical perspective which is centrally concerned with the construction of meaning through experience. In other words, pragmatism is a philosophy which asserts that meaning and knowledge can only be defined in terms of their role in experience. It emphasizes experiences, experimental inquiry and truth as criteria for evaluating consequences. In other words, pragmatism is the 'position in philosophy that defines meaning and knowledge in terms of their function in experience, with reference to adjustment and the resolution of problematic situations'.[8]

Pragmatism is a modified form of positivism. Like positivism, pragmatism advocates the use of scientific method. The only difference being that this movement tries to find solutions to human problems.[9] The proponents of pragmatism use value-based scientific methodology (incorporating human attitudes, beliefs and norms) to solve the practical problems of the society and to ascertain the geographic reality. In other words, it is action-oriented, user-oriented and extends the experimental method to include evaluation and implementation. Research is undertaken for the purpose of solving an immediate problem and the results are a means to an end for some target population. The researcher gives directives for action and serves as an 'action agent' in the implementation of the results.

In geography, this approach is distinguishable as planned action rather than thought for planning. Because evaluation and implementation stages are involved, we must deal with the way things are.[10] Providing means to ends, which involve human activities and welfare, include values which are an integral part of the reality. Being action-oriented, lobbying, persuading and other actions are very important in pragmatism.

Pragmatism was developed in America after the civil war and brought about intellectual and social changes up to the Second World War. Its proponents in substantial numbers are found in West European countries also.

A common position attributed to this philosophy is one of dealing with practical problems; thus, the emphasis of pragmatism has been on the practical. A pragmatist believes that the 'concrete' or 'particular' situation is important in obtaining scientific knowledge and for understanding the world. Accordingly, 'abstract' or 'general' laws and theories are also important and useful as 'guiding principles' in any scientific inquiry. In other words, the pragmatist deals with theoretical notions as well as practical situations. This is exactly what many geographers have been supporting and that needs focus for our discipline. For example, one geographer proposed geographic strategies which include 'organization', 'persuasion', and 'action' to facilitate needed societal transformations.[11]

The major attributes of pragmatism are:

1. *The imperfection of reality*: The pragmatists believe that the current reality is composite of knowledge and error.

2. *The fallibilistic views of knowledge*: Because of the changing nature of reality (world) and the mind's view of it, it is impossible to guarantee an expected result from a specific experiment. Past successes do not guarantee future successes. They argue, therefore, that when prediction fails, the underlying assumptions and hypotheses should be revaluated and modified.
3. The scientific method and the hypothetico-deductive models are the best modes of investigation found to date and should be adhered to.
4. Logic should be used as a problem-solving device. The problems should be practical and used for the promotion of human welfare.[12] Owing to these approaches they reject the positivistic viewpoint of value-free research.

Pragmatism in geography led to the development of applied geography. Experts and thinkers agree that human attitudes, interests, desires, prejudices and group values differ in space and time. Policy based on applied geography, whether it be the modification of an environment, the removal of inequality in housing, educational or medical facilities, or the preservation or cultural landscape, include researcher and client values, which may vary substantially from the values of other sub-populations, especially those directly involved. The researchers' recommendations can also have long-term impact. In this research, while doing empirical work, value judgements are involved.[13]

For the pragmatist geographers, spatial laws are valid, and they provide the framework for hypotheses formulation and data collection. Furthermore, hypotheses about spatial structure can be formulated, tested and modified in the light of empirical evidence. The pragmatist geographers have a strong belief in the resolution of geographic problems by constant adjustments and modifications of the hypotheses, in the light of empirical data.

The aim of pragmatism is to emphasize the human element; 'our thoughts determine our acts, and our acts determine the previous nature of the world'.[14] Here, man is central. This view is similar to those expressed by Vidal de la Blache and the French School of Geography (Chapter 8). In humanistic geography, man and science are reconciled. The principal aim of modern humanism in geography is the reconciliation of social science and man, to accommodate understanding and wisdom, objectivity and subjectivity, and materialism and idealism.[15] From the above discussion, some of the elements of pragmatic geography can be identified as under:

1. Geographic space is a composite of knowledge and error.
2. Geographic space is changeable as our knowledge of it changes and the scale of measurement becomes more refined.
3. Geographic space is a manifestation of the 'human element' through time.
4. Geographic space is structured and restructured as a result of solution to practical human problems.
5. Spatial reality is a composite of human experience.
6. Spatial laws are useful for hypothesis formulation but the hypothesis may be modified in the light of our knowledge.
7. Geographical studies are concerned with the practical problems of man in space, and they can be studied using the scientific method.

FUNCTIONALISM

The definition of functionalism has varied over time and across disciplines. The word 'function', which is the key ingredient to functionalism, has been interpreted in the following five major ways:[16]

1. It refers to a public gathering for a specific ceremonial purpose.
2. In political science, it refers to the duties associated with a job that involves the exercise of authority.
3. In mathematical sense, it refers to the relationship between a variable and another.
4. In sociology and biology, it refers to the process which contributes to the maintenance of organism.
5. In geography, it is synonymous with occupation.

The diversity of definitions of function has resulted in diversity of meanings of functionalism within a discipline and in the various social sciences. It is, however, a viewpoint that investigates functional linkages with emphasis on the goals, the needs and links between role and actor. In simple words, functionalism is concerned with functions (occupations) and the analysis of the functions of a society.[17] It is a perspective which views the world as a set of differentiated and interdependent systems, whose collective actions are instances of repeatable and predictable regularities in which form and function can be assumed to be related, and which explains these form-function relations in terms of their role in maintaining the continuity of the systems.

The basic principles of functionalism are as under:

1. Societies should be examined holistically in an interrelated system framework.
2. Causation is reciprocal and, in many instances, multiple.
3. Social systems are generally in a state of equilibrium.
4. The functionalists are less interested in the history of a society, but more concerned with social interaction.
5. The functionalists attempt to find the interrelationships between the compounds of social structure.

Functional approach in geographical research can be seen in the writings of French scholars like Jean Brunhes and his contemporaries. The French scholars of the late 19th and early 20th century argued that culture is an indivisible wholeness. 'Region' was considered as a functional unit – an 'organism' which was more than the sum of its parts.

At present, functionalism is very popular in geography. Mumbai (Bombay), Tatanagar and Gulmarg can be explained in terms of their functions – as a chief port, an iron and steel manufacturing centre, and a centre of tourism, respectively. Moreover, smaller towns can be explained in terms of their function in a central place hierarchy. Each town has two types of functions, i.e. manifest and latent. For example, manufacturing of iron and steel is the manifest function of Tatanagar but the latent function is the exchange of ideas and social meeting of people with each other. Thus, a place may be explained in terms of its manifest and latent functions.

Functionalism has been criticized on both conceptual and methodological grounds. Functionalism conceives the geographical realities as states of equilibrium (a status quo) and the assumption of total social (spatial or global) integration on the methodological or logical grounds. The basic criticism is that of teleology.[18]

Critics of functionalism, on conceptual grounds, contend that their preoccupation with society as a system does not allow functionalists to study or focus on such contemporary problems as poverty, war, disease and racism. Functionalism has been accused of advocating social control rather than social change. It has a 'static bias' and, therefore, is incapable of accounting for social change.

Another criticism is that many functional explanations are structural in nature, i.e. the explanation of an observed pattern does not make reference to the underlying motives or processes; it is largely in terms of subsystem interrelationships. Moreover, there is an absence of definitional clarity.

On logical and methodological grounds, one of the major criticisms against functionalism is that of teleological explanation. Teleological explanation explains a given situation 'not by reference of causes which 'bring about' the event in question, but by reference to ends which determine its course'.[19] For example, vultures were created by nature in order to get rid of corpses.[20] In this example, the implication is that vultures are indispensable for the specific function they perform. There are alternatives that would equally fulfil these functions. The functions of vulture can be efficiently performed by others such as foxes, hyenas and men. This is known as the 'principle of functional substitutability'. The substitute, however, must be from the same ecosystem otherwise it would affect changes in the ecosystem and may damage it. For example, the introduction of snowmobiles and firearms in the life of the Eskimos has resulted in upsetting the delicate ecological balance between Eskimos and the Arctic wildlife.

The above discussion reveals that functionalism has six interrelated concepts which are used by geographers: (i) functions, (ii) functional substitutability, (iii) goals, (iv) pattern maintenance, self-regulation/status quo, (v) adaptation, and (vi) integration.

In geography, underlying the notion of functional region is the assumption that the region functions as a unit in order to maintain the existing intense inter-subsystem and intra-subsystem interaction that is essential for meeting a need or needs.

EXISTENTIALISM

Existentialism is a philosophical view that man is responsible for making his own nature.[21] It lays stress on personal freedom, personal decision and personal commitment. It emerges to challenge and even abandon the purely objective, quantitative and deterministic analysis. It urges a concern for human values, quality, subjectivity and spirituality. In existential geography, a central concept is that of existential space. According to Samuel, it is the 'assignment of space'.[22] Such an assignment is a result of human reality. Existentialism has been considered as an endeavour, on the one hand, to restore the concrete, immediate experience of existence *in situ* to the realm of knowledge, and on the other hand, to bridge the logical gap that separates subjective from objective, idealism from materialism, and essence from existence. It is based on 'existence comes before essence'. This phrase means that 'man first of all exists, encounters himself, surges up in the world, and defines himself afterwards'. This also means that to understand man we must first of all 'begin with the subjective life' and 'man is nothing else but that which he makes of himself'.[23] The first principle of existentialism is 'once thrown into the world (man) is responsible for everything he does'. Existence comes before essence, as it were, because man is free. In brief, existentialism is a philosophical view declaring that man is responsible for making his own nature. As said earlier, it lays stress on personal freedom, personal decision and personal commitment. It emphatically argued, 'we are along with no excuses'. The essential argument rejects any sentiment that would abandon man to nature, making him the product of his environment. As such, existentialism provides a firm foundation for the philosophy of human geography.

The operative principle of any existential logic is that man 'expresses himself as a whole in even his most insignificant behaviour'. In other words, there is not a taste, a mannerism, or human act which is not revealing.[24] An existential method is one that endeavours to 'decipher' that total expression, and to do so by 'beginning with the subjective'. In the process '... man's inward attitude, the way in which he contemplates his world and grows aware of it, the essential value of his satisfaction are the origins of what he does'.

An existential method, in short, begins by examining the centres, people (and, in particular, individuals) occupy and the way they define their relations with the world. It begins with an analysis of existential space, by first analysing the partiality that people project into their situations. More specifically, the method entails 'existential psychoanalysis'.[25]

Some scholars have considered existentialism as a 'biography of landscape'. An analysis of the landscape biography may be either backward from an already given landscape to particular individuals or groups, or forward from the latter towards landscape articulation.[26] In both cases, the concern is to identify the source of a landscape and the meaning that landscape conveys.

The scale of existential space may vary, ranging from individuals to groups/societies. In an existential analysis, the researcher can look to individuals or specific groups (large or small) for the biographies of landscapes. Though with this approach the researcher is constrained to be consistent in the scale of analysis, yet he can investigate the 'great' figures and civilizations or any number of 'lesser' individuals and communities. Every landscape is someone's existential space.

In the light of the above discussion, it may be said that existential geography is a study of the biography of landscape. In other words, existential geography is a type of historical geography that endeavours to reconstruct a landscape in the eyes of its occupants, users, explorers and students in the light of historical situations that condition, modify, or change relationships.[27] Thus, the essence of the existential geography is that for every landscape or every existential geography there is someone who can be held responsible and accountable. The responsibility may lie on individuals or groups, whether sane or insane, rational or irrational, well intended or demonic, the fact is they make their choices and their landscapes. To summarize, existential geography places the greatest emphasis on the human core of existence. The main advantage of this philosophy lies in the fact that it is anthropocentric which should help in the expansion of the horizon of geographical research.

IDEALISM

This is a view that reality is mental or mind-dependent. In a philosophical sense, idealism is the view that the activity of mind is the foundation of human existence and knowledge.[28] Idealism is in opposition to the proponents of naturalism and materialism. The essence of idealistic philosophy is that mental activity has a life of its own which is not controlled by material things and processes, and the world can only be known indirectly through ideas. According to this view, all knowledge is ultimately based on an individual's subjective experience of the world, and comprises mental constructs and ideas.[29] There is no 'real' world that can be known independently of mind.

Guelke – the most celebrated advocate of idealism in geography – argued that we have developed methods of entering the minds of our subjects so as to think their thoughts and justify their expectations, methods that will determine human intentions and understanding of our role in changing the earth. The idea that human behaviour is largely controlled by mental activity is the basis on which idealists insist that the social science and history are logically separate from the natural sciences. The logical positivists' idea of a social science has its own approach and methods. Although human behaviour cannot be treated as a material process in the normal (natural) scientific way, the rational character of human thoughts makes it possible for one to understand deliberate activity in a way that it is not possible to comprehend material processes.[30] It is because of this fact that a number of idealist philosophers have developed distinctive methodologies for the social sciences and history on the assumption that human activity must be understood in terms of thought.

The rethinking or reconstruction of human activity in order to discover what really happened is called *verstehen*. This is against the nomothetic approach of positivists or the natural scientists.

The followers of idealism maintain that it is possible to reconstruct and understand logical sequence of thought of another person in a way that it is not possible, for instance, to re-experience emotional life. Geographers are concerned overwhelmingly with activities which are the outcome of deliberate rational actions. When people grow crops, build houses and exploit resources, their actions are the result of rational thought and are, therefore, in principle, open to be understood in terms of the idealistic method of rethinking.

In human geography, man-environment (resource) relationship is studied. Geographers should try to find out what were the reasons for clearing of forests, different field patterns and settlement types. In fact, the patterns of any of the phenomena of cultural landscape are not arbitrary but reflect the thinking of the people who created them. In many situations it will involve historical research, because the purpose that, say, a building or road was originally designed to serve might no longer be related to the landscape of the present. Many of the churches of the Victorian era in England have been converted into places of social gathering in which libraries and things related to other cultural activities have been installed. The old forts of India have now become points of attraction for tourists while in the past they were places of rulers, administrators and defence personnel.

Another characteristic of idealism is that the idealist geographer recognizes the importance of studying human activity on the earth in relation to the overall cultural context. It is an idiographic approach which means that the validity of the generalizations will be limited to regions and peoples of broadly similar culture.

A general criticism of the idealist approach is that one can never know for sure whether one has actually succeeded in providing true explanation. This allegation seems to be genuine, but on a closer examination it loses much of its force as an argument for not taking the approach seriously. Although one can never know with certainty that an idealist explanation is true, the same objection is applicable to all empirical, interpretative and theoretical work. The theoretical physicist can never be certain of his theories. Indeed, 'the history of natural science is largely a history of abandoned theories'.[31] Yet progress has been made, because with the failure of old theories, new more powerful ones have emerged. In social sciences, an idealistic interpretation will also be challenged in terms of new evidence and new argument. In the process of reinterpretation of old theories and new evidence, a more accurate and powerful account of 'what really happened' will gradually emerge.

An important implication of the adoption of an idealist approach to explanation is the methodological separation of human and physical geography. This dichotomy of dividing the discipline into physical and human geography, however, does not imply that human geographers do not need to consider the physical environment or that physical geographers can ignore human activity. The human geographers will consider the physical environment mainly in terms of the way people of different cultures and circumstances have used their resources. As human ideas on technology, institutions and social priorities have changed so has the relationship between man and his environment. The physical geographer, on the other hand, is basically concerned with man as an agent of landscape change.[32]

The idealist approach ideally suits the study of regional geography. In the delineation of regions the idealist geographer will attempt to group together people who share a common culture or worldview, because such views will largely shape geographical behaviour. In order to ascertain a more reliable picture the macro regions will be divided into meso and micro regions. The sub-divisions, however, would not be identical for each geographical region, owing to the variations in resource potential. This potential will vary from place to place in relation to technical, social, political and economic factors.

Many of the geographers who build their model on rational principles (empirical scientific models) have used hypothetical situations in terms of rational principles. For example, the Von Thunen model of land use and crop intensity is developed on a model of rational man. The propounder of this model comprehends the behaviour of hypothetical farmers at various distances from the isolated city in terms of what makes rational economic sense. This understanding is close to the idealist method of rethinking, and the idealist geographer has no objection to the use of simplified models as aids to understanding how human activity on the earth's surface might be affected by certain factors under ideal conditions. Thus, the idealist approach is competent enough to offer an insight into man's activities on the face of the earth, in their social and cultural context.

Geographers during the last two decades have realized that the application of the methodology of the natural sciences has caused adverse results in the study of regional geography. Regional geography conducted along the lines of natural sciences encouraged its practitioners to investigate external relationships between such things as soils and settlements, climate and crops. In the absence of general theories and laws such relationships were basically descriptive instead of explanatory. As a result, regional geography was reduced to systematic inventory and description.[33] These problems can be avoided with the idealist method which permits the researcher to progress beyond a description of events and relationships to the meanings expressed in them.

The idealist approach is quite distinct from the other major approaches accepted by geographers. It contrasts with logical positivists. The idealists emphasize the autonomy of social science which is particularly useful in the study of human and regional geography. The methodology of idealism is seen as providing an appropriate analytical tool for gaining fundamental understanding of human behaviour similar to the fundamental understanding that theoretical knowledge gives us of the physical world.

REALISM

Realism is the view that reality exists independent of the mind; it is not mind-dependent. It is antithetical in many respects to idealism. Gibson suggests realism as a viable alternative explanation to idealism.[34] The basic philosophy of realism is that facts speak for themselves and explanation is logical and inductive. Realism advocates the use of theories and models in geographical explanation. It is very close to the objective philosophy of positivism but has different methodology of explanation.

Historically, realism by the Platonic-Socratic thought was used in opposition to nominalism* for the doctrine that universal and abstract entities have real objective existence. At present, it is, however, used in opposition to idealism. In contrast to other philosophies of science such as naturalism, positivism, and idealism, realism is based on the doctrine that human science is an empirically-based rational enterprise which explains observable regularities by describing the hidden, but 'real' structures that casually generate them.[35]

In his theory of ideas, Plato asserted that the forms we see, touch, taste and smell in time and space do not exist and are not knowable with our senses.[36] A particular phenomenon is only appearance which shall disappear in due course of time. For example, a specific mountain like the Himalayas does not exist, it will be worn down to the ocean floor over geological time. Contrary to this, the general and universal term 'mountain' is rigid and fixed. Just opposite to it, the nominalists, notably, Aristotelians denied the existence of an ideal mountain. For the early nominalists 'mountain' was a mere term. What is real is a particular mountain that we can all see and touch.[37]

The battle between the realists and nominalists over the existence of abstract entities or problematic entities, carried over the medieval period. During the medieval period, the Platonic-Socratic thought came to be known as scholastic realism. The main proponent of scholastic realism was John Scot. In his essay 'On the Division of Nature', he reasoned that the divisions of the physical world all signify something hidden. In themselves they are not real. The cyclic process of the physical world – seasonal and astrological cycles for instance – all proved for Scot the existence of a divine order, a harmony and a law. They proved that in the ordinary sense world is not real.

In the 19th century, 'realism' took the shape of 'direct' or 'naive' realism which was a polemic against idealism. Cook Wilson was the founder of 'naive realism'. He and his pupils denied the existence of any problematic or abstract entity – a denial which of course runs counter to Plato. For direct realist:

> Nothing existed that was not observable in time and space. From this ontology they developed a logic of perceiving the world, a commonsense logic which argued that our views of the world are constructed in the mind by an interaction, through society, with the physical world.[38]

This direct naive realism has had a sustaining influence on geography, especially on commercial and military geography, since the Victorian period. For the naive or commonsense geographers, the mind grasps the world in a

---

* A philosophy that believes that only particulars are real and that universals or general terms, are only names.

simple effortless process, something we do all the time. Geographical facts of observed phenomena and changes within them can be objectively established and any question of unseen entities, problematic, abstract forms or subjective impressions is irrelevant. With due precautions, by simple commonsense, we can know the reality of a place, as the topography looks, as the soil feels, as the water tastes, etc. The objectives of 'naive realism' are common to social reforms or national surveys. Stamp, with this objective, conducted the land use survey of Britain and suggested significant changes in the utilization of land. It gave more job opportunities and greater respectability to geographers. The sixties can be called as a period of extreme of naive realism, which has also been termed as a phase of 'quantitative revolution' in geography. This revolution was based mainly on a move toward philosophical 'positivism'.

Realism has taken the shape of 'new or critical realism' in the last few decades. It was propounded by T.P. Nunn.[39] The essence of the concept of new realism is that anything we experience depends for its existence upon the fact that it is experienced. In other words, the objects of our perception are actual properties of the physical world. The proponents of this philosophy asserted that nothing exists excepting that which is experienced.

The basic difference between positivism and realism can be summarized as under:

1. The positivist asks 'how' a pattern is produced while the realist asks 'why'. For the realist to answer the question why, it is necessary to go beyond the logical argument of the positivist, to go to a description of the causal processes generating the observed patterns of physical and social regularities.
2. The positivists speak of logical (mathematical) necessity of the regularity, while the realists speak of natural necessity. In other words, the realists seek to discover the rational cause and effect in nature and society.
3. For the positivists theory could be a logical conjecture, an abstraction, e.g. a mathematical law, while for the realists there is greater emphasis on 'commonsense' argument following from analogy.[40] In other words, positivist theories come from theoretical laws, whereas realist theories come from problematic entities with assumed ontological status.

In brief, realists believe in 'objects' or physical world which exists independent of our thought about them – a physical world that science can genuinely discover. They lay stress on the use of theories and models in geographic explanation.

MARXISM

Marxist geography is a perspective within 'historical materialism' which is centrally concerned with the ways in which the production of space, place, and landscape is implicated in the reproduction of specific 'social formations'. It developed against capitalism. The historical materialism of Marx is based on dialectic – the perpetual resolution of opposition (rich and poor, developed and undeveloped, abundance and scarcity, etc.), in which each resolution produces its own contradiction. Its main concern is uneven development, social justice and regional inequalities as a result of division and exploitation of labour resources and environment, through surplus value. Marxist geography insists on value-free inquiry. In other words, it does not give any importance to social values (a set of beliefs and ideas, e.g. religion, which inform about our assessments of worthiness). Marxist geography discloses 'social processes which originate spatial appearance' and which in turn 'forms an input into the continuing social process'. It presumes that 'space' and 'society' interact.

Marxism is a form of realism, which seeks to relate the empirical world of appearances to a set of infrastructural determinants – economic processes. Marxism and Marxist geography attempted to provide 'a powerful theoretical and political base for resistance' to the dominance of the capitalist imperatives on individual action. Its goal is based on Marx's humanism. He argued that people are alienated by the capitalist system; in particular the proletariat is exploited and has its human dignity removed through the process of selling its labour. To restore this dignity and to give individuals full control over self and destiny, capitalism must be overthrown and replaced by communism. The argument is that truly human relationships can be achieved only when everyone can take responsibility for the conditions of their own lives and when there is freedom from the ideologies and actions of bourgeois professional class. The Marxist and related realist works therefore suggest that the objectives of human geography should be:

1. to explain and interpret the patterns of spatial organization and of society-environment relationship. These patterns can be understood only by examining the economic processes;
2. that the economic processes cannot be understood directly, but can be appreciated through the development of theories of superstructure (religion and legal system);
3. that the economic processes are continuously changing, and therefore universal laws of superstructure cannot be derived;
4. that class struggle (bourgeoisie vs. proletariat) is central to the economic processes;
5. that any attempt to retain the present superstructure can only help the present unjust system (capitalism) to survive; and
6. that the objective of human geography should be to bring social change, to overcome the problems of exploitation of man and environment (resources).

Thus, the Marxist geography emerged as a critique of quantitative revolution which made geography as a regional science which promotes capitalism.

Marxists argued that positivist spatial analysis was flawed in three basic ways: (1) Insofar as existing geographical realities were treated as spatial than social patterns. In their opinion geographers might map urban segregation according to class and race, but never interrogate the political and economic process that led to such unequal geographies. (2) Spatial science sought to identify the most efficient locations for factories, supermarkets and social services. (3) Universal spatial laws of the sort sought by positive spatial analysis are a misnomer, and very different spatial arrangements are obtained in different societies.

The system of thought developed by Marx propounds that the state, through history, has been a device for the exploitation of the masses by a dominant class and that class struggle has been the main agent of historical change.[41] In Marxist philosophy, economic classes and private property are the main cause of historical change. These two factors determine the man and environment relationship also. The Marxist geographers conceive that territorial structures essentially reflect the prevailing socio-spatial dialectics. Marxist geography analyses the dialectical relationships between social processes, natural environment and spatial relations.

The essence of Marxist philosophy is positivistic approach which puts emphasis on materialism. Marx writes: 'It is not consciousness (ideas) that determines life, but life that determines ideas'.[42] It is not the ideas that change the world, but the development of actual reality (space and place) which changes the ideas. Marx believed that society develops in stages in accordance with the development in the factors of production (materials and instruments of labour). In other words, consciousness develops with productivity, increase in needs, and with increase in the number of people. It emerges as a dynamic process because in transforming the bonds of our experience we transform ourselves. It develops specially with the division of labour, particularly the division between material and mental labour. From this point, consciousness may proceed to the formation of 'pure' theory, theology, philosophy, morality, etc. Hence, the key to the understanding of the structure of social life and consciousness is the mode of production of the material basis of that life and consciousness.[43] Marx refused to accept that the scientific laws of society were eternal. This view contrasts sharply with the claim of positivists who assert that scientific laws are universal and eternal in space and time. Engels pointed out that to us the so-called economic laws are not eternal laws of nature; they are like historical laws which appear and disappear. A given society or a cultural landscape is organized on the basis of several modes of production. The concept of 'social formation' thus designates a social whole composed of distinct but interrelated 'instances' of the whole. Social formations are made up of these instances: those of its modes of production or the economic structure – forces and relations of production – and those of superstructures corresponding to those modes-politics legal (law and state) and cultural-ideological (religion, ethics, law, politics etc.).[44]

According to Marx, the man and environment relationship or the man and space relationship changes with the change in the mode of production. For example, from the stage of nomads, mankind turned to the stage of settled living, hunting, fruit gathering and sheep rearing gave way to domestication of plants and animals (agriculture). Then came the urban culture and a rich diversity of vocations. For the first time economic classes within human societies were distinctly recognizable, and the relationship between them formed the basis of their belief systems, social hierarchy, codes of behaviour, rules of punishment for crime and misconduct and institutionalized arrangements for the same, worship, recreation, family obligations and ties, the apparatus of ruling authority, etc. These constituted what is called the cultural landscape and the cultural pattern of society.

The division of people into economic classes had the effect of initiating class struggle. Polarized class interests could only set class against class. Thus, there was a class of employees, another of employed. In familiar communist terminology, this would be called exploiters versus exploited.

Like many other things, property is anathema to the communists. Class and property are the principal twin evils which man in his folly nourished for his own undoing. What the communists object to the private ownership of property, as such ownership hurts the general good of the community at large. In particular, property which qualifies as primary means of production such as land, forests, mines, factories, mills, etc. is not permitted to be privately owned. Instead, these units should be owned by the community at large. This is because they produce goods of basic necessity for the whole community. In private hands, they yield profits which go into the pockets of individuals or groups of individuals. These individuals are at liberty to utilize their profits according to their will and pleasure. Where such profits are large and far beyond reasonable needs, they confer considerable power on their owners and such power is invariably exercised irresponsibly and anti-socially. Private wealth has often been expanded on personal luxury and pleasure and seldom on the relief of distress. Had rich men spent their wealth mostly to help the poor and needy and had this been the rule and not the exception, there would have been strong presumption in favour of private ownership.

The question is: to what extent has private ownership of property affected human life? The answer is: very greatly indeed. For it always influenced normative ideas and beliefs, religion, customs, law and tradition in all societies. Moreover, private ownership engenders pride, vanity, extravagance, fear, jealousy, envy, hate, corruption and crime.

As stated in the preceding paras, the man and environment relationship, and Marx's philosophical interpretation is through dialectical materialism. According to dialectical materialism, in the world there exist entities which are antithetical such as day and night, white and black, right and wrong, male and female, positive and negative, productive and unproductive, hot and cold, elevation and depression, wet and dry, torrid and frigid, rich and poor, etc. Numberless examples of this coexistence of opposites can be cited.

The Marxian premise is that because classes exist as differential units with different functions to perform, their economic interests are necessarily hostile and impossible to reconcile. For that reason collision and clash must occur. High and low income groups are under some dire necessity to cut each other's throats. Put bluntly, it is like the robber and his victims having contradictory interests.

Moreover, Marx and Engels proclaimed that ownership of private property is the cardinal sin of mankind. In order to remove the private ownership of means of production, there is a need of a minority of secretly trained and armed agitators. It may include terrorists, saboteurs, gangsters, guerrillas, secret agents, hardened criminals, disgruntled youngsters and the smooth-checked politicians who know how to deceive as long as it is unsafe to come out in the open. The only thing that unites them is a common bond in absolute loyalty to the communist party.[45]

In explaining the man and environment relationship, the Marxist geography has also failed. The results no longer deniable are:

1. Totally failed agriculture with dwindling production and chronic shortages which are met by massive imports.
2. Industrial production for consumer goods is woefully deficient.
3. The upper layer of state functionaries enjoys life better than the rich landlords of the past.
4. There are many environmental hazards like that of Chernobyl.
5. Workers do not enjoy a better standard.
6. There are vast black markets at the outskirts of large cities like that of Moscow and Leningrad.

To develop economy and society and to maintain ecology and environment, the Marxist approach could not bring the desired results and has been criticized by the scholars all over the world. The disintegration of Soviet Union has proved this point beyond any doubt.

Man-Nature Relationship

In the Marxist philosophy, man is a part of nature, immersed in it, yet also 'apart' as conscious subject. Consciousness itself is natural in the sense that knowledge is not drawn from nature via sense experience, but man's capacity for thought is a product of his nature.

In the materialist conception, the key interaction between man and nature is labour. Man is confronted with a natural world which cannot be transcended and which must be appropriated in order to survive. His mode of appropriation is labour. Labour transforms natural objects into use values in the context of particular social relations. Different modes of production have natural environmental relations which reflect the character of their dominant social relations:

> Thus, under capitalism men struggle with nature in order to satisfy new needs, but they do so in a prescribed way (namely, under conditions of wage labour) that differs profoundly from other modes of production ... the antagonism of a class divided society makes it impossible for men to bring their productive system (of which mastery over nature is a part) under their control.[46]

This idea can be applied to analyse capitalism's relations with natural environment. In the capitalist mode of production the competitive nature of capitalism compels constant economic expansion (i.e. the accumulation of capital).[47] Accumulation at the societal scale, leads to an expansion in the demand for raw material and thus nature is exploited which leads to environmental crisis and ecological problems. Secondly, in order to have the greatest possible accounts of surplus value, labour power is exploited to the greatest possible extent.[48] The characteristics of economic structure – an exploitative, competitive, alienated commoditized population – is necessary to perpetuate an exploitative, competitive, alienating mode of production of commodity. These cultural characteristics of social relations extend into the social formations of environmental relations. Hence, an economic system compelled to expand production by its own inner laws, characterized by aggressive, exploitative social and environmental relations, necessarily comes into a contradictory relationship with a finite, fragile world.[49] In this era of high technology, this contradiction threatens to destroy the natural basis of life.

Marxist science rests on, and is integrated by, its assumptions about the importance of material production in structuring social processes. The Marxist geography shares these assumptions. Geography is that part of the whole science, according to Marx, which specializes in the dialectical relations between social processes on the one hand and the natural environment and spatial relations on the other. It is aimed at changing the fundamental operation of social processes by changing the social relations of production (see Soviet School of Geography: Chapter 6). The main criticism against Marxist geography is that it reduces man as a cog in a machine who is governed by the space and is not able to change the space through his mind and thinking process.

RADICALISM IN GEOGRAPHY

The radical approach in geography developed in the middle of 1960s as a part of an oppositional politics (the 'movement' as it was called). It coalesced around domestic issues like inequality, racism, sexism, environment and opposition to the Vietnam War. It also developed as a reaction to 'quantitative revolution' and positivism which tried to make geography as a spatial science, with great emphasis on locational analysis. It began as a critique within the contemporary liberal capitalistic society but later coalesced around a belief in the power of Marxian analysis. According to radicalists, inequality is inherent in the capitalist mode of production. Redistribution of income through taxation policies will not solve the problems of poverty and inequality, and according to Peet, alternative, environment designs, with removal of central bureaucracies and their replacement by anarchistic models of community control are needed, and geographers should work towards their creation.

As stated above the followers of radical approach in geography mainly concentrated on the issues of great social relevance like, inequality, racism, sexism, crime, delinquency, discrimination against blacks and non-whites, females, exploitation of juveniles and environment resources and the opposition of the Vietnam War in USA. Events of the late 1960s, such as the burning of large cities in the western world, student unrest, worker-uprising in Paris in 1968, massive anti-Vietnam War protest actions and radical cultural reformation exposed the social and political irrelevance of geography as a spatial science and proved the hollowness of locational analysis. It was in this background that the radicalized students and junior faculty members challenged the traditional geography (geography as spatial science) and they started publishing articles with more 'socially relevant' geographic topics in the professional journals. In 1969, *Antipode* – a radical journal of geography was founded at the Clark University in Worcester (Massachusetts), specifically to publish the research papers of the younger geographers with a revolutionary leaning. The young radical geographers published papers in *Antipode* dealing with urban poverty, discrimination against women, coloured people and minority groups, unequal access to social amenities, crimes, deprivation, permissiveness and sexism. They also published articles on underdevelopment, poverty, malnutrition, unemployment and resource misuse in the Third World countries. Thus, the radicalists took the side of the oppressed, advocating their causes and pressing for fundamental social change. In brief, the radical geography was a quest for social relevance of the discipline geography at a time of contradiction and crises in the capitalistic society of the West.

The origin of the radical geography movement can be traced to the late 1960s, especially in the USA with three contemporary political issues:

1. the Vietnam War,
2. civil rights (especially of the American blacks), and
3. the pervasive poverty and inequality suffered by the residents of urban *ghettos* and deprived rural areas all of which were generating increased social unrest and tension.

In the words of Peet (1977), radical geography developed largely as a negative reaction to the established discipline (spatial science). The radical geographers introduced the study of topics such as poverty, hunger, health, and crime to human geographers, who had previously very largely ignored them.

The salient features and objectives of radical geography were:

1. To expose the issues of inequality, deprivation, discrimination, health, exploitation, crime and environmental degradation in the capitalist countries.
2. To highlight the weaknesses of the positivism and quantitative revolution in geography which emphasized on geography as a 'spatial science' with a thrust on locational analysis.
3. To bring a cultural revolution to eradicate permissiveness, sexism and discrimination against females.
4. To remove regional inequalities.
5. Radicalists opposed political centralization and economic concentration. Contrary to multinationals, they favoured small-scale, self-sufficient social units, living in greater harmony with their natural surroundings.
6. They were against imperialism, nationalism, national chauvinism and racism.

7. They opposed the idea of the superiority of the white and the West.
8. According to radicalists the man and environment relationship may be understood through history. In other words, the mode of production in any society determines the economic relation among its people.
9. One of the objectives of the radicalists was to explain not only what is happening but also to prescribe revolutionary changes and solution to the social problems.
10. To develop a more just, equal, tension free, peaceful and enjoyable society.

The development of adequate social theories proved difficult for the radical geographers who were largely trained in the fieldwork traditions. Theoretically sophisticated ideas tended to form in areas of radical geography with clear connections with more heavily theorized stream of thought outside the discipline. There are numerous examples to illustrate the radicalists' ideas about the social issues. Imperialism, women and environment relationship and racism are some of the important issues on which the radicalists concentrated. The work of radical geographers about imperialism, women and environment and racism have been illustrated in the following paras.

Geography and Imperialism

An unequal territorial relationship, usually between states, based on domination and subordination is known as imperialism. Such a relationship does not necessarily imply colonialism, for imperialist control over a subordinate territory's economic and political activities can exist without military intervention and the establishment of a colonial regime.

Imperialism has been considered largely responsible for destructive economy – 'robber economy' or violent attack on nature which is often the main cause of poverty in the Third World countries. Jean Brunhes was against this approach of the imperialists towards the exploitation of natural resources of the colonial and subordinate countries. The involvement of the United States of America in the Vietnam War was considered by the radicals as a device of the imperialists to exploit and destroy the economy of the developing countries. The radicals agreed with the J.A. Hobson (1902) theory of imperialism which was later elaborated by the Russian Marxist, V.I. Lenin. In his 1915 thesis, Lenin argued that both the causes of the First World War and the continuation of capitalism were linked to the main features of imperialism. In his opinion, in the epoch of imperialism, production and capital are concentrated to such a degree they give rise to monopolies, which play the decisive part in the economic life of the capitalist states. Moreover, the process of monopolization brings about the formation of international monopolies which divide the world among themselves economically, and ultimately leads to rivalry, conflicts and wars.

As a criticism to the Vietnam War a series of articles were published against the imperialism, under development in the Third World countries and centre-periphery relations, to provide theoretical insight. In a monumental work, James Blaut (1970), a professor of geography at the Clark University, later at the University of Illinois (Chicago), argued that the conventional western science is closely interwoven with imperialism. He defined imperialism as 'white exploitation of the non-white world'. Imperialism, he asserted, is underpinned by western 'ethonoscience'. Ethnocentrism is a form of prejudice or stereotyping that assumes the superiority of one's own culture or ethnic group – a mild version of racism or xenophobia, which assumes that one's own way of doing thing is the normal or 'natural' way and that other ways are inherently inferior. In his opinion, European ethnoscience contains a set of historical beliefs and social scientific generalizations about the world biased in favour of whites and congruent with the interests of western imperialism. Mc Gee (1991) has criticized geography for its own form of ethnocentrism (a prejudice against everyone who is perceived to belong to a different ethnic group), arguing that the discipline has defined Asia and Africa in Eurocentric terms. For Blaut, the European model of the world has a unicentric form with a distinctive geometry, an inner Europe space originally closed from an outer non-European space. The West has some kind of unique historical advantage (race, ethnicity, culture, mind, spirit, traditions, customs, etc.) which gives it superiority over all other peoples. European civilization is supposedly generated mainly by inner processes. Europe makes history, while non-Europeans play little or no crucial role in epochal events. Rest of the world is traditional. Non-Europeans are characterized as primitive and unprogressive, barbarous, uncivilized, uncultured, heathen, less intelligent and less virtuous than white Europeans. The expansion of the Europeans is thought to be self-generated. Whenever non-Europeans show evidence of progress, this is proportional to the European impact on their society. This amounts to global diffusions model of the belief generalized 'diffusionism' (i.e. cultural process flow from the European centre to the non-European periphery).

The unicentric model of European culture was criticized by Blaut, who advocated the multicentric-ethnocentric model of the Third World. According to multicentric model, centres of development are springing up at strategic points throughout the world. In a Third World list understanding, the multicentred pattern of relatively equal levels of development was disrupted by the European plunder of the New World (the discovery of North and South America and the colonization of the African and Asian countries). The discovery of America and colonization of Afro-Asian countries resulted into the flood of wealth and bullion into Europe, which ultimately led to the development and expansion of commercial, agricultural, industrial, educational, medical, scientific, and technological advancement in Europe. Thereafter, the gulf between development in Europe and underdevelopment in the developing countries widened. From this perspective, Blaut (1976) argues against the notion of the 'European miracle' and 'superiority of white race' by underpinning its more concrete theory of the 'autonomous rise of Europe'. He asserted that (1) Europe was not superior to other regions prior to 1492. In fact, in the past the Egyptians, Babylonians, Indus and Hawang Ho Valley civilizations were much advanced and Europe was inhabited by the white barbaric tribes. In the words of Aristotle the inhabitants of cold countries like that of North and West Europe are courageous, brave but unintelligent, lacking in political organization and capacity to rule their neighbours. Contrary to this, the people living in warm climates of Asia are intelligent but lack in courage and so slavery is their destiny. Kant, the leading German thinker, endorsed the view and stressed that the inhabitants of hot and humid regions are exceptionally lazy and timid but intelligent while people of cold countries are strong but less intelligent and more diligent. The Europeans were in the 'dark age' when the Arabs were the leaders of the world commerce, science and education. In the medieval period, India, Iran and China were having the main centres of education and learning and their handicrafts and artifacts were well known all over the world. (2) Colonialism and the wealth plundered from the Third World countries were the basic processes leading to the rise of Europe. (3) Europe's advantage lay solely in the 'mundane realities of location' that is nearness to the Americas (Blaut, 1994).

The theory of superiority of white race and ethnocentrism of Europe were based on certain prejudices and racism was therefore rejected by the radicalists.

Women and Environment

The radicalists were of the firm opinion that women are oppressed both in the developed and the developing countries. The women's movement of 1960 inspired the radical geographers to explore and make an in-depth study of the man and environment. The female geographers raised the question of women and environment, and the role of females in the decision-making process about the utilization of natural resources. In an atmosphere of social and academic ferment, typical of the early 1979s, geographers began to inquire into relations between women and space (Mackenzie, 1984: 3). At first, work, gender and environment took the form of criticism of the 'invisibility' of women in the geographic literature or the 'gender blindness' of the neoclassical and behavioural models of spatial structures (Brunet, 1973). In the 1970s, most work in liberal feminist geography tried to create a 'geography of women' which documented the disadvantages systematically suffered by women, the constraints of women's activities, and women's inequalities in general. They emphasized on the constraints on women's spatial choices, arguing that problems of access result from constraints of gender role such as the social expectations that women should primarily be involved in house keeping and family care.

In an article, Allison Hayford (1974) argued that women were as invisible in geography as they had been in history. In his opinion, women assumed either to have no role of their own, or to be continuously adjusting to the male determined order. She (Hayford) thought women embody, the means by which people attempt to relieve the tensions, which come from dealing with infinite space in finite time. They are mainly to provide comfort to men and to relieve their physical and mental tension. Thus, women are almost essentially the essence of locality. In traditional and less developed societies, women have main responsibility for types of production (gathering, herding and agriculture) that reinforce locality. In such societies, there is little distinction between public and private spheres of activity. Women, she found are central in space because of their role in the household, the main means people have devised in the tension of space:

> The household was the nodal point, in the spatial network of productive systems and formed the point around which rights to the use of the earth were determined .... At the same the household had tremendous symbolic importance; it contained the iconography of the locality – it was the ultimate and often the complete expression of 'here'. It was the site of the most dependable personal relationship. It was the one place where human obligations were most supportive and most fixed, the one place, where people could spend their weakest and most private moments – sleeping, eating, childhood, adult, mature and old age – in relative security. (Hayford, 1974–76)

The household was also an important means of extending control over space, socializing labour, and allocating resources. As the centre of the household, women had main roles in establishing the linkages through which the earth's surface was manipulated.

However, the growing separation of the public and private spheres of activity created a tension between the household and the larger society. With the development of class society, the separation of economy and politics subjects the household to external power, diminishing the symbolic linking significance of women. Under capitalism, direct and personal organizations with women at their cores are replaced by the impersonal, invisible power of capital. The private sphere of women is replaced in importance by the male dominated public sphere. And, the functions of women in the household are confined to reproduction, care of personal needs and the security of the locality. Capitalism thus changes the position of women from centrality to peripherality. The safe space of household still provides relief from the stresses of involvement in capitalism productive relations, but it is also under pressure from these relations. The separation of work from living space, and of the various other kinds of space from each other, subjects women to continual spatial tensions – for example, between an ideological commitment to the household and an economic need to function in a wider space. Women do not have the same freedom as men to move in space, or organize it, and have no power to change the structures of their lives. Such arguments, Hayford (1974: 17–18) concludes, make it important for geographers to investigate the spatial roles of women, in particular the meaning of the 'continuing transition from centrality to periphery, from being the pivot of society's relations to being nowhere'.

Apart from imperialism and discrimination against females, the radicalists strongly opposed apartheid. The policy of spatial separation of the races was severely criticized by them. In their opinion racial discrimination has resulted in marked inequalities in levels of living, with whites enjoying the highest standards overall, between them, the blacks and coloured experiencing the lowest.

Anarchic Leaning

The early radical geographers took the help of numerous political and social theories including 'anarchism'. Anarchism advocates the removal of the *state* and its replacement by voluntary groups of individuals who can sustain social order without any external authority. Such a social order may emphasize either individualism (and thus is a logical conclusion of liberalism, stressing the importance of individual liberty) or socialism (some versions of which reject private property as well as the state). Among the early proponents of anarchist's communism were Peter Kropotkin and Elisee Reclus, whose geographical writings were rediscovered by some of the advocates of radical geography.

Peter Kropotkin, a leading anarchist theoretician of the late 19th and early 20th centuries thought that we should learn from the greater sweep of history in building an alternative to capitalism. For long periods, humans lived in groups organized around principles of co-operation and mutual support, for it was found that cooperation, and altruism or unselfish devotion to the welfare of others were the only lasting bases of social life. Kropotkin believed that natural co-operativism forms the basis of people's ethical system. He thought that capitalism leads to severe competition which increases the economic disparities and threatens the very survival of human society. Kropotkin thought we must return to societies based on co-operation and mutual aid, principles which have continued to be practised (for example, in the family) and which are propagated still via an underground people's history.

The anarcho-radicalists believe in 'integrated labour' rather than the division of labour as the basis of production. People should perform several different kinds of tasks in free associations with the means of production and products held in common by a decentralized society. Essentially, self-sufficient regions would become 'integrated cells' to equally interchange ideas and products with other regions. In their opinion, production decisions should be democratically made at the grassroot level taking into account the needs of the people and the available resources. Work places and living places should be nearer together, allowing greater integration of various spaces in which life is lived. Believing deeply in such ideals, many radical geographers supported communism which is radically democratic, decentralized society, where people directly control the means of production and make their own spaces.

The radical approach adopted by some of the geographers of the 1970s and 1980s with the set objective to change the theory and practice of geography and to replace the capitalistic social order, however, could not achieve much. The radical approach has also been criticized on several counts. Some of the weaknesses of radicalism are as under:

1. The theoretical base of the radical paradigm was weak. They attempted to borrow theoretically sophisticated ideas from other disciplines without much success.
2. The radicalists tried to draw freely on a variety of sources of political theory including anarchism. With such approach, radical geographers made numerous attempts at developing an anarchist base with a distinct political perspective on alternative forms of society. Their objective was not to reform but to change the society. They were determined to turn the society upside down. They, however, could not change the capitalistic society to make a more socialistic social order and to make life of everyone enjoyable and tension free. Thus the discipline of geography remained as a regional science, dependent on statistical tools to explain regional variations.
3. Radical geography was radical in topics (imperialism, anti-war, discrimination, poverty, crimes, etc.) and politics (anti-capitalist) but not in theory or method of analysis.
4. The radicalist could not develop an appropriate model about population resource in an integrated dynamic way to remove inter-regional and intra-regional inequalities.
5. The radical geographers developed a deep leaning towards Marxism (historical materialism), and gave over weightage to the Marxian analysis to explain the spatial variations of geographical phenomena. They attempted to evict human agency from human history. In other words, men and women are reduced to the passive 'bearers' of historical and structural determination. In fact, man is not the product of history and historical materialism. He is an active agent who created history, affects the historical processes and in turn gets transformed.
6. The radicalists and Marxists gave priority to time over space. Those who criticized radical geography insist that 'just as there is a theory of history, in historical materialism too, there is also a geography' – the space dimension is by no means sacrificed to the time dimension; both are indelibly present. The humanists who attacked and criticized radical geography stressed that we have to liberate ourselves from the chain of spaceless Marxist orthodoxy.
7. The radicalists however, could not develop a theory of uneven development and they followed the Marxist philosophy, emphasizing on 'wage labour' as the key to historical geography of capitalism. The fall of the state apparatus in the erstwhile USSR and eastern Europe during the 1989–91 gave the impression that socialism has fallen and is the final victory for capitalism which is based on locational analysis. Thus, it was established that the socialist governments following the model of Marx were not clean and were marred by bureaucratic and oppressive character of the real existing socialism. Thus, despite Marxian leaning the radicalists could not liberate human beings, especially the oppressed class from the natural and social restraints. In the process of creation of space, and man

and environment relationship, the followers of radicalism in geography ultimately realized that eradication of social injustice and removal of regional inequalities from both the capitalistic and socialistic societies is a gigantic task which demands deep thinking and more empirical research.

In brief, it may be said that geography cannot evolve through positivism (quantitative revolution, regional science) nor through radical methodologies. Instead, geography must return to its roots and revive its concern with the relationship of environment and social concern within place, area, region or context. An amalgamation of the quantitative and qualitative methodologies seems to be necessary for a reliable interpretation of man, space, place and environment relationship.

BEHAVIOURALISM

Dissatisfaction with the models and theories developed by the positivists, using the statistical techniques which were based on the 'economic rationality' of man led to the development of behavioural approach in geography. It was increasingly realized by the geographers that the models propounded and tested with the help of quantitative techniques, provided poor descriptions of geographic reality and man and environment relationship. Consequently, progress towards the development of geographical theory was painfully slow and its predictive powers were weak. Theories such as Central Place Theory, based on statistical and mathematical techniques, were found inadequate to explain the spatial organization of society. The economic rationality of decision-making was also criticized as it does not explain the behaviour of floodplain dweller, who does not leave his place despite the risk of flood. It was a psychological turn in human geography which emphasized the role of cognitive (subjective) and decision-making variables as mediating the relationship between environment and spatial behaviour. The axiom of 'economic person' who always tries to maximize his profit was challenged by Wolpert. In an important paper, Wolpert (1964) showed that, for a sample of Swedish farmers, optimal farming practices were not attainable. He concluded that the farmers were not optimizers but, in Simon's term, *satisficers*.

The objectives of behavioural approach were:

1. to develop models for humanity which were alternative to the spatial location theories developed through quantitative revolution;
2. to define the cognitive (subjective) environment which determines the decision-making process of man;
3. to unfold the spatial dimensions of psychological and social theories of human decision-making and behaviour;
4. to explain the spatial dimensions of psychological, social and other theories of human decision-making and behaviour;
5. to change in emphasis from aggregate populations to the disaggregate scale of individuals and small groups;

6. to search for methods other than the mathematical and statistical that could uncover the latent structure in data and decision-making;
7. to emphasize on processual rather than structural explanations of human activity and physical environment;
8. to generate primary data about human behaviour and not to rely heavily on the published data; and
9. to adopt an interdisciplinary approach for theory-building and problem-solving.

The fundamental arguments of the behavioural geography to achieve these objective are that:

1. people have environmental images;
2. those images can be identified accurately by researchers; and
3. there is a strong relationship between environmental image and actual behaviour or the decision-making process of man.

The behavioural approach in geography was introduced in the 1960s. Its origin can be traced to the frustration that was widely felt with normative and mechanistic models developed with the help of quantitative techniques. These normative and mechanistic models are mainly based on such unreal behavioural postulates as 'rational economic man' and isotropic earth surface. In normative models, there are always several assumptions, and generally the centre of attention is a set of *omniscient* (having infinite knowledge) fully rational actors (men) operating freely in a competitive manner on isotropic plane (homogeneous land surface). Many normative models are thus grossly unrealistic as they ignore the complexities of real world situations and instead concentrate on idealized behavioural postulate such as rational economic man. People behave rationally, but within constraints – the cultures in which they have been socialized to make decisions.

Behavioural geography banks heavily on 'behaviouralism'. Behaviouralism is an important approach adopted mainly by psychologists and philosophers to analyse the man-environment relationship. The behaviouristic approach is largely inductive, aiming to build general statements out of observations of ongoing processes. The essence of behavioural approach in geography lies in the fact that the way in which people behave is mediated by their understanding of the environment in which they live or by the environment itself with which they are confronted. In behavioural geography, an explanation for man-environment problem is founded upon the premise that environmental cognition and behaviour are intimately related. In other words, behavioural approach has taken the view that a deeper understanding of man-environment interaction can be achieved by looking at the various psychological processes through which man comes to know environment in which he lives, and by examining the way in which these processes influence the nature of resultant behaviour.

The basic philosophy of behaviouralism may be summed up as under:

> The behavioural geographer recognizes that man shapes as well as responds to his environment and that man and environment are dynamically interrelated.

Man is viewed as a motivated social being, whose decisions and actions are mediated by his cognition of the spatial environment.

Salient Features

The salient features of behavioural geography are as under:

1. The behavioural geographers argued that environmental cognition (perception) upon which people act may well differ markedly from the true nature of the real environment of the real world. Space (environment) thus can be said to have a dual character:
   (i) as an objective environment – the world of actuality – which may be gauged by some direct means (senses); and
   (ii) as a behavioural environment – the world of the mind – which can be studied only by indirect means.

No matter how partial or selective the behavioural environment may be, it is this milieu which is the basis of decision-making and action of man. By behavioural environment it is meant: reality as is perceived by individuals. In other words, people make choices and the choices are made on the basis of knowledge.

Thus, the view of behaviour was rooted in the world as *perceived* rather than in the world of *actuality*. The nature of the difference between these two environments and their implications for behaviour was neatly made by Koffka (1935–36) in an allusion to a medieval Swiss tale about a winter travel:

> On a winter evening amidst a driving snow-storm a man on a horseback arrived at an inn, happy to have reached after hours of riding over the winter-swept plain on which the blanket of snow had covered all paths and landmarks. The landlord who came to the door viewed the stranger with surprise and asked from whence he came? The man pointed in a direction away from the inn, whereupon the landlord in a tone of awe and wonder said: 'Do you know that you have ridden across the Great Lake of Constance?' At which the rider dropped stone dead at his feet.

This example vividly shows the difference between the 'objective environment' of the ice-covered lake Constance and the rider's subjective or 'behavioural environment' of a wind-swept plain. The rider reacted to the situation by travelling across the lake as if it were dryland – we may safely surmise that he would have acted otherwise had he known!

2. Secondly, behavioural geographers give more weight to an individual rather than to groups, or organizations or society. In other words, the focus of study is the individual, not the group or community. They assert that research must recognize the fact that the individual shapes and responds to his physical and social environment. In fact, it is necessary to recognize that the actions of each and every person have an impact upon the environment, however, slight or inadvertent that impact may be. Man is a goal-directed animal who influences the environment and in turn is influenced by it. In brief, an individual rather than a group of people or social group is more important in man-nature relationship.

3. Behavioural approach in geography postulated a mutually interacting relationship between man and his environment, whereby man shaped the environment and was subsequently shaped by it (Gold, 1980: 4).
4. The fourth important feature of behavioural geography is its multidisciplinary outlook. A behavioural geographer takes the help of ideas, paradigms, and theories produced by psychologists, philosophers, historians, sociologists, anthropologists, ethnologists and planners. However, the lack of theories of its own is coming in the way of rapid development of behavioural geography.

Historical Perspective

In geography, behaviouralism has a long history. Consciously or unconsciously, the behavioural approach has been adopted since the time of Immanuel Kant. In the last decades of the 19th century, Reclus, a French geographer, emphasized the point that in man-environment relationship man is not a passive agent. The landscape school in American geography focused on man as a morphological agent. Similarly, advocates of human geography – as a type of human ecology – owed much to the possibilist philosophical position (French School) that stressed the significance of choice in human behaviour.

Sauer, the leading American historical geographer, also recognized fully the important role played by man in shaping his socio-cultural environment by transforming and utilizing his physical surroundings. In 1947, Wright put emphasis on behavioural approach for the interpretation of man-nature interaction. He proposed that a profitable direction for geography would be to study geographical knowledge in all its forms, whether contained in formal geographical inquiries or in the vast range of informal sources, such as travel books, magazines, newspapers, fiction, prose, poetry and painting. The works of Sauer, White and many others demonstrate that people act according to habits and experience not as rational persons. Wolpert (1964) concluded in his doctoral thesis that farmers face an uncertain environment – both physical and economic – when making land use decisions, which in aggregate produce a land use map. Wolpert decided that the farmers were satisficers and not economic men. They behave according to the available information and their image about the environment and the resource. Subsequently, Kirk (1952–63) supplied one of the first behavioural models. In his model, he asserted that in space and time the same information would have different meanings for people of different socio-economic, cultural and ethnic backgrounds living in a similar geographical environment. Each individual of a society reacts differently to a piece of information about the resource, space and environment. This point may be explained by citing an example. The highly productive Indo-Gangetic plains have different meanings for different individuals belonging to various caste, creed and religion. Jats, Gujjars, Ahirs, Sainis, Jhojas and Gadas living in the same village perceive their environment differently. A Jat farmer may like to sow sugarcane in his field, a Gada and a Jhoja may devote his land to sugarcane, wheat and rice, an Ahir may like to grow fodder crops for the milch animals, and a Saini is invariably interested in intensive cultivation, especially that of vegetables. For a Saini (vegetable grower), even five acres of arable land may be a large holding, while a Jat who uses a tractor considers even 10 acres a small holding. The perceived environment of each of these farmers living in the same environment thus differs from each other both in space and time.

The followers of behavioural geography do not recognize man as a rational person or an 'economic man' who always tries to optimize his profits. Man always does not take into consideration the profit aspect while performing an economic function. Most of his decisions are based on behavioural environment (mental map) rather than on the 'objective or real environment'. The fundamental arguments of behavioural geography are that:

1. people have environmental images;
2. those images can be identified accurately by researchers; and
3. there is a strong relationship between environmental images and actual behaviour.

Figure 12.1 A Conventional Model of Man-Environment Relationship, after Boulding (1956)

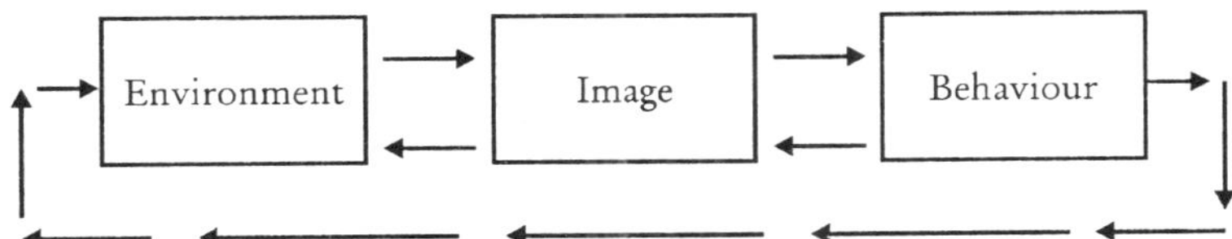

The behavioural paradigm has been shown in Figure 12.1. In this paradigm, man has been depicted as a thinking individual whose transactions with the environment are mediated by mental processes and cognitive representation of external environment. In geographical circles, this concept is derived primarily from the work of Boulding (1956) who suggested that over time individuals' developmental impressions of the world (images) are formed through their everyday contacts with the environment and that these images act as the basis of their behaviour.

The conceptual framework provided by Downs has been illustrated in Figure 12.2. This framework proposes that information from environment (real world) is filtered as a result of personality, culture, beliefs, and cognitive variables to form image in the mind of man who utilizes the environment. On the basis of the image formed in the mind of the utilizer about the environment he takes a decision and uses the resources to fulfil his basic and higher needs. Downs' framework also suggests that there exist an 'objective' and a 'behavioural' environment.

A similar but slightly more complex classification came from Porteous (1977) who recognized the existence of:

1. the phenomenal environment (physical objects);

Figure 12.2 Environmental Perception and Behaviour, after Downs (1970a: 85)

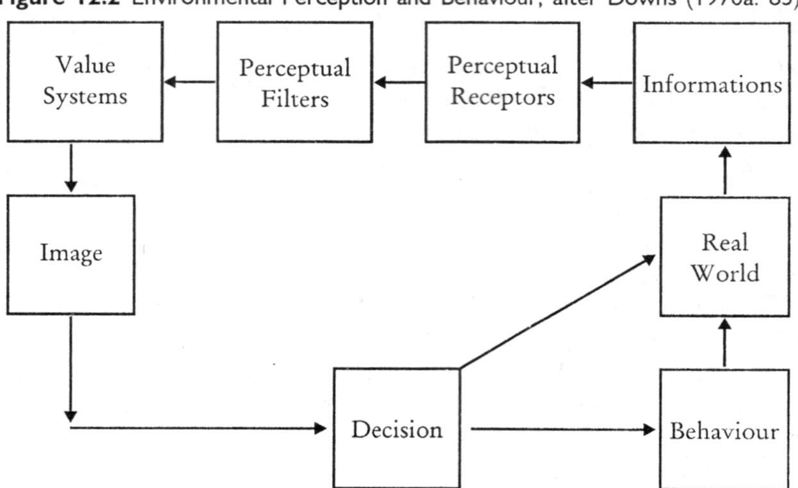

2. the personal environment (perceived images of real phenomenal environment); and
3. contextual environment (culture, religion, beliefs and expectations that influence behaviour).

Sonnenfeld (1972) went even further and proposed four levels at which the environment should be studied. The four-fold environment, advocated by Sonnenfeld, has been given as below:

(a) the geographical environment (the world);
(b) the operational environment (those parts of the world that impinge upon a man, whether or not he is aware of them);
(c) the perceptual (the parts of the world that man is aware of as a result of direct and indirect experience); and
(d) the behavioural (that part of the perceptual environment that elicits a behavioural response).

The behavioural approach in geography is a fruitful one and it helps in establishing a scientific relationship between man and his physical environment. The broad scope of behavioural geography is remarkable even by the standards of human geography. There are, however, overall biases in content towards urban topics and towards developed countries. One of the main weaknesses of behavioural geography is that it lacks in synthesis of empirical findings, poor communication, inadvertent duplication, and conflicting terminology.

In behavioural geography, the terminology and concepts remain loosely defined and poorly integrated, primarily owing to the lack of systematically-organized theoretical basis.

Another shortcoming of behavioural geography lies in the fact that most of its data are generated in laboratory experiments on animals and the findings are applied directly to human behaviour. Koestler (1975: 17) pointed to the danger of this strategy, in that behaviouralism 'has replaced the anthropomorphic fallacy – ascribing to animals human faculties and sentiments – with the opposite fallacy; denying man faculties not found in lower animals; it has substituted for the erstwhile anthropomorphic view of rat, a ratomorphic view of man'. In short, behaviouralist theories are elegant but unhelpful when it comes to understanding the real world man-environment interaction.

Behavioural geography has too often put too much emphasis on ego-centred interpretations of the environment. Specifically, scholars are critical of two assumptions on which a great deal of behavioural research in geography is based. The first assumption is that there exists identifiable environmental images that can be accurately measured. It is not clear whether an environmental image can be extracted without distortion from the totality of mental imagery. Moreover, not enough effort has gone into checking and validating the methods by which images are elicited.

The second critical assumption is that there exists a strong relationship between revealed images or references and actual or real world behaviour. The main objection to this assumption is that it is an unfounded assumption because extremely little research has been undertaken to examine the congruence between image and behaviour.

A more serious criticism of behavioural approach in geography is that it frequently views man as homo-psychologicus and tends to treat environmental behaviour as a non-dimensional phenomenon to the extent that the economic, social and political considerations that act concomitantly with environmental influences are frequently overlooked.

Another significant deficiency in behavioural geography has been the gap between theory and practice. This has been most noticeable over the question of public policy. In fact, behavioural geographers remain observers rather than participants. There is a serious lack of knowledge of planning theories and methods amongst behavioural geographers, which is an impediment to more active involvement. It is a barrier that can be removed only by developing the requisite understanding of the planning processes; it cannot be camouflaged by noble sentiments and moral tone. For instance, it will be only rarely that a small survey carried out upon a sample of students will supply the basis for far-reaching policy recommendations, yet the final paragraphs of many such works contain this seemingly obligatory element. In other words, generalization on the basis of small sample studies should not be made the basis of wider and important policy decision-making. It is, therefore, necessary to conduct research on problems that specifically deal with policy questions, that are well-versed in planning theory and methodology, and communicate the results intelligently to the interested parties. There are signs that such an approach is developing, but the gap is still wide. The future of behavioural geography would be bright only if it could improve its standing in the subject while maintaining its multidisciplinary links.

Despite several constraints and methodological limitations, behavioural geography is now widely accepted within the positivist orientation. It seeks to account for spatial patterns by establishing generalizations about people-environment interrelationship, which may then be used to stimulate change through environmental planning activities that modify the stimuli which affect the spatial behaviour of ourselves and others.

The research methods of behavioural geography vary substantially but the general orientation – inductive generalization leading to planning for environmental change – remains. Eventually, it is hoped, a 'powerful new theory' will emerge. Golledge argued that substantial advances in understanding spatial behaviour have already been made by studying – individual preferences, opinions, attitudes, cognitions, cognitive maps, perception, and so on – what he terms process variables.

HUMANISM

'Humanistic Geography' emerged in the late 1960s as the most sophisticated of a series of critical reactions against logical positivism. Humanistic geography developed due to a deep dissatisfaction with the mechanistic models of spatial science that had developed during the quantitative revolution. The cultural and historical geographers attacked the positivism from the early 1970s. In fact, it was a rejection of the geometric determinism in which men and women were made to respond automatically to the dictates of universal spatial structures and abstract spatial laws. The followers of spatial science (positivists) treated people as dots on a map, data on a graph, and number in an equation. It was at the same time that a claim was made for human geography with the human being at its very centre, a people's geography, about the real people and for the people, to develop human being for all.

One of the first geographers to attract a wide audience with his advocacy of a humanistic approach was Kirk (1951). But, it was Tuan (1976) who argued for humanistic geography. The term 'humanistic geography' was used for the first time by Yi-Fu-Tuan in 1976. The focus of humanistic geography is on people and their condition. For Tuan, humanistic geography was a perspective that disclosed the complexity and ambiguity of relations between people and place (man and environment).

Humanistic geography gives central and active role to human awareness and human agency, human consciousness and human creativity. It is an attempt at understanding meaning, value and the human significance of life events. In the humanistic stand, the intent has been to understand and recognize the dignity and humanity of the individual. Humanistic geography looks at environment and sees place that is of locales in which people find themselves, live, have experiences, interpret, understand, and find meaning. Humanists explain and interpret man and space relationship mainly with historical approach. Humanism does not treat humans as machines. It is a subjective approach which aims at *verstehn*, at an understanding of man in his environment. Humanism is a conviction that men and women can best improve the circumstances of their lives by thinking and acting for themselves, and especially by exercising their capacity for reason (Ralph, 1981).

As stated above, humanism in geography developed as a criticism against positivism and quantitative revolution in geography. The basic objection of humanists against quantitative revolution is that its tools and assumptions do not adequately explain human world and human issues, especially those relating to social institutions, attitudes, morals, customs, traditions and aesthetics.

Humanistic geographers propose that reasoning in humanistic geography should conserve contact with the world of everyday experience and recognize, if not celebrate, the human potential for creativity. The followers of this approach consider geography as 'the study of the earth as the home of man'. Humanistic geography is thus not an earth science in its ultimate aim. It belongs to the humanities and the social sciences to the extent that they all share the hope of providing an accurate picture of the human world. In humanities the scholars gain insight into the human world by focusing on what man does supremely well in the arts and logical thought. In fact, in humanities, knowledge of human world is acquired by examining social institutions. These institutions can be viewed both as examples of human inventiveness and as forces limiting the free activity of individuals. Humanistic geography achieves an understanding of the human world by studying people's relation with nature, their geographical behaviour as well as their feelings and ideas in regard to space and place.

Humanists reject the reduction of space and place to geometrical concepts of surface and point as perceived and preached by the positivists through the methodology of quantitative techniques. Place (landscape, region) is a key concept in humanistic geography. Much humanistic writing is devoted to illustrating and clarifying space. From a humanistic perspective, the meaning of a place (landscape, region) is inseparable from the consciousness of those (men) who inhabit it. The scope of place as a concept varies according to the extension of the thoughts, feelings and experiences that make the consciousness of inhabitants.

The methodology of humanists is characterized with:

(a) A self-conscious drive to connect with that special body of knowledge, reflection and substance about human experience and human expression, about what it means to be a human being on this earth, namely, the humanities.
(b) Its methods are essentially those of *literary criticism, aesthetics* and *art history*. It is essentially based on hermeneutics (the theory of interpretation and clarification of meanings).
(c) Its interest is the recovery of place and the iconography (the description and interpretation of landscape to disclose their symbolic meanings) of landscape. In other words, the interpretation of the landscape is a carrier

and repository of symbolic meaning, widening the traditional definitions of iconography – the study, description, cataloguing and collective representation of portraiture as revealing of the prevailing aesthetic of an age – to include the landscape specifically.

(d) It lays emphasis on *participant observation, interviewing, focus groups discussion, filmed approaches* and *logical inferences*, rather than statistical and quantitative techniques for establishing correlation between people and place (environment).

(e) It is a philosophy which seeks to disclose the world as it shows before scientific inquiry, as that which is pre-given and pre-supposed by the sciences.

(f) Humanists argue that 'objectification' is never the simple exercise which conventional forms of science assume them to be.

Themes in Humanistic Geography

Scientific approaches like positivism, empiricism, and quantification tend to minimize the role of human awareness and knowledge. Humanistic geography, by contrast, especially tries to understand how geographical activities and phenomena reveal the quality of human awareness. Humanistic geography does not consider human being as an 'economic man'. The propounder of humanistic geography (Tuan) explored five themes of general interest to geographers, namely: (i) geographical knowledge (personal geographies), (ii) territory and place, (iii) crowding and privacy, (iv) livelihood and economics, and (v) religion.

Geographical Knowledge (Personal Geographies)

Man is the superior form of life and has special capacity for thought and reflection. The primary task of humanistic geographers, therefore, is the study of articulated ideas (geographical knowledge). In general, broadly conceived knowledge of geography is necessary to biological survival. All animals must have it, and even the migratory birds have a mental map. For example, in the winter season, the Siberian birds migrate and many of them arrive at the Bharatpur Sanctuary (Rajasthan). These birds start their return journey in the ending part of February. These birds have a mental map which helps them to follow a set route of migration. Knowledge of geography in this sense is animal instinct, developed to varying degrees of acuity (sharpness and vision) in the different species. The people (who are not trained in geography) have a broad range of ideas regarding space, location, place and resources. All human groups possess such ideas, though their degree of articulation varies widely from group to group. For example, some primitive people like the Polynesians of the Pacific Islands are able cartographers, whereas materially more advanced people, lack the concept of map and map-making.

Territory and Place

Territory and place is also an important animal instinct. Some species of animals, like honeybee, tiger, lion, etc. defend their living space against intruders. They behave so as they regard certain areas as their own; they appear to have a sense of territory. Human attitudes and attachment to territory and to place bear a clear resemblance to those of other animals. All animals, including human beings, occupy and use space. A song bird, perched high on a tree, is able to survey the entire area that it takes to be its own. Contrary to this mammals living close to the ground cannot survey a whole area. Their whole territory is not bounded space but a network of paths and places. Similarly, the food hunters and gatherers generally do not envisage the boundary of their territory. Territory for them is therefore not circumscribed area, but essentially a network of paths and places. By comparison the shifting cultivators and settled cultivators tend to have a strong sense of property and of the bounded space (territory).

Much more than animals, man develops emotional attachment to place as he satisfies his biological needs (drinks, eat and rest). Moreover, in comparison to animals he has a strong memory. He remembers the past and thinks of the future. It is because of these feelings that he attaches so much importance to events like birth and death. Consequently, man becomes sentimental and attaches more importance to his birthplace. How mere space becomes an intensely human place is a task of human geographers to explore and explain according to the preachers of humanistic geography.

Crowding and Privacy

Crowding of a place leads to physical and psychological stress. It has been observed that the behaviour of animals at a crowded place becomes abnormal. Same is the case with man. Culture, social institutions and infrastructures, however, help in reducing these stresses. For example, people in crowded Hong Kong are no more prone to crime than people living in relatively spacious American, European and Australian cities. Contrary to this, in the Kalahari desert, the Bushmen are crowded by choice, and biological indicators of stress are absent despite the high density at places where water is available.

Similarly, privacy and solitude also affect the thinking process and decision-making of a person regarding space. In solitude a person creates his own world. All people need privacy; the degree and kind may vary. Crowded conditions make it difficult to escape the human gaze, and thereby a developed sense of self. In solitude a person creates his own world; safe from another's gaze he seems to sustain the existence of all that he sees.

Livelihood and Economics

Man sustains himself by doing some economic and social activities. All human activities appear to be economic and functional in the sense that they support the social system outside of which people cannot live. Whether it is worship of the sacred cow or ritual human sacrifice, they may be shown to have important economic consequences, and hence they are not beyond the economic rationale.

While working for his livelihood, man differentiates between life-sustaining and life-destroying activities. Production of armaments, for example, is an economic activity that provides a livelihood for many workers, but its contribution to the survival of the species is in doubt. All people and professional planners plan their economic activities according to their knowledge and technology. To what extent do planners make use of economic theory and facts in reaching the decision? How good are the results? Such questions need to be asked by the humanistic geographers.

Religion

Religion is present at varying degrees in all cultures. It appears to be a universal trait. In religion human beings are clearly distinguished from other animals.

Religion (Latin *religare*) means to bind again, i.e. to bind oneself strongly to a set of beliefs, faith, or ethic. More broadly speaking, the religious person is one who seeks coherence and meaning in his world, and a religious culture is one that has a clearly structured worldview. Since everybody tries to understand cosmos in his own way, everybody is religious. In other words, if religion is broadly defined as the impulse for coherence and meaning, then all human beings are religious. In fact, at individual level, Albert Einstein also was a religious person. The strength of the impulse varies enormously from culture to culture and from person to person. A humanistic approach to religion would require that we should be aware of the differences in the human desire for coherence, and not how these are manifest in the organization of space and time in attitude to nature or physical environment.

Historical Perspective

Although humanism in geography is traced back to Vidal de la Blache's writings, its real beginning is attributed to the Kantian philosophy. Kant asserted:

> History differs from geography only in consideration of time and space. The former (history) is a report of phenomena that follow one another and has reference to time. The latter (geography) is report of phenomena beside each other in space. History is a narrative, geography a description.

Geography and history fill up the entire circumference of our perception; geography that of space, history that of time.

The humanistic approach in geography became popular by French geographers, especially Febvre and Vidal de la Blache. The school of possibilism advocated the view that the physical environment provides the opportunity for a range of possible human responses and that people have considerable discretion to choose between them. The possibilists emphasized that 'nature is never more than an adviser' and that the *milieu interne* revealed the human being as 'at once both active and passive'. Vidal de la Blache's writings, however, bear many of the hallmarks of functionalism and pragmatism, and Vidal de la Blache himself regarded human geography as a natural science. Sauer wrote of the phenomenology of landscape in 1925. In 1936, Wooldridge claimed that historical geography must seek to view the countryside through the eyes of the farmer. In 1947, John Wright introduced the term *geosophy* as part of his contention that geographical knowledge is part of the mental stock of all humans.

In 1939, Hartshorne pleaded the cause of humanistic geography in his book, *The Nature of Geography*. He accepted that geography's basic task was essentially Kantian:

> Geography and history are alike in that they are integrating sciences concerned with studying the world. There is, therefore, a universal and mutual relation between them, even though their bases of integration are in a sense opposite – geography in terms of earth space, history in terms of periods of time. (Hartshorne, 1939)

Subsequently, it was Kirk (1951) and Tuan (1976) who laid a strong foundation of humanism in geography.

The revival of humanism in geography in the 1970s owed much to a deep dissatisfaction with the more mechanistic models developed during the 'quantitative revolution'. For this reason, its early steps were made alongside 'behavioural geography'; but the two soon parted company and humanistic geography came to recognize the essential subjectivity of both the investigator and the investigated.

Over the past decade humanistic geography has moved far from its previous position. It has advanced from its early attack on positivism to make an assault on structuralism (man is bound in socio-economic and political structures). Moreover, it has developed a more incisive and logical methodology for empirical investigation.

The anarchism of Kropotkin and Reclus and their writings were also the typical examples of humanism. The approaches of Fleure and Herbertson were also humanistic.

In humanistic geography, as discussed above, central importance is given to the actor's (man's) definition and behaviour for examining the social world. The researcher needs to discover the actor's definition of the situation, namely, his or her perception and interpretation of reality and how these relate to behaviour. In other words, the researcher must be able to see the world as the actor sees it. This approach has, however, been criticized on more than one grounds as below.

1. A general criticism of humanistic geography is that the investigator can never know for sure whether one has actually succeeded in providing true explanation or not. Undoubtedly, one can never know with certainty that a humanistic explanation is true, the same objection may be raised to positivists, quantitative and theoretical approaches. The theoretical physicist

can never be certain of his theories. In fact, the history of natural science is largely a history of abandoned theories. Yet progress has been made, because with the failure of old theories, new more powerful ones have emerged.

2. The second criticism of humanistic geography is that on methodological grounds it separates physical geography from human geography. In physical geography, the scientific techniques can be applied for theory and models building and hypotheses testing as it mainly deals with non-living objects. Contrary to this, in human geography, such quantitative techniques may not give the authentic and reliable results as the behaviour of man varies in space and time. The dichotomy of physical geography and human geography is thus harmful for the growth and development of the discipline. This dualism has eroded the geographical core of the subject – the unity of the subject.
3. In humanistic geography which is largely based on participant observation, it is difficult to develop theory, abstraction, generalization and spatial geometry. Thus, it has no sound and valid methodological base as it involves more subjective than objective research.
4. There is insignificant emphasis on applied research. For example, it does not give emphasis to applied research or policy related to the location of industry, locational analysis of land use and crop intensity. The indifference towards applied research may destroy the base of the subject. The potential dangers are greater because other disciplines have been more effective at academic imperialism than geography. For example, applied research on the economics of location (economic geography) is in danger of being consumed within economics; research on climatic variability might be swallowed by atmospheric physics; research on slope and soil might be absorbed by engineering soil mechanics; and so on.
5. Humanistic geography does not offer a viable alternative to, nor a pre-suppositionless basis for, scientific geography. Rather, the humanist approach is best understood as a form of criticism (Entrikin, 1976).
6. Humanistic approach is 'methodologically obscure'. The goals of understanding man's meaningful experience seem to lead to a situation in which any method is acceptable. It is not a practical philosophy as it involves thinking rather than practical activity. Its methodology is eclectic and sources of interpretation are numerous and therefore it becomes difficult to ascertain the reality.
7. Humanistic geography has also been criticized on the ground that it is unscientific. The new age sentiment considers it as incapable of producing generalization, theories and laws about human behaviour.

Most of the criticisms of humanistic geography are, however, ill-founded. Is it not a fact that all history is the history of human thought? The geographical reality of a place or region may be appreciably understood through participant observation and social interaction, by giving central and active role to human awareness, and human agency.

GEOGRAPHY AND PUBLIC POLICY

As discussed in Chapters 6 and 7, geography as a university discipline got recognition in the early decades of the 19th century in the German universities and subsequently in the French and British universities. During the period of evolution, geography, like all other sister social science disciplines, faced many philosophical and methodological problems. Geography did not develop as a well-regulated activity. It followed a process of varying tensions in which tranquil periods, characterized by steady accretion of knowledge, are followed by crisis which can lead to upheaval within subject discipline and breaks in continuity. In each phase of tranquillity and crisis, geographical literature was and has been written with changing philosophies and methodologies; the philosophy and methodology being largely governed by the individual beliefs of the author, the political system, the social requirements of the people of the region and its economic institutions.

The last sixty years can be regarded as a period in which enormous geographical literature has been produced. This literature in the shape of books, research papers and monographs pertains to teaching, research, professional employment and pragmatic plans for the public and private bodies. Geography up to the Second World War, however, was regarded as a discipline providing general information about topography, relief features, weather, climate, mountains, rivers, routes, towns, cities and seaports. Geography for most of the people was nothing but general knowledge. In the recent past, geographers have, however, adopted a new strategy in the restructuring of their courses and designed the syllabi around the theme of social welfare, making the subject the principal source of awareness of local surroundings, regional milieu, environmental pollution and world environment. Geographers are venturing into the areas of environmental management and problems of pollution to make the social environment conducive for the proper development of individuals and societies. In order to achieve the welfare target, geographers are attacking social problems and exploring the causes of socio-economic backwardness, environmental pollution, and uneven levels of development in a given physical setting. Now, the main objective of geographical teaching and research is to train students in the analysis of phenomena, so that they can take up subsequently the problems of society as the fields of their research and investigation, thereby helping the local, state and national administration to overcome the regional and intra-regional problems. The social problems are being tackled with approaches ranging from positive to normative, from radicalism to humanism, and from idealism to realism. In brief, geographers are increasingly concerning themselves with the problems of society, conditions of mankind, economic inequalities, social justice, and environmental pollution.

For the reduction of regional inequalities and for the improvement of the quality of life, the main concern of geographers is with what should be the spatial distribution of phenomena instead of with what it is. It is in this context that the spatial inequality in social amenities and living standards is investigated by geographers to trace the origin of disparity rather than to condemn injustice.

Historically, in the initial phases of its development, the main area of employment of geography students in the developed countries was teaching. In the Third World countries, geographers even today are not much actively involved in the process of planning and development. Regrettably, research had less important place in the geographical profession than in many of the social and physical sciences. Moreover, the research done by individuals mainly remained confined to the libraries and has hardly been utilized for the purpose of planning. Unfortunately, the policy-makers in developing countries like India do not seem to be aware of the spatial dimensions of their problems of policies. Another reason is widespread ignorance of and even prejudice against geography particularly among the present generation of decision-makers whose opinions have been shaped by the experience of the previous generation school geography – when geography occupied a low place and was, as a subject, considered to be nothing more than general knowledge. In fact, in most of the social fields, very little contribution had been made by geographers, and in the past they could not significantly suggest alternative strategies for the spatial organization of space. The last three decades have, however, seen some particularly important changes in the subject matter, philosophy and methodology of geography.[50] The major issues on which the geographers are concentrating include poverty, hunger, pollution, racial discrimination, social inequality or injustice, environmental pollution, and use and misuse of resources. Some of the leading works which have been useful in the public policy making are: *Geography of Crimes*,[51] *Black-Ghetto*,[52] and *Geography of Social Well-being*.[53] The quantitative revolution of the 1960s in geography gave to it some kind of intellectual vigour so essential for the rigorous analysis required in any public context and in the formulation of proposals for public policy. It is an encouraging fact that now geographers all over the world are envisaging research on social problems with a welfare theme. They are working with a pragmatic approach to overcome the problems of inequalities. In fact, the objective of welfare geography is the evolution of the social desirability of alternative geographical state.[54]

Scientific revolution entered in geography in the early 1970s. The pragmatists advocated the use of scientific methods (positivism) for finding solutions to human problems. It is with this intention that scholars like David M. Smith had adopted the welfare approach while discussing the problems and prospects of human geography.[55] The welfare geography has been defined differently by different scholars of geography. In the words of Mishan, 'theoretical welfare geography as that branch of study which endeavours to formulate positions by which we may rank, on the scale of better or worse, alternative geographical situation open to society'.[56] While Nath expressed 'welfare geography' as that part of geography where we study the possible effects of various geographical policies on the welfare of society.[57] In the spatial context, Smith defined welfare geography as the study of 'who gets what, where and how'.[58]

The geographical 'state' or situation, in the sense used above, may refer to any aspect of the spatial arrangement of human existence. It may relate to the spatial allocation of resources, income, or any other source of human well-being. It may concern with the spatial incidence of poverty or any other social problem. The expression may also be used in desirable industrial location pattern, the distribution and concentration of population, the location of social service facilities, transportation network, patterns of movement of people or goods and any other spatial arrangement which has a bearing on the quality of life as a geographically variable condition. And beneath them all, in the type of society – the economic, social, political structures that generate the pattern.[59]

The welfare approach, nevertheless, has had different meanings in the different periods of human history. The humanist endeavours in various periods of different nations and societies like Jewish, Christians, Muslims, Confucians, Hellenistic, Scientific, Realists, Marxist and Existentialists, and many other forms of humanism appeared on the map of intellectual history.[60]

The geographers who are mainly concerned with the problems of society and trying to formulate pragmatic proposals for public policy clarify the description and explanation of the phenomena. On the basis of such analysis they evaluate their plans and prescribe suitable strategies for balanced development. Description involves the empirical identification of territorial levels of human well-being – the human condition. This is a major and immediate research area in which surprisingly little work has been done in India and in other developing countries. Explanation covers the how .... It involves identifying the cause and effect links among the various activities undertaken in society, as they contribute to determining who gets what and where. This is where the analysis of the kind of economic, demographic and social patterns mentioned above logically fits into the welfare structure. Evaluation involves making judgement on the desirability of alternative geographical states and the societal structure from which they arise. To say that one spatial pattern of human well-being is preferable to another is to say that a higher level of welfare is attached to it. Such judgements must be made with reference to equity as well as the efficiency criteria with which the geographer is more familiar. Geographical patterns of all kinds can be judged with respect to their profit maximizing and cost minimizing criteria. Prescription requires the specifications of alternative geographical state, and alternative societal structures designed to produce them. Prescription involves answering the ethical question: who should get what, where? Implementation is the final process replacing the state deemed undesirable by something superior. It covers the question of how, once it has been decided who should get what, where. Just what role should be adopted by geographer *qua* geographer in a changing world.[61]

In the contemporary world, there is a growing awareness among geographers that all physical development has a potential income redistributive impact. Any development proposed at any time in space has the capacity to benefit some people in some places more than others. It would be very difficult to construct anything anywhere which would be of equal benefit to every citizen. This is because of this situation that the benefits of government developmental policies in developing societies do not percolate down to the lowest strata of these societies. Geographical distance and accessibility mean that some people will be better placed to enjoy the advantages or disadvantages, whether the structure is hospital, school, road, railway, community hall, cinema, theatre, park, recreational place or sewage works. Therefore, location decisions and plans for spatial allocation of resources must be made with utmost care, if the benefits and penalties are to be proportional among the population in a predictable and equitable manner. In such public policy decisions, geographers' role becomes imperative as they have the basic training in the spatial and temporal analysis of phenomena.

Spatial allocation problems are associated with identification of priority areas, planning routes, location of factories or other sources of employment, spatial arrangement of infrastructure, facilities providing medical care, housing complexes, shopping centres and allocation of land for different agricultural-urban and recreational uses. Each of these decisions can be made in a number of ways, and each decision can have a different impact. Geographers by their training can build up more sophisticated knowledge of the process of development. This involves disentangling complex networks of economic, social and cultural relationships and also the ecological relationships in a balance, so easily disturbed by ill-conceived 'development' projects. Geographers by allocation, analysis and synthesis of space can contribute, successfully, meaningfully and effectively to the formation of public policy.

In developing countries like India there is a high degree of internal inequality. In the Third World nations wealth and power are still largely in the hands of a small urban elite or big landlords. The most obvious example is South Africa. In India also, nearly 50 per cent of the population is below the poverty line while over 50 per cent of the total national assets are in the hands of only about two dozen families. Moreover, in India, most of the economic activity is concentrated in metropolitan cores, though still about 70 per cent of the total population is residing in the rural areas (2011). The urban biased industrial and social infrastructural policy adopted by planners is widening the gap between the rich and the poor on the one hand and rural and urban population on the other.

The highly advanced countries like the USA, Canada, Japan, Germany, France, and Australia also have spatial variations in levels of human well-being. In the United States, the general material standard of living is higher than anywhere else in the world. Yet, millions of Americans, especially Negroes, live in poverty and social deprivation in ghettos – city slums. In parts of the rural south of USA (Texas, Georgia, etc.) people can be found living in conditions as bad as anywhere in South Africa. In these urban slums, the rate of crimes and drug addiction is fairly high. The persistence of widespread poverty in American slums – the most affluent society in the world – is a contradiction which underlines the failure of economic growth under a capitalist system to uplift the lives of all people to current standard of decency. In 1976, according to the US Census Bureau, about 12 per cent (26 million) Americans have income below the officially recognized poverty line.

One of the arguments put forward by the capitalist for the existing regional and intra-regional inequalities is that peoples are not born equal and they cannot be equal in their societies owing to the unequal distribution of the means of production. In fact, the chance of birth into a particular family or group in a particular locality, immediately constrains a child's opportunity. This situation gets further aggravated if the socio-political and economic organization is planned with an urban-biased or rich people-oriented policy. The planners in consultation with geographers can construct general social amenities which can benefit all sections of the society. Geographers, however, cannot be a panacea to all the ills, inequalities and socio-economic imbalances. They know it much better than any other experts that they cannot make all deserts fertile, eliminate drought and create mineral resources where none exist in nature. There are physical limitations in the development of societies living in harsh environment. Such people, however, can have better chances of development if their resource base and needs of society will help to highlight fundamental issues of choice, efficiency and equity. Moreover, it would be useful in the provision of public services and other aspects of local life quality.

Geographers have the ability to analyse the spatial dimension of environmental problems and more particularly, to handle, analyse and interpret spatially distributed data. This awareness of and facility of handling the spatial dimension, which is a major ingredient of all problems of environmental and resource management, is something not generally provided by those in other disciplines and tends to be overlooked if a geographer does not provide it. A welfare society needs better allocation of commodities, better distribution of commodities and better allocation of means of production among individuals (groups or classes) and among places. All these things are more easily achievable if geographers who deal with the man-environment interaction and examine the spatial distribution of phenomena are actively involved in the process of planning and formulation of public policies at the local, regional, national and international levels. In countries like Sweden, Norway, the Netherlands, Israel, Denmark, USSR, France, New Zealand and Australia where geographers in collaboration with other scientists design public policies the use and beneficial effects of resources are reaching all sections of the societies. Geographers in India can provide pragmatic proposals for solving the various socio-economic and employment problems facing the rapidly increasing population. By their efforts geographers can consider the causal relationships between inequity, the spatial organization of society and social structure. Public policies about reorganization and redistribution can be designed through planning by the experts who have expertise in man-environment interaction and spatial analysis of phenomena. For this purpose, geographers have to assert themselves through their applied and utilitarian researches.

Applied Geography

The application of geographical knowledge and skills to the solution or resolution of problems within society is the main concern of applied geography. The geographers for quite sometime were occupied with the study of the production and distribution of goods and the exploitation of natural resources, while ignoring important conditions of *human welfare* and *social justice*. Applied geographers argued that:

> Research should highlight *particulars*, and teaching should place emphasis on *a man in harmony with nature* rather than master of it, on *social health* rather than *economic health*, on equity rather than efficiency, and on the quality of life rather than the quantity of goods. (Adams, 1976)

About applied geography, the first major statement was made by L.D. Stamp (1960) in *Applied Geography*, which presented the geographer's unique contribution as the holistic approach in which he sees the relationship between man and his environment (resources) with its attached problems, as a whole. The relationship between man and environment can be discerned by: (i) survey in the field, and (ii) gathering of facts (data) systematically and objectively. The data thus obtained is to be plotted on maps and studied cartographically. Such surveys and analyses were perceived by Prof. Stamp as extremely relevant to many of the world's most pressing problems, such as population pressure on land, economic development, inequalities in the standard of living and regional planning. He conducted the first Land Use Survey of Great Britain and published the book, *Land of Britain its Use and Misuse*.

Stamp presented the geographer as an information gatherer and synthesizer, who could stand outside the political processes within which the formulation and pursuit of planning goals were set. It was because of this direction to geography by Stamp that many trained geographers were employed as central and local government planning officers in the 1940s, 1950s and 1960s, where their skills were relevant to the focus on land use survey and planning. Geographers were also employed by national government agencies, notably in wartime when information about environment was needed as part of military intelligence. They were also involved in the interpretation of aerial photographs, out of which has grown the discipline's expertise in remote sensing, and GPS which are now the most important tool of geographers for survey and mapping work.

From the 1960s onwards, developments in geographical approaches and techniques like quantitative revolution and later, Geographic Information Systems (GIS), computer aided cartography enhanced the range of available applied contributions and led to employment of geographers for private as well as public sector work. Quantitative models were developed not only to describe but also to predict, as in the study of traffic flows. The Entropy Maximizing Models were later adapted to provide more comprehensive information.

After Stamp's land use studies geographers focused on society-environment interaction. The geographers also concentrated on the problems of environmental hazards and rehabilitation. They founded the Journal of Applied Geography in 1981 to concentrate on such type of issues. Subsequently, applied geographers like (Coppock, 1974) took up the theme of 'geography and public policy'.

Throughout the 1980s there was increasing pressure on geographers to become more involved in applied work. This reflects growing political requirements on universities and other institutions of higher learning to increase their contributions to tackling society's problems and to earn larger proportions of their incomes from such research and consultancy activity.

In order to increase the 'marketable skills' some geographers stressed on the saleability of the technical advances - a characteristic of modern geography. It was with this objective that the applied geographers utilized the tools of GIS and remote sensing.

Applied geography as practised by geographers working outside academia is more apparent in North America than in European countries. There is a much greater recognition of a graduate profession of geography in North America. The Association of American Geographers has a very large and active Applied Geography Speciality Group.

In the capitalist economies, many geographers have worked in the field of urban and regional planning. More recently, geographers have been obtaining contracts and employment from private sector firms in a wide variety of business. In socialist countries, especially those of eastern Europe and the erstwhile USSR, much of the work of academic geographers (between 1950 and 2000) was directed by the state apparatus (of which they were a part) towards the solution of economic and environmental problems and applied geography was volumetrically greater there than in most of the capitalist world.

Not all geographers, however, have accepted this call for a particular form of applied geography, perceiving it as a narrow presentation of their discipline's expertise (especially the emphasis on technical skills). Some recognize three types of geography, each of which has its own applied programme: (i) that form of geography, such as *spatial analysis*, which adopts the tenets of *positivism* and seek technical solutions to problems within the applications of GIS to location and allocation issues; (ii) *Humanistic Geography*, which gives central role to human agency; and (iii) *Radical Geography* - the goal of which is to emancipate people by clarifying from them the nature of the society in which they live and thereby enabling them to participate in its restructuring. Stoddart (1987) has criticized the work of applied geographers by stating that they are concentrating on 'trivial issues'. He stressed that geographers should instead 'do some real geography' and focus on the large issue of people-environment relationship.

Postmodernism

Postmodernism is a recent movement in humanities, philosophy, arts and social sciences. It developed in reaction to historicism in modern geographic thought. Historicism gives emphasis on biography (chronological description of individual and collective events). Consequently, it (historicism) neglects spatiality. In the opinion of Soja (1989) historicism is based on an overdeveloped historical contextualization of social life and social theory that actually submerges and marginalizes the geographical or spatial imagination. This results into subordination of space to time that obscures geographical interpretation of the changeability of social world.

The term postmodernism has been used differently by different authors in different disciplines. Postmodernism in geography, however, stresses on openness in social and geographical enquiry, and artistic experimentation and political empowerment. In fact, the difference between modern and postmodern is not at all clear. In use postmodernism is a shorthand for a heterogeneous movement which has its origin in architecture and literary theory. Postmodernism has, however, a wide range of explicit and implicit meanings and its core is hard to identify. In the opinion of Dear (1994: 3):

> Postmodernity is everywhere, from literature, design, art, architecture, philosophy, mass media, clothing style, to music and television. Postmodernism raises urgent questions about place, space and landscape in the production of social life.

Those who supported postmodernism argued that social and historical processes are differently constituted in different places/regions, and therefore, the historical flow is not the same everywhere. For example, postmodern novels have an apparently chaotic structure when they try to represent different things happening simultaneously in different places, and postmodern architecture lacks a clear, functional structure.

The problem of synchronicity has long been recognized by geographers. Darby pointed out:

> A series of geographical facts is much more difficult to present than a sequence of historical facts. Events follow one another in time in an inherently dramatic fashion that makes juxtaposition in time easier to convey through the written word than juxtaposition in space. Geographical description is inevitably more difficult to achieve than is historical narrative.

Although it is difficult to define postmodernism Dear (1986) usefully classified postmodernism into three components: (i) postmodern style, (ii) postmodern method, and (iii) postmodern epoch.

(i) Postmodernism as Style

Postmodernism as a style originated in literature and literary criticism, and spread to other artistic fields such as design, film, art, photography, and architecture, the general trend involved the promotion of difference and the lack of conformity to over-riding structural imperatives.

The postmodern style of architecture has been criticized for its attention to facade variation, a diversity of colour, design elements and iconography which is no more than a superficial gift wrapping. However, this criticism is incomplete, for style is centrally implicated in the constitution of meaning and identity.

(ii) Postmodernism as Method

Postmodernism as method is, according to Dear, likely to be the most enduring of the three main trends. It eschews the notions of universal truths and meta-theories which can account for 'the meaning of everything'. No portrayal can claim dominance over another; separate theories are incommensurable and so cannot be evaluated: 'even the attempt to reconcile or resolve the tensions among competing theories should *a priori* be resisted'. Deconstruction is a principal strategy, a mode of critical interpretation which seeks to demonstrate how the (multiple) positioning of an author (or a reader) in terms of culture, class, gender, etc. has influenced the writing (and reading) of a text. Deconstruction is essentially a destabilizing method, throwing into doubt the authority claims of preceding traditions, and seeking to praise loose alternative readings of texts. In human geography, Olsson (1980) was the earliest exponent of deconstruction and remained its most innovative and skilful practitioner.

(iii) Postmodernism as Epoch

Postmodernism may be thought of as an epoch, a historic era in which changes in culture and philosophy are themselves located in the evolution of a global economy and geopolitics. Thus, postmodernism is the culture of *late capitalism*. Postmodernism as epoch portrays current developments within society as a major radical break with the past hence use of the term 'postmodernity' to contrast it with the modernity of the previous epoch. These 'new times' are characterized by difference, so that study of the postmodern epoch involves:

> ... grappling with the fundamental problem of theorizing contemporaneity, i.e. the task of making sense out of an infinity of con-composed, of obsolete, current, and emergent artifacts; but how do we begin to codify and understand this variety?

This emphasis on 'heterogeneity, particularity and uniqueness' (Gregory, 1989a: 70) undoubtedly attracted some human geographers to postmodernism – or, as Dear (1994: 3) expressed it in a quote from the *New York Times*, 'the great lesson of the twentieth century is that all the great truths are false'. Human geographers under the sway of modernism emphasized order in their promotion of spatial science, when their empirical observations (as their critics pointed out: see p. 184) could really only identify disorder, which suggested the absence of generally applicable theories and universal truths (Barnes, 1996). Postmodernism gave them a philosophical hanger, recognizing (Gregory, 1989a: 91–92) that:

> ... there is more *disorder* in the world than it appears at first sight. It is not discovered until that order is looked for ... we need, to go back in parts on the question of areal differentiation but armed with a new theoretical sensitivity towards the world in which we live and to the ways in which we represent it.

In human geography, postmodernism is, in a very real sense, 'post-paradigm', i.e. postmodern writers are immensely suspicious of any attempt to construct a system of thought which claims to be complete and comprehensive. The paradigms like environmental determinism, possibilism, positivism, structuralism, behaviouralism, humanism and system approach have been rejected by those who believed in postmodernism. In brief, postmodernism presents a substantial critique to the approaches which dominated geography between 1950s and 1980s.

Postmodern writers are also hostile to the totalizing ambitions of the conventional social sciences and humanities. They reject the notion that social life displays what could be called a 'global coherence' or the structure of a society regulates its everyday life in some automatic, pre-set fashion. They opposed the idea of structuralism and it is largely through this opposition that postmodernism is sometimes called as 'post-structuralism'. However, postmodernism is not another humanism.

One of the distinguishing features of postmodern culture is its sensitivity to heterogeneity, particularity and uniqueness. Thus, it resulted in a remarkable return to *areal* differentiation. But it is a return with a difference.

Time Geography

The time geography was developed by the Swedish geographer Torsten Hägerstrand and his associates at the University of Lund (The Lund School). Hägerstrand conceives of time and space as providers of 'room' for collateral processes. In the opinion of Hägerstrand, 'every situation is inevitably rooted in past situations'. All human beings have goals. To attain these, they must have projects, series of tasks which act as vehicle for goal attainment and which, when added up, form a project. The time geography is based on *naturalism* (the thesis that there is or can be an essential unity of method between the natural and the social sciences).

Time geography emphasizes the continuity and connectedness of sequences of events, which necessarily take place in situation bounded in time and space, and the outcome of which is thereby mutually modified by their common localization.

The concept of time geography is analogous to Kant's view of history and geography as the architects of 'physical' rather than 'logical' classifications. Kant considered that knowledge could be classified in two ways: either logically or physically. The logical classification collects all individual items in separate classes according to similarities of morphological features; it could be called something like an archive and will lead, if pursued to a natural system, e.g. rocks in geology, plants in botany and animals in zoology. The physical classification, in contrast, collects individual items which belong to same time or the same space. Geography and history fill up the entire circumference of our perceptions: geography that of space, history that of time (cited in Hartshorne, 1939).

According to Hägerstrand, time and space are resources that constrain activity. Any behaviour requiring movement involves individuals tracing a path simultaneously through space and time (Figure 12.3). In the Figure 12.3, the movements along the horizontal axis indicate spatial traverses and those along the vertical signify the passage of time. All journeys, or lifelines, involve movement along both and are displayed by lines that are neither vertical nor horizontal; vertical lines indicate remaining in one place; horizontal lines are not possible for people, though they are (or virtually so) for the transmission of messages.

Figure 12.3 The Time-Space Prism. In this simple example, a person cannot leave a place until time I and must return there by time II; the prism between these two times indicates the maximum available spatial range (in miles/km or some other metric).

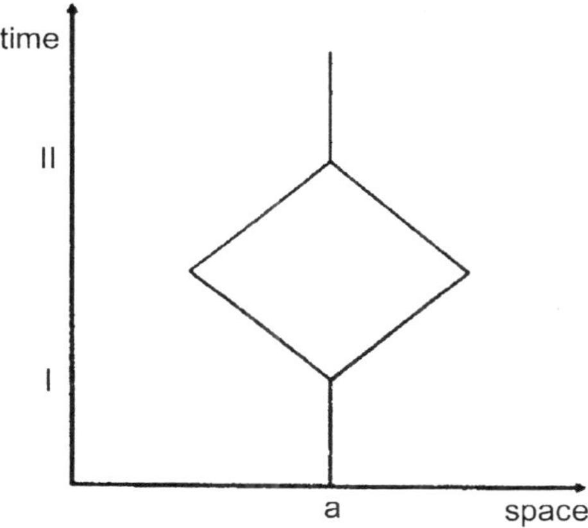

Hägerstrand developed an elementary time-space notation from standard Lexis-Becker diagrams used in demography. His basic framework can be represented as a web model (see Figure 12.4) spun across four basal propositions:

Figure 12.4 Time Geography (Hägerstrand's Web Model)

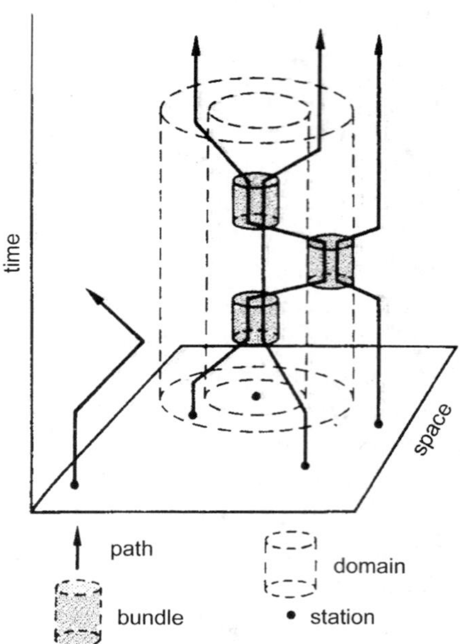

(a) Space and time are resources on which individuals have to draw in order to realize projects.

(b) The realization of any project is subject to three constraints as under:

1. *Capability constraints*, which limit the activities of individuals through their own physical capabilities and/or the facilities which they can command. Over time, these include the biological need for about eight hours sleep in every 24 hours, whereas movement across space is constrained by the available means of transport. The individual prism contains a set of feasible time-space paths (lifelines). These paths are succession of situations traced by individuals flowing through a constellation of accessible stations, e.g. farms, factories, schools and shops.
2. *Coupling constraints* require certain individuals and groups to be in particular places at stated times (e.g. teachers and pupils in schools), and thus limit the range of mobility during 'free time'. Coupling constraints define time-space bundles.
3. *Authority* or steering constraints may preclude individuals from being in stated places at defined times.

(c) These constraints are interactive rather than additive, and together they delineate a series of possibility boundaries which mark out the paths available for individuals or groups to fulfil particular projects.

(d) Within these structural templates, *competition* between projects for 'free paths' and 'open space-times' is the 'central problem for analysis' and is mediated by specific *institutions* which seek to maintain an essential time-space coherence (Hägerstrand, 1973, 1975).

Geographical study of such situations has traditionally involved the concept of 'landscape', devised to represent the 'momentary thereness and relative location of all continuants'. Hägerstrand claimed that this concept insufficiently incorporates the human body subjects, the keepers of memories, feelings, thoughts and intentions, and initiators of projects, and preferred the concept of diorama, normally used to denote static museum display which depicts people and animals in their usual environments. The concept implied to Hägerstrand that 'all sorts of entities are in touch with each other in a mixture produced by history, whether visible or not, ... (we) appreciate how situations evolve as an aggregate outcome quite apart from the specific intentions actors might have had when they conceived and launched projects out of their different positions'.

Hägerstrand's seminal papers were written before the 'information technology' which enabled the almost instantaneous transmission of information to multiple sites around the globe and promised virtual reality, whereby people in one place could operate as if in another. The time geography, as advocated by Hägerstrand, was appreciated by many geographers. In the opinion of Baker (1981), time geography could be of value in a reorientation of geographical work.

The time geography and its methodology have been appreciated by many geographers. Its main criticism is that most of the empirical work carried out under the aegis has been illustrative and confined to small-scale, short-term and essentially individual level. It gives little importance to the institutional factors which mould the personality of individuals and influence their decision-making capacity for their economic activities and paths and projects.

Postmodernism and Feminism

In addition to race, and ethnicity, gender is one of the main concerns of postmodern literature. Feminist geography traces the interconnections between all aspects of daily life in the economic, social and cultural geography. In other words, feminist geography 'emphasizes questions of gender inequality and the oppression of women in virtually all spheres of life', and its goal includes uncovering and countering such inequality and discrimination within the geographical profession itself.

Feminist geography, according to Johnson (1989), involves recognizing women's common experience, and resistance to, oppression by men, and a commitment to end it 'so that women can define and control themselves'. Evaluation of geographical practice will demonstrate that it is sexist, patriarchal and phallocentric and will open the way to emancipation, by providing a guide to political practice.

Feminists like Rose (1993) asserted that:

1. the academic discipline of geography has historically been dominated by men;
2. within the profession, women have been patronized, harassed and marginalized;

3. feminism remains 'outside the project' of geography; and
4. the domination of geography by men has serious consequences both for what counts as legitimate geographical knowledge and who can produce such knowledge .... [Men] have insisted that geography holds a series of unstated assumptions about what men and women do, and that the discipline concentrates on spaces, places and landscapes that it sees as men's.

The discipline of geography is thus mainly 'masculinist' in which the concerns of women have been ignored. Moreover, gender differences are human creations.

Postmodern or postrational feminism argues that treatment of women as a single category involves linking together very different groups with separate experiences and needs. The postmodern geographers argued that powerful groups within societies have imposed their interpretations of landscape and nature. They no longer focus solely on male-female differences but incorporate race, class and sexual orientation.

The postmodern human geographers outlined the following main themes to be explored as:

1. *moral philosophy, moral geographies and the geographer's morality* – stressing the need to downplay the dominant economic focus of geography and to replace it by the moral frameworks which form life;
2. *processes of social differentiation* – involving greater appreciation to race, ethnicity, class, sexuality, age, health, etc. which have largely been taken for granted in discussions of spatial differentiation;
3. *constructions and boundaries of the self* – how individuals define themselves and relate to others within the context of various categories used within society, which involves interrogating psychoanalytic literature, something not previously undertaken by geographers;
4. *globality and territoriality* – the location of individuals and groups in spaces and places and cultural practices involved; and
5. *society, culture and natural environment* – addressing the social construction of 'nature' and 'environment' their importance to approaches to the resolution of environmental problems.

Notes

1. Harvey, D., 1969, *Explanation in Geography*, p. 6.
2. Ibid.
3. Harvey, M.E. and Holly, B.P., 1983, *Themes in Geographic Thought*, London.
4. Holt-Jensen, A., 1981, *Geography: Its History and Concepts*, London, pp. 76–82.
5. Ibid.
6. Van-Valkenberg, S., 1952, 'The German School of Geography' in *Geography in Twentieth Century*, ed. G. Taylor, London, pp. 91–117.
7. Holt-Jensen, A., op. cit., p. 80.

8. Beck, R.N., 1969, *Perspectives in Philosophy*, New York.
9. Harvey, M.E. and Holly, B.P., op. cit., p. 35.
10. Fraizer, J.W., 1983, 'Pragmatism: Geography and the Real World', in *Themes in Geographic Thought*, op. cit., p. 97.
11. Morril, R.J., 1970, 'Geography and Transformation of Society', Pt.II, *Antipode*, 2, pp. 4–10.
12. Aune, B., 1970, *Rationalism, Empiricism and Pragmatism: An Introduction*, New York.
13. Ibid.
14. James, W., 1932, *A Pluralistic Universe: Hibbert Lectures on the Present Situation in Philosophy*, Longmans, New York.
15. Ley, D. and Samuels, M.S. (eds), 1978, *Humanistic Geography: Prospects and Problems*, Chicago.
16. Harvey, M.E., op. cit., pp. 73–98.
17. Ibid.
18. Ibid.,p.84.
19. Hempel, C.G., 1959, 'The Logic of Functional Analysis' in L. Gross (ed.), *Symposium on Sociological Theory*, pp. 271–307.
20. Goode, W.J., 1973, *Explanations in Social Theory*, Oxford University Press, New York.
21. Harvey, D., op. cit., p. 65.
22. Samuel, M.S., op. cit., pp. 22–40.
23. Sartre, J.P., 1966, *Being and Nothingness*, New York.
24. Ibid.
25. Ibid.
26. Samuel, M.S., 1983, 'An Existential Geography' in *Themes in Geographic Thought*, London, pp. 115–132.
27. Ibid., p. 129.
28. Acton, H.B., 1967, 'Idealism' in P. Edwards (ed.), *The Encyclopedia of Philosophy*, New York, Vol. IV, p. 110.
29. Guelke, L., 1983, *Idealism in Themes in Geographic Thought*, op. cit., p. 133.
30. Ibid., p. 134.
31. Ibid., p. 139.
32. Ibid., p. 140.
33. Collingwood, R.G., 1956, *The Idea of History*, New York.
34. Gibson, E.M.W., 'Realism in Geography', op. cit., p. 148.
35. Ibid.
36. Russell, B., 1959, *Wisdom of the West*, Doubleday, Garden City.
37. Gibson, op. cit., p. 150.
38. Ibid., 152.

39. Passmore, J., 1966, *A Hundred Years of Philosophy*, New York.
40. Hosse, M., 1966, *Models and Analogues in Science*, Norte Dame III.
41. Harvey and Holly, op. cit., p. 207.
42. Marx, K. and Engels, F., 1976, *The German Ideology in Marx and Engels*, collected works, Vol. V, New York.
43. Peet, R.J. and Lyens, J.M., 'Marxism: Dialectical Materialism, Social Formation and the Geographical Relation', in *Themes in Geography Thought*, London, p. 190.
44. Godelier, M., 1977, *Perspectives in Marxist Anthropology*, Cambridge.
45. Jawad, A., 1992, *Development and Social Order*, Anmol Publications, New Delhi.
46. Leiss, W., 1974, *The Domination of Nature*, New York.
47. Marx, K., 1967, *Capital*, 3 Vols., International Publications, New York.
48. Ibid., p. 331.
49. Poet, R., 1977, 'The Development of Radical Geography in the United States' in R. Poet (ed.), *Radical Geography*, Chicago, pp. 6–30.
50. Chisholm, M., 1975, *Human Geography: Evolution or Revolution*, Penguin, Harmondsworth.
51. Harris, K.D., 1974, *The Geography of Crime and Justice*, McGraw Hill, New York.
52. Rose, H.M., 1971, *The Black Ghetto: A Special Behavioural Perspective*, McGraw Hill, New York.
53. Smith, D.M., 1972, 'Towards a Geography of Social Well-being: Inter-state Variations in the United States', R. Peet, editor, Geographical Perspectives on American Poverty, *Antipode*, Monograph in Social Geography, I Worcester, Mass, pp. 17–46.
54. Henderson, J.M. and Quandt, R.E., 1958, *Micro-economic Theory: A Mathematical Approach*, McGraw Hill, New York.
55. Smith, D.N., 1977, *Human Geography: A Welfare Approach*, London.
56. Mishan, E.J., 1964, *Welfare Economics: Five Introductory Essays*, New York.
57. Nath, S.K., 1969, *A Reappraisal of Welfare Economics*, London.
58. Smith, D.M., op. cit., p. 7.
59. Ibid., p. 10.
60. David, Ley and Samuel, M., 1978, *Humanistic Geography*, Chicago.
61. Smith, D.M., op. cit., pp. 9–10.

13

Feminist Geography

Feminist Geography is a branch of Human Geography, concerned with the spatial differentiation and organization of female activity and their use of the physical environment (resources). In other words, it deals with the females' relations in the space and the spatial structures that underpin those relations.

Feminist geography emphasizes the oppression of women and gender inequality. It focuses on how gender and geographies are mutually produced and transformed, and the ways in which gender differentiation permeates social life. Feminist geographies perspectives draw on feminist politics and theories to explore how gender relations and geographies are mutually structured. It developed after 1970s, drawing inspiration from the women's movements of the 1960s.

The present chapter looks at feminist thought in geography in the context of broader economic, political, philosophical and social-theoretic tendencies. Feminist theory grew from ideas generated by the political practices of social movements. Feminism is a set of political and social philosophies and practices that centre on the role of gender in the constitution of social life. The main focus is on the critique of hierarchical construction of binaries associated with masculinity and femininity. In other words, feminism is a system of thought radicalized by women's experiences of multiple forms of oppressions. These oppressions include gender, class, race, religion, and region. Feminist geography bears the marks of women's exclusion from the upper echelons of the organized discipline and from a research agenda which long ignored their existence. In addition, feminist geography has had a sophisticated radical view since its first appearance in the middle 1970s.

Women's Movements – Feminist Thought

As stated above, feminist theory begins with a history of women's movement, usually divided into first, second and third waves. Each wave has counter currents. The struggles occur over women's rights, gender, class, and race differences. All over the world, strategies vary between groups; the tensions are such that it is difficult, if not impossible, to speak of a single women's movement. The significant features of the three waves of women's movements are given below:

First Wave

In the United States, the 1948 Seneca Falls Convention and its Declaration of Rights and Sentiments, shaped after the Declaration of Independence, called for equality in marriage, for women's property rights and the right to earn equal wages, rights to have child custody, to make contracts, testify in court, inherit property, and the right to vote. The Convention organized by Elizabeth Cady Stanton and Susan B. Anthony saw achievement of women's political suffrage as the crucial goal. With the vote, women would gain entrance into public life and through it, legal equality, a gateway to liberation from social, political and economic oppression. Moreover, suffrage would unite all women, not just the middle class, because the root of oppression was the same: male domination (Balser, 1987: 53–55).

The International Women's Suffrage Movement was influenced by the Second Socialist International and world's Woman's Christian Temperance Union (WCTU). This movement aimed, in part, at dissolving the distinction between a private women's sphere and a public male arena (Nolan and Daley, 1994: 17). It used militant tactics, civil disobedience, and mass demonstrations, all met with arrest and forced prison feeding, which 'intensified women's militance around the world' (DuBois, 1994: 266).

Women's movements in Asia, Latin America and the Middle East developed after 1920s, in connection with the larger working class movements. In general, women challenged men's exclusive control of political rights and reaffirmed difference according to women's special, positive abilities and biologically-based distinctions (Lavrin, 1994: 198–199). Eventually, the achievement of suffrage in most western countries, an expanding economy, and the appeal of mass consumption demobilized the first wave of feminist movements.

Second Wave

The second wave of feminism began with the radicalization of women during the anti-Vietnam War and civil rights movements of the 1960s. While women and men participated equally in these political movements, women were often denied leadership positions, and were silenced when they tried to raise issues of women's rights, for example, within the activist organization, Students for a Democratic Society (SDS). As one SDS organizer said, 'women made peanut butter, waited on tables, cleaned up and got laid. That was their role' (Evans, 1979: 160). Partly in response, women began organizing consciousness-raising groups which became critical, strategic components of feminist struggles. The notion that 'the personal is political' questions oppression in women's lives. The National Organization of Women (NOW) formed in 1966, connected liberals with socialists. This was a time of optimism, debate and publication of feminism's formative literature. There was, however, a decline of the second wave of feminism by the 1980s.

Third Wave

A number of factors shaped the development of third phase, often labelled as 'postfeminist'. Feminism increasingly became an academic critique of knowledge. In this phase, women became critical for a more progressive education in colleges and universities. A number of discourses concerning abortion, welfare, work, crime (especially drug related), and immigration, coalesced around women's questions.

Contemporary feminist practices, whether in North America, Europe, or the Third World, among academics or activists, now reflects debates begun by women of colour and lesbians in the late 1970s. Poets and writers criticized the women's movements for downplaying sexual, racial, and class differences.

Women in the Third World organize around economic, environment, legal, military, cultural, and physical threats, and resistance to dictatorship, militarism, fundamentalism, economic dependence, and violence against women. Women's movements are not necessarily organized around feminist agendas but promote women's perspectives: examples include the Chipko Movement of Garhwal, Uttarakhand (India), the Green Belt Movement in Kenya, the Self-Employed Women's Association in India, and the Mothers of the Disappeared Movements in Latin America. Increasingly, groups posit culturally specific feminism as a political base. Yet, feminists also around the world unite on the issues of economic justice, human rights, and degradation of environment, the idea being unity through diversity. The main points of the three waves of women's movements have been given as under:

Topical Focus	*Theoretical Influences*	*Geographical Focus*
Wave One: The Geography of Women (mid to late 1970s)		
description of the effects of gender inequality	welfare geography, liberal feminism	constraints of distance and spatial separation
Wave Two: Social Feminist Geography (early to late 1980s)		
explanation of gender inequality, and relation between capitalism and patriarchy	Marxism, socialist feminism	spatial separation, place, and localities
Wave Three: Postcolonial Feminist Geography (late 1980s to present)		
the construction of gender identities, differences among women, gender and nature, gender and nationalism	cultural, post-structural, postcolonial, psycho-analytical theories; and writings of women, lesbian, gay-man, women from 'developing' countries	micro-geographies of the body, distance, separation, place, imaginative and symbolic spaces, environment

Feminist Geography: The Basic Theoretical Questions

The impetus for contemporary feminist theory begins in a deceptively simple question: 'And what about the women?'. In other words, where are the women in any situation being investigated? If they are not present, why? If they are present, then what exactly are they doing? How do they experience the situation? What do they contribute to it? What does it mean to them?

Over thirty years of posing this question has produced some generalizable conclusions: Women are present in most social situations. Where they are not present, it is not because they lack ability or interest but because there have been deliberate efforts to exclude them. Where they are present, women have played roles very different from the popular conception of them (as, for example, passive wives and mothers). Yet, though women are actively present in most social situations, scholars, publics, and social actors themselves, both male and female, have been blind to their presence. Moreover, women's role in most social situations, although essential, have been different from, less privileged than, and subordinate to those of men. Their invisibility is only one indicator of this inequality.

About the differences among women, it has been argued by the sociologists that the women's lives are profoundly affected by a women's social location, that is, by her race, ethnicity, class, religion, age, affectional preference, marital status, and global location.

Contemporary Feminist Theory

The typology of feminist theory is organized around feminism's most basic question, 'And what about women?'. The pattern of responses to this question generates the main categories of our classification given in Table 1.

Table 1 *Overview of Varieties of Feminist Theory*

Basic varieties of feminist theory – answer to the descriptive question, "What about the women?"	Distinction within theories – answers to the exploratory question, "Why is women's situation as it is?"
Gender differences	
Women's location in, and experience of, most situations is *different* from that of men in the same situation.	Cultural feminism Biological Institutional and socialization Social-psychological
Gender inequality	
Women's location in most situations is not only different but also less privileged than or *unequal* to that of men.	Liberal feminism
Gender oppression	
Women are *oppressed*, not just different from or unequal to, but actively restrained, subordinated, molded, and used and abused by men.	Psychoanalytic feminism Radical feminism
Structural oppression	
Women's experience of difference, inequality, and oppression varies by their social location within capitalism, patriarchy, and racism.	Socialist feminism Intersectionality theory
Feminism and postmodernism	

Source: Ritzer, G., 2000, *Sociological Theory*, New Delhi, Tata McGraw-Hill Edition, p. 450.

Gender Inequality

The following four themes are important in gender inequality:

1. First, men and women are situated in society not only differently but also unequally. Specifically, women get less of the material resources, social status, power and opportunities for self-actualization than do men who share their social location – be it a location based on class, race, ethnicity, religion, occupation, education, nationality, or any other socially significant factor.
2. This inequality results from the organization of society, not from the significant biological or personality differences between women and men.
3. Although individual human beings may vary somewhat from each other in their profile of potentials and traits, no significant pattern of natural variation distinguishes the sexes.
4. All inequality theories assume that both women and men will respond fairly easily and naturally to more egalitarian social structures and situations. In other words, it is possible to change the situations.

The gender differences, whatever be their cause, are more durable, more penetrative of personality, and less easily changed.

Jaggar, a professor of Philosophy at the University of Colorado, outlines the following conceptions of women's liberation: Liberal, Radical, Socialist and Feminism in Postmodernism.

Liberal Feminism

The major expression of gender inequality theory is liberal feminism, which argues that women may claim equality with men on the basis of an essential human capacity for reasoned moral agency, that gender inequality is the result of a patriarchal and sexist pattering of the division of labour, and that gender equality can be produced by transforming the division of labour through the repattering of key institutions – law, work, family, education, and media.

Social forms of labour determine the fundamental features of society and its inhabitants. Hence, women's nature is formed by the interaction between praxis (customs, practice and traditions), biological constitution, and physical and social environment. Understanding women in a specific society basically means examining its social relations of production. Engels argues that early societies had a sexual division of labour, with men producing means of subsistence and women working in the household. Development of forces of production in the male sphere gave men social dominance. Wanting to control the inheritance of wealth, men instituted monogamy as an instrument of economic dependence and subordination of women. For Engels, women's subordination results from the institution of class society and is maintained because it serves the interest of capital. In brief, traditional Marxist feminists want radical women to adopt the standpoint of the working class.

According to the supporters of liberal feminism, liberal theory basically states that human beings have intrinsic values as rational agents, with rationality being the mental capacity for moral reasoning and calculation. Liberal feminism is committed to neo-positivist conceptions of objectivity which supposedly eliminate social interests, values and emotions from consideration. Theories produced by the rational, detached observers are universally applicable. In other words, liberal feminists claim no special privilege for women, but only equal rights and opportunities for all.

The contemporary liberal feminism's explanation of gender inequality rests on the interplay of four factors: (i) the social construction of gender, (ii) the gender division of labour, (iii) the doctrine and practice of public and private spheres, and (iv) the patriarchal ideology.

The sexual division of labour in modern societies divides production in terms of both gender and spheres denoted as 'public' and 'private'. In this type of division of labour, women are given primary responsibility for the private sphere, while men are given privileged access to the public sphere. The access to public sphere rewards men in the form of money, power, status, freedom, opportunities and self-worth.

Thus, liberal feminism is based on neopositivism, which is objective and free from values and emotions. The liberal feminists claim no special privilege for women, but equal rights and equal opportunities for all.

Radical Feminism

Radical feminism developed as a critique of patriarchy. It is based on two emotionally charged central beliefs: (i) that women are of absolute positive value as women, a belief asserted against what they claim to be a universal devaluing of women, and (ii) that women are everywhere oppressed – violently oppressed – by the system of patriarchy.

Radical feminism generated by the women's liberation movement of the 1960s consists of a series of positions united by commitment to eradicating the systematic cause of women's oppression. Explanation initially emphasized the sex roles played by men and women, the identities assumed by people playing these roles, and the need for androgyny, or elimination of social and psychological distinctions between genders. In the second line of explanation, the discovery of universal and trans-historical male privilege is interpreted in biologically materialistic terms, as with notions of the male as a natural predator (a person who exploits others) and the female as a natural prey (victim or an animal hunted and killed by another for food).

Shulamith Firestone's *The Dialectic of Sex* (1970) adapts Marxism in theorizing procreative reproduction, rather than productive labour, as the moving force of history. In his opinion, the biological family is the base of society and economy part of the super structure. Class struggle occurs between two distinct biological and reproductive classes, i.e. women who are weaker due to reproductive physiology and stronger men with dependent families. The mother-child interdependency shapes the psychology of all females (and infants) in each society while biologically-based social institutions reinforce male domination. According to Firestone, development in the forces of reproduction (contraception, test-tube babies) will now make it possible to transform the biological state of women's subordination in the sense that genital sex distinctions would no longer have the same cultural significance.

Other radical feminist theories link women's special powers to female biology, or in some versions, to relations with nature. Susan Griffen (1980: 226), for example, says that 'we are nature'. Women have special ways of conceiving the world which emphasize feelings rather than men's reasoning. Closeness to nature becomes the main source of women's strength and power. The human ideal is the women with empathy, intuition, and protectiveness, an emotional as well as a rational being. In the place of the biologically based would come consciously designed and chosen social reproductivity. In brief, radical feminist knowledge is created directly from the particular experiences of women and is different from that of men.

Radical feminists see in every institution and in society's most basic structures – heterosexuality, class, caste, race, ethnicity, age, and gender – systems of oppression in which some people dominate others. Of all these systems of domination and subordination, the most fundamental structure of oppression is gender, the system of patriarchy. Within patriarchy, men see and women learn what subordination looks like. Patriarchy creates guilt and repression, sadism, masochism, manipulation and deception, all of which drive men and women to other forms of tyranny. Patriarchy, to radical feminists, is the least noticed and yet the most significant structure of social inequality.

The patriarchal violence may not always take the form of overt physical cruelty. It can be hidden in more complex practices of exploitation and control; in standards of fashion and beauty; in tyrannical ideas of motherhood, monogamy, chastity, and heterosexuality; in sexual harassment in the workplace; in practice of gynecology, obstetrics, and psychotherapy; in unpaid household drudgery, and underpaid wage work (MacKinnon, 1979).

Violence as physical cruelty lies at the heart of radical feminism's linking of patriarchy to violence. These are rape, sexual abuse, enforced prostitution, spouse abuse, incest, sexual molestation of children, hysterectomies, and other excessive surgery, the sadism in pornography, the historic and cross-cultural practices of witch burning, the stoning to death of adulteresses, the prosecution of lesbians, female infanticide, Chinese foot-binding, the abuse of widows, and the practice of clitoridectomy.

Patriarchy exists as a near universal social form because men can muster the most basic power source, physical force, to establish control. Men create and maintain patriarchy not only because they have the resources to do so but also because they have real interests in making women serve as compliant tools. Women are uniquely effective means of satisfying male sexual desire. Their bodies are essential for the production of children. Women are useful labour force. They can be ornamental signs of male status and power. As carefully controlled companions to both the child and adult male, they are pleasant partners, source of emotional support, and useful foils who reinforce the male's sense of central social significance. These useful functions mean that men everywhere seek to keep women compliant. Radical feminists give us both an explanation of universal gender oppression and a model for understanding cross-cultural variations in this oppression.

The question is how is patriarchy to be defeated? Radicals hold the view that this defeat must begin with a basic reworking of women's consciousness so that each woman recognizes her own value and strength, and works in unity with other women to establish a broad-based sisterhood of trust, support, appreciation, and mutual defense. With this sisterhood in place, two strategies suggest themselves: (i) a critical confrontation with any facet of patriarchal domination, and (ii) a degree of separatism as women withdraw into women-run business, households, communities, centres of artistic creativity, and lesbian love relationship. Lesbian feminism, as a major strand in radical feminism, is the practice and belief that 'erotic and/or emotional commitment to women is part of resistance to patriarchal domination' (Phelan, 1994). According to radicalists, if these steps are taken together, then the male domination in society can be reduced substantially.

Socialist Feminism

Socialist feminism is committed to the Marxist notion of the historical and social creation of human nature, a process which includes gender, race, ethnicity, and other distributions, as well as class. In socialist feminism, causal emphasis is placed on the sexual division of labour, or on different types of social praxis (customs and practices), as bases of physical and psychological differences between men and women. The central idea of social feminism is that women are constituted by the social relations they inhabit and the types of labour (work) they perform. Beginning with the Marxian notion of production for satisfaction of physical needs, socialist feminism argues that needs for bearing and raising children are equally fundamental, as well as needs of sexual satisfaction and emotional nurturing, all of which require labour (usually female). Girls achieve femininity by being like their mothers, retaining their capacity for empathy, and developing super egos open to persuasion, and vulnerable to the judgment of others. Boys grow into achievement-oriented men, adapted to work outside the home; and girls grow into women adapted to emotional work inside or outside the home. Gender struggles over reproductive activity are fundamental, yet often ignored by traditional Marxist theory.

Various socialist feminist theories elucidate several implications of this basic position. Nancy Chodorow (1978) argues that the individual's most intense early relationships involve love of mother, the female reproductive worker. To become masculine, boys must separate themselves via hostility towards their mothers, generating lifelong contempt for women, denying their emotional needs, and creating rigid and punitive super egos. Girls achieve femininity, as stated above, by being like their mothers. Socialist feminists, in general theorize procreative activities and public sphere production as mutually interdependent, and not independent.

Feminism as Postmodernism

Postmodernism refers to both the thought of the period following (or at the end of) modernism and to the theory that is opposed to the thought associated with modernity, particularly the rigid certainties of forms of science, planning and structuralism. It developed after modernism. Modernism is a general term for modern thought and practice. It is often associated with the application of rationality and abstraction that is applicable anywhere.

The two most well-known books in geography which concern postmodernity are David Harvey's *The Condition of Postmodernity* and Edward Soja's *Postmodern Geographies*. These authors describe postmodernism in the cultural sphere as a product of features of late capitalism. In this argument, capitalism has indeed changed from a form of accumulation based on mass production to a much more flexible, footloose, and just-in-time mode. This, the argument goes, was accompanied by a fragmented cultural sphere.

Postmodernist theory does not offer an answer to the fundamental question of feminist scholarship, 'And what about the women?'. Instead, postmodernism would respond with the counter question, 'How are you constructing the category or concept of 'women'?'. Thus, sociologists argue that postmodernism is important to feminist theory primarily as 'an oppositional epistemology'.

Postmodernist theory begins with the observation that we (that is, those of us living at the run of the century) no longer live under conditions of modernity but of 'post modernity'. This postmodern world is produced by the interplay of three major changes: (i) an aggressively expansive stage in global capitalism, (ii) the weakening of centralized state power (with the collapse of the old imperial systems, the fragmentation of the communist bloc, and the rise of ethnic politics within nation states), (iii) the patterning of life by an increasingly powerful and penetrative technology that controls pro movement based not in class but in other forms of identity – nationalism (the revolutions of formerly colonial states), race (the African, American civil rights movement), gender (feminism as a global movement), sexual orientation (gay rights), and environmentalism. Liberationist movements may have been the most important of these developments producing the postmodern challenges to modernist theory.

Postmodernism means changing or breaking away from modernity. It is a perspective that says that everything is contextual in this world. There is no single truth and that world phenomena is multi-layered in nature. Postmodern feminism finds modern reason to be normalizing, western, masculine prejudice whose 'enlightenment' embodies a colonizing scientific rationalism. Postmodern feminists think that western reason makes oppressive, universalizing and dogmatically assumed truth claims by opposing the masculine knowing subject to a known conquered (often feminized) object.

Feminist Geography

Feminist geography was initially closely related with radical geography and went through similar phases of development. As stated above, the feminist geography dates to mid-1970s. The theoretical positions have opened up geographical enquiry to domains that had previously been largely unexplored. These include: (i) geographies of fear, (ii) geographies of nature, (iii) geographies of mobility, and (iv) geography of development. Some common covers of feminist geographies are given below:

Feminist Geographies of Fear: Crime has long been an area of research in social geography. Susan Smith suggested in 1987 that women experienced crime and the fear of crime, in a very different way than men. She revealed how women lived in constant fear of crime in urban areas. In addition, women were more likely to not use the spaces of the city night and to lock themselves into their homes (Smith, 1987). The public space is as dangerous to women and private space is often described as a safe sanctuary for women. But, a vast majority of violence against women, and particularly rape, occurs in private space – at home. The reason for this is that public acts of violence are reported in the media, while private acts of violence are unlikely to be reported in the media. This leads to women excluding themselves from public space yet continuing to believe that the space of home will be safe. This mismatch between the geography of crime and the geography of fear leads to the reproduction of gendered geographies of violence and fear.

Feminist Geographies of Nature: Geography has often been described as the study of human environment relationship. There has been a long tradition of associating women with nature in western thought since Renaissance. Gillian Rose charts how dualistic thought has divided western thought in general and geography in particular into two, unequal halves:

Culture	Nature
Man	Woman
Masculine	Feminine
Reason	Emotion
Public	Private
Space	Place

This kind of list can be extended almost infinitely. Broadly, feminists have taken two approaches between the feminine and nature. One has been to embrace it and to use it as a mirror for those who are out to damage both women and nature. Under the banner of eco-feminism, many have used the association between women, and particularly women's bodies, and the natural world to critique seemingly cold, unnatural, rational masculinity (Merchant 1989). Consider this example:

> Nothing links the human, animal and nature so profoundly as women's reproductive system which enables her to share the experience of bringing forth and nourishing life with the rest of the living world. Whether or not she personally experiences biological mothering, it is in this that woman is truly a child of nature and in this lies the well-being of her strength. (Collard and Contrucci, 106)

Feminist Geographies of Mobility: Geography of female mobility is an interesting area of research in feminist geography. Feminist geographers are pursuing research on movements of the body to daily travel patterns, seasonal and temporary to permanent migration. Daily travel patterns for men and women in family units may be different. Men generally going to the place of work and returning back, while women have to perform several things while going out of the house. Thus, the urban mobility of males and females are different within an urban place.

At the level of the body, the work on the gendered geography of fear indicates that simply walking through the city can be a very different experience for men and women. These differences are further marked by differences in skin colour or of physical ability. Again, when mixed with other social markers such as race, this becomes even more complicated. Some of the earliest work in feminist geography considered these differences and how they challenged the assumptions behind spatial science's rational 'man'. Thus, the geography of work and travel reflects the spatiality of patriarchy and structural racism.

Feminist Geography of Development: Up to 1970s, development theorists of all theoretical persuasions tended to talk about development in a general way, paying no attention to the different ways men and women might experience the process of development. This was challenged by Ester Boserup in her book, *Women's Role in Economic Development* in 1970. In her book, Boserup used United Nations' statistics to show how women played a much more significant role in developing economies than had hitherto been recognized. Economic development she argued, led to an increased specialization of labour moving away from the family as the main unit of production and consumption to a specialized economy based on market. This led to further research in feminist geography especially in the following areas:

(i) Women's oppression in society, (ii) Sexism within geographical institutions, (iii) Interpretations are context bound and partial, rather than detached and universal, (iv) Interconnections between all aspects of social life (Economic, Social and Cultural Geography).

The first task of feminist geographers was to make women visible by developing a geography of women. In order to make women visible in geographic literature, two points were made: (i) women's experiences and perceptions often differ from those of men, and (ii) women have restricted access to a range of opportunities from paid employment to services. This is largely an empirical tradition, loosely influenced by liberal feminism and welfare geography. It has tended to focus on individuals, documenting how women's roles as care givers and 'housewives', in conjunction with existing urban spatial structure, housing design and policy, and patterns of accessibility to transport and other services such as child care, conspire to constrain women's access to paid employment and other urban resources.

This geography of women approach has been criticized because gender inequality is typically explained in terms of the concept of gender roles; especially women's role as housewives and mothers, in conjunction with some notions of spatial constraint.

In 1984, a Cooperative of feminist scholars-practitioners in the Women and Geography Study Group of the Institute of British Geographers published *Geography and Gender*. This followed earlier theoretical articles on women (Hayford, 1974), empirical studies of women and space (Tivers, 1978), etc. But, the cooperative study was the first sustained effort at outlining a critique of human geography and an alternative feminist approach. For the Women and Geography Study Group, geographical works portray humanity so much in terms of man and his environment that women's existence is denied. The group advocated change not simply by adding women to geography, but by developing 'an entirely different approach to geography as a whole' (Women and Geography Study Group, 1984: 21). Feminist geography, they say, looks at how socially created gender structures form and transform space in a project dedicated to ending gender inequalities through social change. The Cooperative chose three topics as examples of feminist approaches to mainstream geography:

1. *Feminist urban geography*: This branch is interested in conventionally neglected aspect of land use, especially the effects of separation of public sphere of waged work from the private sphere of family and home. The main objective of feminist urban geography is the separation and creation of a new form of built environment. This should be a feminist priority.
2. *Women's increased participation in the public sphere*: This area deals with the composition of workforce, pattern of women's employment, regional inequalities, and topics which they find are best examined by attending to women's everyday lives.
3. *Studies of the effects of the separation between home and waged work*: Females are not getting equal access to facilities and public services. The gender-based inequalities should be removed. Feminist geography needs to reconsider women's activity in changing the developed world.

Basically, the Cooperative argued that feminist geography must concern itself with changing gender relations, within and outside the discipline, by doing research not on people, but with them. Feminism challenges us to examine our personal, political and academic lives (Women and Geography Study Group, 1984: 146). With this, feminist geography entered as fully theorized, socially engaged dimension of a changing discipline.

The period of the middle to late 1980s saw two kinds of divide opening in feminist radical geography: between feminist and masculinist radical geography; and between perspectives within feminist geography, especially as post-structural and postmodern theory entered the discussion. A number of competing perspectives emerged within the feminist geography, which are categorized below.

Approaches to Feminist Geography

The geography of gender involves spatial variations in gender relations, the social construction of gender identities in particular environment; and the nature is related to gendered distinctions, and similar issues. According to Harding (1986), these approaches are considered as trinity of feminist geography. Feminist approaches within geography, in the opinion of McDowell, address three central concerns: (i) *space*, (ii) *place*, and (iii) *nature* (environment). A brief description of these approaches of feminist geography has been given in the following:

1. *Feminist empiricist geography:* This approach emphasizes on the unfairness of the exclusion of women from the discipline of geography, emphasizing equal rights for women. Feminist empiricist geography maps women's subordination and documents variations in the social status of women in the different parts of the world. It also examines the differences in poverty rates among black-female-headed households. Thus, feminist empiricism objects to women's exclusion from the public arena justified by 'natural' association with the home, childbirth and unpaid reproductive labour.

   According to empiricists, human geography excludes pressing questions related to daily life, household and domestic activities. The females are spatially restricted than men, because environment is 'man-made' in terms of architecture and planning. The problems faced by women in urban areas, like the housing problems of single women or women's fear in cities or the effect of urban design and layout on women have not been given adequate emphasis in human geography written by male geographers.
2. *Feminist standpoint theory:* The followers of postmodern approach want to map tensions. The feminist standpoint theory investigates about women making their own geographies by examining their responses to landscape. Anti-rationalism or radical feminism rejects the rationalist vision of equality and valorizes instead female difference and female knowledge.
3. *Postmodern or post-rational feminism:* The followers of postmodern or post-rational feminists opine that the very notion of gender is implicated in a disastrous, oppressive fiction of 'women' which rides roughshod over differences between and within women, that is, gender ignores other subjugated arenas of difference. From this view, both the rationalist and anti-rationalist feminist paradigms should be rejected and the subject 'women' should be replaced by decentred, partial and fractured identities.

Gender and Space

Feminist social theory has long shown considerable interest in the relations between gender and space. At present, feminist geographers are increasingly interested in the relations between gender and space. The spatial arrangements determine the status of women and men. The women's knowledge is reduced and that produces and reproduces power for men. Some feminists argue that women attempt to take up as little space as possible in the public sphere by positioning their bodies in unobtrusive ways. In other words, women situate themselves in space in more remote ways than men. In fact, different groups of women have different relations to space. In the early part of the feminist geographic research and writing, in the late 1970s and early 1980s, feminist positions were modified either from political philosophies such as Marxism, established traditions of geographic thought such as urban geography, or by combination of the two.

The initial impetus for work in feminist geography came from the political demands of women's movements – putting women first, making women visible, and changing the content and direction of research. During the 1980s, theoretical debates began to outline the other feminist-geographic positions, drawing more freely from the broader reach of feminist social theories.

Feminist Research and Database

According to the International Labour Organization, women comprise about 50 per cent of the world's population, do two-thirds of world's work hours, receive only 10 per cent of world's income and own less than one per cent of world property. But, reliable data on their work and wages is not available. This is the biggest hurdle in the research proposals of women research analysis.

Feminist geography is concerned with the spatial differentiation and organization of feminine activity and with human use of physical environment. The human geographers are increasingly interested in developing suitable research methods, appropriate research problems and other methodological issues. Women were excluded from geographical research because of methodological factors, like the non-availability of data on women's waged work and unpaid (household) labour. Human geographers either focus on men to the exclusion of women or drop women when problem of analysis becomes methodologically intractable.

In the absence of reliable and adequate data on women, it is appropriate to generate primary data in the form of structured questionnaires and schedules. Thus, collaborative relationships with participants in their research employing qualitative methodologies of some kind are: in-depth interviews, participant observation, and ethnographic research at micro level. In other words, inter-subjectivity rather than objectivity is the ideal method of data collection and research analysis.

Feminist geography has differentiated itself from masculinist geography through fierce, profound critique. The women geographers were of the opinion that the existing masculinist geography erases women's existence.

Collection of primary data on women is a sensitive issue. There are many limitations in the participatory collection of data also. Methods like participant observation, empathy and ethnographic research also as participatory research are complicated, restrictive, expensive and time consuming. For the determination and ascertaining of inequality between men and women, a qualitative-cum-quantitative approach is more appropriate.

A group of geographers contemplate their experiences with feminist participatory research. These methods demonstrate solidarity with oppressed people. Most of the researchers opine that participatory research is complicated and restrictive, involving reflexive as part of research process, and that processes of marginalization are embedded in the academy.

The recent history of feminist thought and feminist geography is rapidly growing and has enhanced political awareness. The scholars belonging to the oppressed groups have special responsibility to collect more information and to do more analysis on the issues pertaining to females and their problems. The feminist geography has to produce conceptions of space and nature (environment) that are its own, that drive from women's experiences. Although little has been published, yet feminist geographers are expanding their consideration of geography to include environmental concerns. This area is likely to receive much more attention in the coming years.

Gender Geography

It is generally accepted that sex is biologically determined, while gender is the role fabricated by society, which constructs appropriate behaviour for each sex. When space is constructed by men, they make women feel unsafe (secluded woodlands, dark alleyways, ill-lit roads and multi-storey car parks). They (women) are much more aware of their vulnerability and lack of physical strength to combat the male in the men created social environment.

Geography is considered as the mother of all sciences. The study of the various ways in which genders and geographies are mutually constituted is the main area of gender geography. The spatial variation of gender and its significance at all scales from the human individual to world economy is missing from geographical literature. In the history and geography of humanity, women's subordination is omnipresent. No significant society has constituted gender as to produce male subordination. The worldwide theme of geography is female subordination. The forms of subordination differ greatly, but, all over the world, women's work tends to be defined as of less value than men's and women tend to have far less access to all forms of social, economic and political power.

Gender geography deals with the spatial distribution of females and the processes which make them less adventurous than men. It is the study of various ways in which genders and geographies are mutually constituted. Gender geography emphasizes on how gender relations are constructed in all spheres of life.

As stated above, 'gender' is usually contrasted with 'sex' in an effort to remove women from 'nature or physical environment' and place them within culture as constructed and self-constituting social subjects. The treatment of gender in human geography is missing. In an effort to theorize 'patriarchy', Foord and Gregson (1986) attempt to identify the necessary relations that constitute gender relations. They reason that two genders, male and female, are the basic characteristics of gender relations. In order to theorize the necessary relations between these basic characteristics, they ask: 'Under what conditions do men and women require each other's existence?', to which they answer, 'for biological reproduction and the practice of heterosexuality'. This gender relationship was seriously criticized, because it made it difficult to theorize how capitalism structures gender relations and for its biologism, especially in terms of its portrayal of heterosexuality as biological or psychologically fixed.

Explanations of Geography of Gender

There are significant variations in the socio-economic condition of the females and their political empowerment. Current explanations of regional patterns in the constitution of gender range from a focus on reproduction to a focus on production. All seek to explain both the subordination of women and the varying construction of gender in time and space.

Force

Sexual asymmetries are nearly universal among primates. Among non-human primates, for instance, it is typically the male which can approach, threaten and displace the females. Among humans, it is typically the male who can coerce the behaviour of females (for some, this is a definition of patriarchy).

According to J.S. Mill, the subordination of women rested solely on the law of the strongest. This subordination has had originally been achieved by force and remained a kind of slavery, distorting female personality so that no one could know that the capacities of women might be in a free and equal society. The maintenance and variety of subordination were social, although the origin was physical, and technical change could free women.

Collins (1971–1975) presents a sophisticated sociological explanation incorporating male strength, used to control women by violence in the interest of male sexual gratification and satisfaction. In this scenario, changing political and economic conditions (over time and space) greatly alter female bargaining power, since there are changes not only in female access to resources, but also in the role of force in society.

Children as Labour Power

For many anthropologists, whole societies are shaped by the institutions which expropriate the fecundity of women and allocate their progeny. Some experts describe the 'hidden' subordinate role of women's as 'always linked to exploitation of women's reproductive functions'. A women's value is in her reproductive capacity; control over this is control over the rights of children. From these institutions, other aspects of gender and indeed of society are shaped. On this model, women's procreative capacity is more important and more stringently controlled in agrarian societies of low technology and therefore, heavy demand for labour. The spatial transfer of females to the family of the male is the rule for most human societies and is an important cause of female powerlessness.

Children as Genetic Survival

According to Engels (1884), 'the family is based on the supremacy of the man, the express purpose being to produce children of undisputed paternity'. For sociologists, the prime aim of all parties is genetic survival (survival through descendants). For example, Dickemann examines stratified societies in which women seek to marry 'up' into families of higher standing. And the assumption is that the chances of reproductive success are better for the better off, since the progeny will be guaranteed better access to resources, thus women's family has a direct, biological stake in her marrying up. Since access to wealthy family is competitive, they have a stake in her reputation and eligibility, which will normally require her chastity. Since competition can, in this case, not only require chastity but exaggerated behaviour over it, so biological interest may intensify seclusion, *purdah* and other restrictions on women. Different genetic survival strategies may create whole cultures.

Marriage Transfers

Cultures may also be derived from institutions, such as the forms of transfer of wealth on marriage. Goody (1976) has argued that the social effects of these systems are extremely pervasive, and that there are strong regional patterns of marriage transfers. As the rate of awareness and education has increased in the developed countries, the females are trying to be at par with the male members of the society. It is because of this factor that the rate of divorce is significantly high in the developed countries of Europe, especially those of the Second World countries like Russia, Belarus, Estonia, Latvia, Lithuania, Hungary, Romania, and Ukraine.

Sexual Division of Labour

The sexual division of labour emerges as an expression of women's roles in both production and reproduction; these roles in turn are derived from those social relations which regulate the formation of families and the bearing and rearing of children as well as those social relations which govern the production of goods and services.

Many writers place sexual divisions of labour at the centre of the social construction of gender. Early writers placed much more emphasis than is now used on male strength for explaining division of labour. A more usual focus now is to examine the compatibility of task with simultaneous responsibilities for childcare, particularly breastfeeding. Lactating women will tend to avoid tasks that are distant and /or dangerous, leaving the men to do the early stages of production (extraction of raw materials and clearing the land for agriculture) while women work in secondary processing and harvesting. The more mobile males are assumed to drive dominance from this situation. In many societies, lactating women do work in 'incompatible' activities. For instance, most women's paid work is in activities incompatible with breastfeeding. Neither male strength nor the requirements of lactation, however, can adequately explain sexual divisions of labour as they actually exist.

Engels presents a historical development from communal, egalitarian societies through the rise of private property and the family of exploitative class societies. Women in this scenario were initially dominant, but became subordinate with the appearance of means of production which could be privately owned, specifically domesticated animals. When men held private property in productive assets and an exchangeable surplus, women came to work for their husbands and families instead of for society. Hence, the division of labour between the sexes became exploitative. With the rise of capitalism, production and reproduction became spatially separated. Home and workplace are no longer the same. As a result, home becomes the locale of reproduction.

Geography of Gender Roles in 'Economic Activity'

In all parts of the world, nearly all men aged between about 25 and 55 are 'in labour force' or economically active. The pattern is for men to spend their adult years in labour market activities. Women, on the other hand, may spend all or part of their adult lives as 'economically inactive home workers' or may be 'economically active but invisible'. According to official figures, the above tendency in advanced industrial economies (First World) and centrally planned economies (Second World) is far more than the third. In developing market economies (Third World), the proportion is usually much less, being the lowest in certain Islamic countries of North Africa, the South-West Asia and South Asia, and the highest in South-East Asia. The most sharply varying patterns are found in Africa, where Angola reports only 9 per cent of women aged 15 to 64 are in labour force; and the corresponding figures for nearby Botswana is 73 per cent.

Women in the Third World Countries

The Third World countries are largely agrarian in their economies. Reliable national data on the discrepancies between men's and women's wages for the same type of work is not available. Data is also lacking in many other important aspects of gender role. For example, abortion may be legal, but is it available (or rather, are qualified doctors available for abortions?). How immediately relevant is divorce legislation in a country where the majority of poor women do not marry and those who cannot afford divorce proceedings?

Education: In most Third World societies, literacy is an important measure of life option. The females are more illiterate as compared to men in the same age group. The women's ability to control important events in their lives is limited. Moreover, they do not have the same life options as men in the same age group. Their choice of food, education and training, occupations, remuneration, occupational advancement, migration, use of time and leisure, land and property ownership are largely controlled by the male members of the society.

For intra-household variables, only survey research will elucidate international contrasts in women's power, i.e. their ability to decide about their own activities and those of others, their control over subsistence decisions and resources, the amount of help received in reproductive activities, their membership of group outside the household, and their role in the formation and dispersal of the household or other productive group. Only at the micro level, it is possible to assess women's informal power, which may be very significant in many societies.

In addition to the above, phenomena like widowhood and marital breakdown also must be studied in their regional context. In some parts of South Asia, an ideal has been that the bride should be one-third the age of her husband, ensuring early widowhood for many, which for widows of the higher Hindu castes has meant social death. Low-caste South Asian women can marry, although they may then lose property rights and access to their children. Self-immolation was expected only of high-caste women, but widowhood has been particularly unhappy state for most. The worst outcome is possible for the childless widow, so the incentives for child-bearing are high. Widowhood is a widespread phenomenon, since many societies have a large age difference between spouses; where tradition has commonly enforced some sort of provision for them, but this is often disrupted by urbanization and economic change.

Marital breakdown is also an intractable phenomenon. In societies with many informal or consensual unions, the breakdown, like the union, may go unrecorded. In Islamic societies, divorce rates have been relatively high but male kin have vowed full support to a divorced woman. It appears that this obligation is becoming less effective; in the Middle East, North Africa, Malaysia and Muslim Indonesia, divorced women are increasingly entering the labour market. At the other extreme, in Latin America, divorced women have been few, ostracized and rejected; they are increasing in number, and they too appear in the labour market.

The proportion of all households in the world headed *de facto* by women has been estimated as being as high as a quarter to a third. Some female-headed households have arisen from male-headed households through death, marital breakdown or migration; some are 'naturally' female headed, either because the woman has no permanent partner or because there are other wives in other dwellings and settlements (West Africa).

Migration is an important component of geography of gender and an important factor in women's life in the Third World. Marriage migration is an important component of migration in both Africa and Asia, and women's subsequent isolation may be a major factor in their lives.

Female Migration

A temporary, periodic, or permanent move to a new location is known as migration. Migration takes place because of numerous physical, socio-economic and political causes. Migration has been enormously influencing in determining cultural and social change at all scales, especially in the socio-economic status of females. It is however, not clear how sex acts as a basis for selective in-migration. The female migration patterns change in space and time. For example, in the 19th century, females of England migrated more than males, especially over short distances. Grigg (1977) attributes this to the lack of employment for women in rural areas, the demand for domestic servants, and the tendency for women to move after marriage. Men, it is alleged, move when the economic risks are greater, and their wives join them later when prospects become less uncertain. In recent years, in the western world, generally males are now more migratory than females.

In the developing countries (Third World), men far outnumber the females in rural to urban migration. There are, however, exceptions. For example, in Latin America, the majority of rural to urban migrants are women, and a similar situation is to be found in the Philippines. In Asia and Africa, however, such migration is typically more masculine than feminine. It is usual for women in Africa to take responsibility for working the land while men move to town on a permanent or temporary basis. Thus, it is the young males who dominate most in African migration streams, and when women do migrate, it is usually to join their husbands already working in town. At present, in Africa also there is an increasing trend of females migrating to urban places.

In India, there is scarcity of reliable data on internal migration. The Census of India provides general information for 'streams of migration' and reasons for internal migration. Prior to the Second World War, the internal migration in India was free and people voluntarily moved from state to state on a large scale in search of employment.

At present, the pattern of internal and international migration has appreciably changed and that has affected the pattern of female migration also. Males and females are migrating towards the metropolitan and megacities in search of employment and other economic opportunities.

How Women Vanish: A Hidden Geography

How can half of humanity be so consistently overlooked by scholars and planners, particularly in the world of work. Many working women are obscured because they are 'economically inactive homemakers' who only produce food for home consumption, or they are too young or too old to be recorded. This is a familiar concept. But, Boulding (1983) has shown that remarkable numbers are not accounted for at all. In the Third World, women suggest that the poor majority work all the time. Statistics of female labour (the percentage of women participating in the labour force) are thus of very poor international comparability.

Moreover, inconsistencies in the recording of women's roles in direct production appear in part-time work and in informal sector activities. For example, domestic service is commonly female, but in India and some countries of Africa, it is partly male and partly female.

Meanwhile, the 'unpaid family workers' remain in such an awkward category that there are some excuses for countries which refuse to enumerate them. In fact, they form an important fraction of the informal sector. By definition, unpaid family workers work in non-domestic activities for at least one-third of the normal hours.

Time is an increasingly scarce resource for many women in the developing world. Many Third World women carry a heavy burden of a double or even a triple workload involving a very long working day. Females in the rural areas put in a much longer day than men in gathering fuel wood, fetching water, and carrying out housework, child care, food preparation and much of the agricultural labour.

In order to give due place to females in geographic literature, there is a need of more research at the household and individual level into the motivations and commitment of women workers. This must be set in the context of broader structural research which analyses the changes in production relationships in both rural and urban areas, and in particular, the growth of mass market in Asian and African countries.

The features unique to humanity among primates are our elaborate division of labour by sex, our reliance on sharing, the existence of authority among us, and our remarkable ingenuity. As a result, 'our species possess the capacity to carry sexual inequality to its greatest known extremes, but we also possess the potential to realize an unusual social equality between the sexes should we choose to exercise that potential'. Our biological capacity is for a great range of social constructions of gender.

The New International Division of Labour

In the new 'world market factories', production processes are standardized, repetitious, call for very little modern knowledge, and are highly labour intensive. The majority of the labour force are young women. It has been argued convincingly that girls are already trained by their experience of domestic labour in the 'manual dexterity' which so many of these jobs require and that their subordination as gender renders them docile, subservient and cheap. These continuities are encountered through a wide range of relations of production from home based to factories with attached hostels.

Other features are more specific to particular activities, as with health in electronics. In electronics, there are widespread reports of stress, fatigue, eyesight deterioration, and overall health problems. It has been reported that more workers of electronic goods producers than clerical workers suffer from complaints of eye, ear, nose, throat, musculoskeletal system, and sleep disorder. In the case of eye trouble, the incidence is double among those using microscopes, etc. Those using chemicals regularly are more prone to menstrual complaints, pregnancy problems and infertility. Other health problems are related not to the technical conditions of production but to the shift work, which appears to produce sleeping problems, gastrointestinal problems, central nervous system complaints, and psychological complaints.

The introduction of female workers in electronics is a new feature of international division of labour. At present, the division of labour between en and women in electronic industry is sharp. For example, management, supervision and maintenance are overwhelmingly male, while tasks directly related to production are female, since 'women are delicate by nature'. This gender ideology compares directly with the preference of manufacturers for women in task requiring 'dexterity' and the electronic workers are again seen as docile, as second earners in their household.

Regional Geography of Gender

The above discussion clearly shows that females are invisible in geographic literature, almost all over the world. The construction of gender in time and space is therefore, necessary to do justice with the fair sex. In order to overcome the problem of invisibility of women, the geographers should develop a regional geography of gender. At one level, we need to know how certain aspects of people's lives vary around the world (often we have reasonable information for males but poor information for females). At another level, we need to know how different countries are incorporated into the world economy, and what are the links between the form of their incorporation and their socio-economic structure, including the construction of gender. We need micro-locality studies, description at the level of household and the individual to develop the macro studies of gender geography. It is therefore, imperative to move towards a regional geography of gender not only to describe the regional variations in the social status and political empowerment of females but also to explain the causes of such variations.

References

Atkinson, T., 1974, *Amazon Odyssey*, New York: Link Books.

Balser, D., 1987, *Sisterhood and Solidarity: Feminism and Labor in Modern Times*, Boston: South End Press.

Bardhan, P.K., 1984, *Land Labour and Rural Poverty*, Delhi: Oxford University Press.

Bernal, J.D., 1971, *Science in History*, Cambridge: MIT Press.

Blaut, J., 1970, 'Geographic Models in Imperialism', *Antipode 2*, pp. 65–85.

Blunt, A. and Dowling, R.M., 2006, *Home*, London: Routledge.

Bondi, L., 1990, 'Feminism, Postmodernism, and Geography: Space for Women?', *Antipode* 22, pp. 156–167.

Bondi, L., 1992, 'Gender and Dichotomy', *Progress in Human Geography*, 16, pp. 98–104.

Boserup, E., 1986, *Women's Role in Economic Development*, London: Earthscan.

Brownmiller, S., 1975, *Against Our Will: Men, Women and Rape*, New York: Simon and Schuster.

Bunge, W., 1966, *Theoretical Geography*, 2nd edn., Lund Studies in Geography Series C, No.1, Lund: C.W.K. Gleerlup.

Collard, A.F.E.E. and Contrucci, J., 1983, *Rape of the Wild: Man's Violence against Animals and the Earth*, London: Women's Press.

Collins, R., 1971, 'A Conflict Theory of Sexual Stratification', *Social Problems, 19*, pp. 3–21.

Collins, R., 1975, *Conflict Sociology: Toward an Exploratory Science*, New York: Academic Press.

Cresswell, 2013, *Geographic Thought: A Critical Introduction*, Wiley: Blackwell.

Daly, M., 1992, *Pure Lust: Elemental Feminist Philosophy*, San Francisco: Harper San Francisco.

DuBois, E.C., 1994, 'Women Suffrage Around the World: Three Phases of Suffragist Internationalism'. In C. Daley and M. Nolan (eds), *Suffrage and Beyond: International Feminist Perspective*, New York: New York University Press, p. 252.

Elson, D. and Pearson, R., 1981, 'The Subordination of Women and the Internationalisation of Factory Production', in Young et al. (eds).

Engels, F., 1884, *The Origin of the Family, Private Property and the State* (English Translation of the 1884 German Original), New York: International Publishers.

Eisenstein, Z., 1981, *The Radical Future of Liberal Feminism*, New York: Longman.

Evans, S., 1979, *Personal Politics: The Roots of Women's Liberation in the Civil Rights Movement and New Left,* New York: Vintage.

Foord, J., and Gregson, N. 1986, 'Patriarchy: Towards a Reconceptualization', *Antipode 18 (2),* pp. 186–211.

Griffin, S., 2000, *Women and Nature: The Roaring Inside Her*, San Francisco: Sierra Club Books.

Grigg, D.B., 1977, 'E.G. Ravenstein and the Laws of Migration', *Journal of Historical Geography*, 3, pp. 41–54.

Goody, J., *Production and Reproduction*, Cambridge: Cambridge University Press.

Hayford, A. 1974, 'The Geography of Women: A Historical Introduction', *Antipode 6*, pp. 1–19.

Haraway, D., 1991, 'Gender for a Marxist Dictionary: the Sexual Politics of a Word'. In her *Simians, Cyborgs, and Women: the Reinvention of Nature,* New York: Routledge, pp. 127–148.

Herbert, D.T., 1982, *The Geography of Urban Crime*, New York: Longman.

Hrdy, S.B., 1981, *The Woman that Never Evolved*, Cambridge: Harvard University Press.

Huntington, E., 1915, *Civilization and Climate*, New Haven: Yale University Press.

Husain, M., 2012, *Human Geography*, 4th edn., Jaipur: Rawat Publications.

Husain, M., 2014, *Geography of India*, Delhi: Tata Mcgraw-Hill.

ILO, 1983, *Year Book of Labour Statistics*, Geneva: ILO.

Jaggar, A., 1983, *Feminist Politics and Human Nature*, Totwa, NJ: Rowman.

Johnson, R.J., 1980, *Geography and Geographers: Anglo-American Human Geography since 1945,* New York: Wiley.

Johston, R.J., et al. 1994, *The Dictionary of Human Geography*, 3rd edn., Cambridge: Blackwell, pp. 214–215.

Kitchin, R., (ed.) 2008, *International Encyclopedia of Human Geography*, Durham, NC: Duke University Press.

Lavrin, A., 1994, 'Suffrage in South America'. In M. Nolan and C. Daley (eds), *Suffrage and Beyond: International Feminist Perspectives,* New York: New York University Press.

Lee, D., 1985, *Women and Geography*, A Bibliography Transaction.

McDowell, L., 1992, 'Doing Gender: Feminism, Feminists, and Research Methods in Human Geography', *Transactions of the Institute of British Geographers NS,* 17, pp. 399–416.

MacKinnon, C., 1979, *Sexual Harassment of Working Women*, New Haven: Yale University Press.

Marx, K., 1963, *The Poverty of Philosophy*, New York: International Publishers.

Marx, K. 1970, *A Contribution to the Critique of Political Economy,* Moscow: Progress Publisher.

Marx, K., 1976, *Capital: A Critique of Political Economy*, Vol. 1, trans. B. Fowkes, Harmondsworth: Penguin.

Massey, D., 1973, 'Towards a Critique of Industrial Location Theory', *Antipode* 5, pp. 33–39.

Massey, D., 1974, 'Social Justice and the City: a Review', *Environment and Planning*, 6, pp. 229–235.

Massey, D., 1984, *Spatial Division of Labour*, New York: Methuen.

Merchant, C., 1989, *The Death of Nature: Women, Ecology and Scientific Revolution*, New York: Harper & Row.

Messey, D., 1984, *Spatial Division of Labour: Social Structures and the Geography of Production,* New York: Methuen.

Messey, D., 1993, *Flexible Sexism, Society and Space*, 9, pp. 32–57.

New Internationalist, 1985, *Women: A World Report*, London: Methuen.

Nolan, M. and C. Daley, M. Nolan and C. Daley (eds), *Suffrage and Beyond: International Feminist Perspectives*, New York: G. Brazilier.

Peet, R., 1977a *Radical Geography*, Chicago: Maaroufa Press.

Peet, R., 1998, *Modern Geographical Thought*, authorized reprint, 2011, Jaipur: Rawat Publications.

Phelan, Shane, 1994, *Getting Specific: Postmodern Lesbian Politics*, Minneapolis: University of Minnesota Press.

Pratt, G. and Hanson, S., 1988, 'Gender, Class and Space, Environment Planning and Development', *Society and Space, 6*, pp. 15–35.

Ritzer, G., 2000, *Sociological Theory*, 5th edn., New Delhi: Tata McGraw-Hill Education Private Limited.

Rose, D., 1984, 'Rethinking Gentrification: Beyond the Uneven Development of Marxist Urban Theory, Environment, Planning Development', *Society and Space, 1*, pp. 47–74.

Rose, G., 1993, *Feminism and Geography, 14*, pp. 317–328.

Royal Geographical Society (with the Institute of British Geographers), Women and Geography Study Group 1977, *Feminist Geographies: Explorations in Diversity and Differences*, Longman: Harlow.

Seager, J. and Olson, A., 1986, *Women in the World: An International Atlas*, London: Pluto.

Smith, S., 1978, 'Fear of Crime: Beyond a Geography of Deviance', *Progress in Human Geography*, 11, pp. 1–23.

Soja, E.W., 2000, *Postmetropolis: Critical Studies of Cities and Regions*, Oxford: Blackwell.

Stoddart, D., 1986, *On Geography and Its History*, Oxford: Blackwell.

Tinker, I., et al. 1976, *Women and World Development*, Washington DC, Overseas Development Council et al. (eds).

Tivers, J., 1978, 'How the Other Half Lives: the Geographical Study of Women', *Area 10*, pp. 302–306.

Tivers, J., 1985, *Women Attached*, London: Croom Helm.

Townsends, G. et al. 1987, *Geography of Gender in the Third World*, Allony NY: State University of New York Press.

Wekerle, G., 1984, 'A Women's Place is in the City', *Antipode 16 (3)*, pp. 11–19.

Women and Geography Study Group of Institute of British Geographers 1984, *Geography and Gender: An Introduction to Feminist Geography*, London: Hutchinson.

Whyte, R.O. and Whyte, P., 1982, *The Women of Rural Asia*, Boulder, Colorado: Westview Press.

Youssef, N.H. and Hetler, C.B., 1983, 'Establishing the Economic Condition of Women-Headed Households in the Third World', in *Buvinic*, M. et al. (eds), pp. 216–243.

Zelinsky, W., Monk, J. and Hanson, S., 1982, 'Women and Geography: a Review and Prospectus', *Progress in Human Geography, 6*, pp. 317–366.

Zipf, G.K., 1949, *Human Behaviour and the Principle of Least Effort*, Cambridge: Addison Wesley.

A Chronology of Geography

Historical Events	*Geographers' Lives*	*Geographical Publications*
1861–65 US Civil War 1867 *Capital* Vol. 1 K. Marx Meiji restoration, Japan	1861 b. H.J. Mackinder 1863 b. E.C. Semple 1864–5 Kropotkin's Siberian expeditions 1869 Powell's I Grand Canyon expedition Ratzel attends Haeckel's course in Jena	1859 *Origin of Species* C. Darwin 1864 *Man & Nature* G.P. Marsh 1869 Suez Canal opens
1870–71 Franco-Prussian War, Paris Commune 1872 Yellowstone National Park designated 1875 Bell patents telephone 1878 1st telephone exchange	1871 E. Reclus captured fighting for Paris Commune & exiled to Switzerland 1872 Vidal de la Blache appointed at Nancy 1873 Ratzel travels to USA 1874 Prussia decrees geography a university discipline 1876 Kropotkin flees from Russia 1878 W.M. Davis moves to Harvard	1875 F. Galton's I weather map published in a newspaper 1876 Laws of migration, E. Ravenstein 1877 *Physiography* T.H. Huxley
1882 Britain occupies Egypt 1884 Greenwich Mean Time established as world standard 1887 Queen Victoria's golden jubilee 1888 William II, emperor of Germany	1883 Boas's expedition to Baffin Island 1885 c. 94 geographical societies around the world 1886 Ratzel to chair at Leipzig 1887 Mackinder got readership at Oxford 1888 geography at Cambridge National Geographical Society, USA 1889 b. C.O. Sauer 1889 Cycle of erosion proposed by W.M. Davis	1881 *Anthropogeographie* Vol. 1 F. Ratzel 1885 *What Geography Ought To Be*, P. Kropotkin 1886 *Elementary Lessons in Physical Geography* A. Geikie 1887 *On the Scope & Methods of Geography*, H. J. Mackinder 1887 *The Study of Geography*, F. Boas

Historical Events	*Geographers' Lives*	*Geographical Publications*
1893 World's Columbian Exposition, Chicago 1895 Roentgen discovers X-rays, Lumiere brothers develop cinema 1896 H. Becquerel discovers radioactivity 1898 start of German naval build-up	1890s Ratzel active in Pan-German league *Annales de Geographie* founded, Paris Semple studies under Ratzel in Leipzig 1893 *The Geographical Journal* founded 1898 Vidal de la Blache to chair at Sorbonne 1899 Mackinder climbs *Mt. Kenya* 1899 Oxford, 1st UK geography department	1893 *Frontier in American History* F. J. Turner 1894 *Universal Geography* E. Reclus *Morphology of the Earth's Surface* A. Penck 1896 *The Territorial Growth of States*, F. Ratzel 1899 *Handbook of Commercial Geography* G.C. Chisholm
1st pan-African conference Planck's quantum theory 1901 T. Roosevelt becomes American President 1st radio signals across Atlantic G. Marconi 1903 Wright brothers, 1st flight 1905 Special theory of relativity, Albert Einstein Japanese defeat Russian army & navy 1909 Model T. Ford	1900 Mackinder became Director of London School of Economics 1901 E. Huntington travels Euphrates on a raft 1903 Chicago, 1st US geography graduate program 1904 d. F. Ratzel Association of American Geographers founded F. Younghusband expedition to Tibet 1907 T.G. Taylor lectures on geography in Sydney 1909 K. Haushofer sent to Japan 1909 A. Weber's industrial location theory	1900 *The Interpretation of Dreams* S. Freud 1902 *Systematic Geography* W.M. Davis 1903 *Tableau de la geographie de la France* P. Vidal de la Blache *Geographic Influence in American History* A.P. Brigham 1904 *The Geographical Pivot of History* H. J. Mackinder 1905 *The Major Natural Regions* A. Herbertson 1907 *The Pulse of Asia* E. Huntington
1911 Atomic structure explained by Rutherford, Amundsen reaches South Pole 1912–13 Balkan Wars 1914 Panama canal opens 1914–19 First World War	1911 *Annals of the Association of American Geographers* founded 1912 W.M. Davis organizes Transcontinental Expedition, USA K. Haushofer fights on western front E. Huntington in military intelligence	1911 *Influences of Geographic Environment* E. Semple *Les genre de vie dans la geographie humaine* P. Vidal de la Blache 1914 *Mutual Aid* P. Kropotkin

Historical Events	*Geographers' Lives*	*Geographical Publications*
A. Einstein's general theory of relativity 1917 Russian Revolution Versailles Peace Settlement	1915 1st Canadian geography course, U. of British Columbia 1918 d. P. Vidal de la Blache	1915 *The Origin of Continents and Oceans* A. Wegener 1915 *Cities in Evolution* P. Geddes 1916 *Regional Environment, Heredity & Consciousness* A. Herbertson 1919 *Human Regions* H. J. Fleure *Climatic Cycles & Evolution* T.G. Taylor
1923 France occupies Ruhr 1927 Stalin comes to power, USSR 1929–33 Great Depression	1921 Semple 1st woman president of AAG Haushofer appointed to Munich 1st full geography course in Canada, U. of Montreal 1922 d. P. Kropotkin Michigan Land Survey 1923 Sauer to Berkeley 1925 *Economic Geography Journal* founded 1927 The first Geography Department in Asia established at the A.M.U. Aligarh 1928 T.G. Taylor leaves Australia because of his views on environmental determinism	1922 *Ulysses* J. Joyce, *The Waste Land* T.S. Eliot, *Economy & Society* M. Weber 1922 *Principes de la geographie humaine* P. Vidal de la Blache *A Geographical Introduction to History* L. Febvre 1923 *Geography as Human Ecology*, H. Barrows 1925 *The Morphology of Landscape*, C.O. Sauer 1926 *A Plea for the History of Geography*, J.K. Wright 1927 *Geography: Its History, Its Nature and Its Methods* A. Hettner 1929 *Sequent Occupance*, D. Whittlesey *Retail Gravitation Model*, W. J. Reilly

Historical Events	*Geographers' Lives*	*Geographical Publications*
1930 Pluto discovered Japan occupies Manchuria 1932 splitting of atom, Cockroft & Walton 1933 Hitler named German chancellor 1933–45 F. D. Roosevelt, US President	1931–15 British Land Utilization Survey under L.D. Stamp 1932 D.E.C. Semple Schaefer flees Germany 1933 Institute of British Geographers founded A. Weber attacked by Hitler youth, A. Hettner prevented from publishing by Nazis I. Bowman chair of National Research Council, USA	1931 *The Logic of Scientific Discovery* K. Popper 1933 *Central Places in Southern Germany* W. Christaller *The British Isles* L. D. Stamp & S. H. Beaver
1936–39 Spanish Civil War 1937 Japan invades China	1934 d. W. M. Davis Tuan's family flee Japanese invasion of China 1938–39 Hartshorne in Germany & Austria	1934 *Geography in Relation to the Social Sciences* I. Bowman *Habitat, Economy & Society*, C.D. Forde 1936 *The General Theory of Employment, Interest & Money* J. M. Keynes, Television broadcasts, UK 1939 *The Nature of Geography* R. Hartshorne
1939–45 Second World War 1945 United Nations charter 1945 International Monetary Fund & World Bank 1947 Indian independence 1948 Apartheid in South Africa 1949 Division of Germany People's Republic of China formed	Geographers recruited to war effort for intelligence, training, air photography, meteorology, etc. Christaller works on planning for occupied Poland 1945 Kirk at fall of Mandalay, Burma (Myanmar) campaign 1945–46 Glacken works in Korea 1946 Haushofer commits suicide 1947 D.H.J. Mackinder 1948 Harvard geography closes AAG & American Society for Professional Geographers merge 1950 d. I. Bowman	1940 *Intervening Opportunities* S. Stouffer 1941 *A Theory of Location for Cities* E. Ullman *Foreword to Historical Geography*, C.O. Sauer 1942 *Human Adjustment to Floods* G.F. White 1945 *The Nature of Cities* C.D. Harris & E. Ullman 1947 J.Q. Stewart's first paper on social physics 1948 *Historical Geography of the United States* R. Brown 1949 *A Sand County Almanac* A. Leopold *Principle of Least Effort*, G.K. Zipf

Historical Events	*Geographers' Lives*	*Geographical Publications*
1950–53 Korean War Mt. Everest summit reached McCarthy investigations, USA 1954 start of Vietnam War 1954–62 Algerian War of independence 1954 FORTRAN devised	1951 Canadian Association of Geographers founded L.D. Stamp appointed director of World Land Use Survey 1953 d. F.K. Schaefer	1951 *The Spirit and Purpose of Geography* S. Wooldridge & G. East *The Inadequacy of the Regional Method* G. Kimble 1952 *The Dynamic Basis of Geomorphology* A. Strahler 1953 *Exceptionalism in Geography* F. K. Schaefer *Innovation Diffusion as a Spatial Process,* T. Hägerstrand 1954 *The Economics of Location* A. Losch, Eng. trans. *Fundamentals of Ecology* E. & H. Odum
1956 Hungarian uprising put down by Soviet troops, Stalin denounced in USSR 1957 Treaty of Rome leading to the European Economic Community Sputnik in orbit	1955 Start of seminars in mathematical statistics under E. Ullman & W. L. Garrison at U. of Washington, Seattle 1958 R. Chorley to Cambridge NATMO established at Kolkata (Calcutta) B.J.L. Berry to Chicago Institute of Australian Geographers founded 1959 Hägerstrand visits U. of Washington, Seattle	1956 *Man's role in changing the face of the earth* W.L. Thomas
	1960 Anuchin's *Theoretical problems in geography* creates fierce debate in Soviet geography	1960 *The Stages of Economic Growth* W. Rostow 1960 *Methods of Regional Analysis* W. Isard
1961 Vostok 1, first manned space flight, Berlin wall built	1961 IBG creates study groups AAG High School Geography Project	1961 *Megalopolis* J. Gottmann *The City in History* L. Mumford

Historical Events	*Geographers' Lives*	*Geographical Publications*
1962 Cuban Missile Crisis, The Beatles *Silent spring*, R. Carson 1st geodetic satellite.		1962 *Theoretical Geography* W. Bunge
1963 President J.F. Kennedy assassinated	1963 d. S. Wooldridge 1st Madingley conference on Frontiers in Geographical Teaching	1963 *Identification of Some Fundamental Spatial Concepts* J. Nystuen *Problems of Geography* W. Kirk
1964, USA 1st weather satellite		1964 *The Decision Process in Spatial Context* J. Wolpert
1965 1st mini-computer	1965 L. Dudley Stamp knighted	1965 *Locational Analysis in Human Geography* P. Haggett
1966 Cultural Revolution, China	1966 P. Haggett to chair at Bristol	1966 *Central Places in Southern Germany* W. Christaller, English translation
1967 Six Day War, Middle East 1967–70 Biafran War, Nigeria	1967 AAG project in remote sensing *Regional Studies* founded	1967 *Traces on the Rhodian shore* C.J. Glacken *Behaviour & Location* A. Pred *Models in Geography* R. Chorley & P. Haggett *Spatial Diffusion as an Innovation Process* T. Hägerstrand
1968 Civil unrest in Paris, Chicago, Baltimore, Mexico City and elsewhere Tet Offensive, Vietnam		1968 *Acadia* A.H. Clark

Historical Events	*Geographers' Lives*	*Geographical Publications*
1969 Apollo 11, first human on moon De Gaulle resigns as French President R. Nixon, US President	1969 *Geographical Analysis, Antipode & Environment & Planning* a founded 1969 d. J.K. Wright, W. Christaller D. Harvey to Baltimore Detroit Geographical Expedition organized by W. Bunge	1969 *Explanation in Geography* D. Harvey *Behavioural Problems in Geography* K. Cox & R. Golledge *On the Environment as Perceived* H. Brookfield
1970 Earth Day, April 22		1970 *Geographic Perspectives on Urban Systems* B. Berry
1971 Indo-Pakistan war leading to secession of Bangladesh (1972) Aswan High Dam opened	1971 social relevance theme of AAG annual conference The Socially and Ecologically Responsible Geographer founded by W. Zelinsky and others AAG committee on status of women geographers	1971 *Spatial Organization* R. Abler *Physical Geography: A Systems Approach* R. Chorley & B. Kennedy *Pivot of the Four Quarters* P. Wheatley
1972 UN Conference on the Environment, Stockholm *Limits to growth*, Club of Rome		1972 *Geography: A Modern Synthesis* P. Haggett
1973–74 Oil Crisis 1973 UK joins EEC Yom-Kippur War, Middle East	1973 Toronto Expedition organized by W. Bunge	1973 *Social Justice and the City* D. Harvey
1974 Nixon resigns as US President	1974 Relevance & public policy theme of IBG annual conference, Norwich	1974 *Space and Place in Humanistic Geography* Y-Tuan *Urban & Regional Modelling in Geography & Planning* A.G. Wilson

Historical Events	*Geographers' Lives*	*Geographical Publications*
1975 Vietnam war ends	1975 d. C.O. Sauer *Journal of Historical Geography* founded	1975 *Spatial Organization of Society* R. Morrill
1977 Queen Elizabeth II Silver Jubilee Terrorist attacks in Italy & Germany	1977 *Progress in Geography* splits to become *Progress in Physical Geography & Progress in Human Geography*	1977 *The Choreography of Existence* A. Pred *Human Geography: A Welfare Approach* D.M. Smith *The Urban Question* M. Castells, English translation
1978 Camp David peace accord Jonestown, Guyana, mass suicide of religious cult		1978 *Humanistic Geography* D. Ley & M. Samuels *The Environment as Hazard* I. Burton, R. Kates & G. White *Orientalism* E. Said *Ideology, Science & Human Geography* D. Gregory *Paradigms and Revolution or Evolution* R. J. Johnston
1979 M. Thatcher elected as British Prime Minister Shah of Iran deposed Soviet invasion of Afghanistan	1979 AAG 75th Anniversary	1979 *In What Sense a Regional Problem?* D. Massey
1980 R. Reagan elected as US President Iran-Iraq war starts Mount St. Helens erupts, Washington, USA	1980 *Urban Geography* founded	1980 *Conceptions of Space in Social Thought* R. Sack *Geography Without Man* D. Ley *Civilizations: Organisms or Systems?* K. Butzer *Geography, Marx & the Concept of Nature* N. Smith & P. O'Keefe

Historical Events	*Geographers' Lives*	*Geographical Publications*
1981 Columbia space shuttle launched	1981 U. of Michigan closes geography department Women and Geography Study Group of IBG founded	1981 *Dictionary of Human Geography* R. Johnston et al. *Institutionalization of Geography and Strategies of Change* H. Capel *Human Geography & Human Agency* D. Gregory
1982 British-Argentina war over Falkland Islands	1982 *Political Geography Quarterly* founded	1982 *Diorama, Path & Project* T. Hägerstrand *Musing on Helicon* A. Buttimer
1983 HIV identified US invasion of Grenada	1983 Society & Space founded	1983 *The City & the Grassroots* M. Castells
1984 Ethiopian famine Assassination of Indira Gandhi. Bhopal gas disaster, India	1984 25th conference of the International Geographical Union, Paris 50th anniversary of IBG	1984 *Discovering the Vernacular Landscape* J.B. Jackson *Spatial Divisions of Labour* D. Massey *Geography & Gender* Women and geography study-group, IBG *A Woman's Place?* D. Massey & L. McDowell
1985 Gorbachev general secretary, Soviet Communist party Mexico City earthquake		1985 *The Urbanization of Capital* D. Harvey *Social Relations & Spatial Structures* D. Gregory & J. Urry *Political Geography* P.J. Taylor
1986 Chernobyl Anglo-French agreement to build Channel Tunnel	1986 BBC Domesday project GIS	*On Geography and its History* D.R. Stoddart

Historical Events	*Geographers' Lives*	*Geographical Publications*
1987 Palestinian intifada Montreal protocol on ozone	1987 National Center for Geographic Information & Analysis, USA The earth as transformed by human action conference, Clark University National Geography Awareness Week, USA	1987 *Place, Practice & Structure* A. Pred *The History of Cartography* Vol. 1 J.B. Harley & D. Woodward 'Localities debate' *Antipodes*
1988 Australia's bicentenary Intergovernmental Panel on Climate Change	1988 H.C. Darby knighted Social and cultural geography study group, IBG	1988 *Iconography of landscape* D. Cosgrove & S. Daniels *The Postmodern Challenge to Human Geography* M. Dear
1989 Berlin Wall comes down Downfall of Ceausescu regime, Romania Tiananmen Square uprising, China	1989 d. C. J. Glacken	1989 *Postmodern Geographies* E. Soja *The Condition of Postmodernity* D. Harvey *Areal Differentiation and Postmodern Human Geography* D. Gregory *Marxism, Culture & the Duplicity of Landscape* S. Daniels
1990 German reunification Iraq invades Kuwait Nelson Mandela released, South Africa		1990 *The Making of the American Landscape* M.P. Conzen *The Geographer's Art* P. Hagget
1991 Gulf War COMECON & Warsaw Pact disbanded 1991 End of USSR War in Yugoslavia Maastricht Treaty	1991 Geography part of UK national curriculum New Words, New Worlds Conference, Edinburgh Environmental Change Unit, U. of Oxford	1991 *The Production of Space* H. Lefebvre [English translation.] *Situated Knowledges* D. Haraway *A View on the GIS Crisis in Geography* S. Openshaw

Historical Events	*Geographers' Lives*	*Geographical Publications*
1992 UN Conference on Environment & Development, Rio North American Free Trade Association Los Angeles riots & earthquake Discovery of 'cosmic ripples'	1992 J. Patten, MP, former geographer, becomes UK Education Secretary George Perkins Marsh Institute, Clark University 1992 d. R. Hartshorne	1992 *Writing Worlds* T. Barnes & J. Duncan *Geography as a Science of Observation* G. Rose *The Americas Before & After 1492* K. Butzer et al.
1993 Moscow rebellion Mississippi & Missouri floods	1993 Geographical Association centenary	1993 *Geography & the Human Spirit* A. Buttimer *Eco-socialism* D. Pepper *Geography & Feminism* G. Rose
1994 South African elections Israel-PLO agreement	1994 D.J. Gottmann *Ecumene and Gender, Place and Culture* founded National Geography Standards Established. Replacing the five themes of geography, a set of eighteen learning standards was created by the National Council for Geographic Education called the National Geography Standards.	1994 *Geographical Imaginations* D. Gregory *Geography & Empire* A. Godlewska & N. Smith *Technopoles of the World* P. Hall & M. Castells
1995 Kobe earthquake	Royal Geographical Society & Institute of British Geographers merger	1995 *Mapping Desires* D. Bell & G. Valentine *Gender, Work & Space* S. Hanson & G. Pratt

Some Recent Developments

1997 Hong Kong returned to China		1998 *Modern Geographical Thought* Richard Peet
1999 Euro established as currency in 11 European countries Macau returned to China	1999 d. Pierre Gourou, French geographer	

Historical Events	*Geographers' Lives*	*Geographical Publications*
2000 U.S. President Clinton orders GPS Selective Availability turned off, instantly making GPS more accurate	2000 End of Selective Availability: On May 1, 2000, the White House announced the end of Selective Availability. Selective Availability (SA) was the practice of distorting GPS signal locations which reduced the spatial accuracy for civilizations using GPS units. The end of SA opened up GPS use to the commercial market.	2000 *Cultural Geography: A Critical Introduction* D. Mitchell 2000 *Alternative Geographies* J.R. Short
2001 New millenium began Southern Ocean established by the International Hydrographic Organization 2002 East Timor gains independence		2002 *Hybrid Geographies: Natures, Cultures, Spaces* S. Whatmore 2004 *Geography and Geographers 6th Edition: Anglo-American Human Geography since 1945* R.J. Johnston and J.D. Sidaway 2005 *All Possible Worlds: A History of Geographical Ideas* Geoffrey J. Martin 2005 *Nature* N. Castree 2006 *A Geography of Big Things, Cultural Geographies* J.M. Jacobs 2007 *Non-representational Theory: Space, Politics, Affect* N. Thrift
	2009 d. David Wright, a geographer and author whose work inspired children around the world.	2009 *Geography: History and Concepts* Arild Holt-Jensen
		2010 *The Geographical Review* publishes its 100th volume.

Historical Events	*Geographers' Lives*	*Geographical Publications*
	May 10, 2013 Australian Curriculum: Geography endorsed. Ministers of Education representing the States and Territories of Australia endorsed the Australian Curriculum: Geography from Foundation (5 year olds) to Year 10 (16 year olds). The first time Australia had a national approach to teaching the discipline of Geography. Karl Zimmerer is awarded the Melamid Medal in recognition for his pioneering research on cultural ecology.	
	2014 d. David Ross Stoddart, geographer and coral-reef scientist	

* Abridged from: A Chronology of Geography 1859–1995 by Alisdair Rogers in *Human Geography* ed. by J. Agnew et al. 1996, London, Blackwell.

Biographic Notes on Selected Geographers

Ackerman, Edward A. (1911–1973), American geographer; Ph.D. Harvard, 1939; taught at the University of Chicago.

Adams, Cyrus C. (1849–1928), American geographer; Editor, Bulletin of the American Geographical Society, 1908–1915.

Al-Balkhi (850–934), Published first climatic atlas in 921.

Al-Biruni (972–1050), Wrote *Kita-al-Hind* (Geography of India) in 1030.

Alexander the Great (356–323 BC), Pupil of Aristotle; king of Macedonia; led Greek army eastward to the Indus River where Alexander defeated Pores – the ruler of west India.

Al-Idrisi (1099–1180), A leading geographer and cartographer with Roger II at Palemo; wrote book on world geography.

Al-Mamun (786–833), Son of Khalifa Haroon-al-Rashid who developed the *Baitul-Hukma* (Academy of Sciences) at Baghdad.

Al-Maqdisi (945–985), Delineated climatic region based on latitudes, and east-west position relative to land and water.

Al-Masudi (896–956), Arab scholar who described the monsoons and the relation of evaporation to rainfall.

Al-Rashid, Harun (763–809), Ruler of Baghdad who established the *Baitul-Hukma* (Academy of Sciences) to promote sciences. Experts were invited at this academy from the distant and neighbouring countries to train the Arab scholars and to translate the available literature, especially the Greek literature into Arabic.

Amundsen, Roald (1872–1928), Norwegian explorer who discovered the South Pole in 1911.

Anaximander (610–547 BC), Disciple of Thales in Miletus (present Turkey). He introduced the *gnomon* to the Greeks; drew map of the known world to the scale.

Anuchin, D.N. (1843–1923), Russian geographer who founded the Department of Geography of the Moscow in 1887.

Aristotle (384–322 BC), Greek philosopher who believed the world was developing toward final perfection. He taught the theory of natural places for earth, air, fire and water. He was the founder of the Lyceum at Athens.

Awad, Mohammad (1896–1967), Egyptian geographer; completed Ph. D. from the University of London. He became the first professor of Geography Department, University of Cairo in 1936, when graduate studies were introduced in the Cairo University.

Baker, Oliver E. (1883–1949), British specialist on the history of geography. He was a teaching faculty at Oxford from 1923 to 1962.

Baranskly, N.N. (1881–1963), Soviet geographer, friend of Lenin, who helped to establish the role of geography after the October Revolution of 1917.

Barrows, Harlan H. (1877–1960), One of the leaders in the development of American geography between 1919 and 1942, when he was the chairman of the Department of Geography at the University of Chicago.

Bartholomew, John (1831–1893), Scottish cartographer.

Bartholomew, John Geonge (1860–1929), Scottish cartographer; head of map publishing firm.

Berg, L.S. (1876–1950), Soviet geographer. He headed the Department of Physical Geography at Leningrad from 1925–1950.

Bjerknes, V.F.K. (1862–1951), Norwegian meteorologist who developed new model of atmospheric circulation. Between 1895 and 1932 he taught at Stockholm, Oslo, Leipzig and Bergen.

Bonpland, Aime (1773–1858), French botanist who travelled with Humboldt in North and South America and assisted in writing some of the reports including *Kosmos*.

Bowman, Isaiah (1878–1950), Director American Geographical Society.

Brigham, Albert Perry (1855–1932), Obtained the degree of M.A. from the Harvard University. He taught at the Colgate between 1892–1925 and was a member of the Association of American Geographers.

Brunhes, Jean (1869–1930), A student of Vidal de la Blache. He was the Professor of Human Geography at the College de France between 1912–1930.

Buchanan, Keith (1919–2013), First Professor of Geography at Victoria University, Wellington (New Zealand).

Bunge, William W. (1928–), An American geographer. He was an outspoken advocate of the greater use of theoretical models and quantitative techniques. He worked on the social problems of inner cities.

Burton, Ian (1935–), Canadian geographer, taught at Toronto (Canada).

Chorley, R.J. (1927–2002), British geomorphologist. He got his education at the Oxford University of Oxford.

Christaller, Walter (1893–1969), A German geographer who postulated the Central Place Theory in 1933.

Clark, Andrew H. (1911–2013), He was a historical geographer; taught geography at the Wisconsin University.

Cole, J.P. (1928–), A British geographer who got his Ph.D. from the Nottingham University in 1962; a faculty member at Nottingham.

Coleman, Alice M. (1923–), A disciple of Prof. L.D. Stamp who taught at the King's College, University of London.

Columbus, Christopher (1451–1506), Born in Genoa, he went on voyages for the Portuguese; with the aid of Ferdinand and Isabella of Spain. He made four voyages to America.

Comte, Auguste (1798–1857), A leading thinker who advocated the school of positivism.

Cook, James (1728–1779), A British naval captain, who made three voyages to the Pacific Ocean. He finally established the outlines of land and water around the oceans.

Copernicus, Nicolaus (1473–1543), A Polish astronomer who published his concept of heliocentric universe in 1543.

Cosmos (522–547), Wrote the book the *Christian Topography*.

Cressey, George B. (1896–1963), A product of Clark University who obtained his Ph.D. from the Chicago University. He was a specialist on Asia and published the *Geography of 500 Million People*.

Daly, Reginald A. (1871–1957), American geologist and authority on the origin of glaciers. He taught at the Harvard University from 1912 to 1942. He opposed Charles Darwin interpretation of coral reefs.

Darwin, Charles (1809–1882), A British naturalist and author of *Origin of Species.*

Davis, William Morris (1850–1934), Father of geomorphology who introduced professional geography into the United States of America. He taught at the Harvard University from 1878–1912. He was the founder of the Association of American Geographers.

De Geer, Sten (1886–1933), Swedish geographer; son of Baron Gerard De Geer, who pioneered in population mapping and urban studies.

de Martonne, Emmanuel (1873–1955), He was a student and son-in-law of Vidal de la Blache; taught physical geography at the Sorbonne (1909–1940). He taught Physical Geography at the Sorbonne University of France.

Demangeon, Albert (1872–1940), A student of Vidal de la Blache who taught at Sorbonne. He published one of the first French regional studies.

Dickinson, Robert E. (1905–), A British geographer who obtained his Ph.D. from the London University in 1932. He was an expert on Urban Geography. He taught at Syracuse, Leeds and Arizona Universities.

Dokuchaiev, V.V. (1846–1903), Russian geographer and Professor at St. Petersburg in 1884. He was the pioneer in the development of zonal soils.

Drake, Sir Francis (1540–1596), A British naval officer. He was the second captain to sail around the world. Drake Passage south of South America was named after him.

Dutton, Clarence E. (1841–1912), American geologist who explored the Colorado Plateau region.

Edrisi or Al-Idrisi (1099–1180), Geographer with King Roger II at Palermo. He was a leading cartographer, who wrote a book on world geography.

Einstein, Albert (1879–1955), German physicist who formulated the general theory of relativity.

Eratosthenes of Cyrene (273–192 BC), First to use the word 'Geography'. He measured the circumference of Earth scientifically and accurately. He was appointed the chief librarian at Alexandria.

Flamsteed, John (1646–1719), He was the first director of the Greenwich Observatory (1675).

Fleure, Herbert J. (1877–1969), British anthropogeographer who taught at Manchester.

Galileo (1564–1642), Galileo was the first to formulate the concept of universal mathematical order and to demonstrate correctness of the ideas of Copernicus.

Gama, Vasco da (1469–1524), Portuguese captain; first to sail around Southern Africa to India.

Geddes, Patrick (1854–1932), Scottish biologist, bio-geographer and town planner.

Gerasimov, I.P. (1905–1985), Soviet physical geographer; appointed Academy of Sciences.

Gottman, Jean (1915–1994), French geographer; got his education at Paris. He was the director at *Ecole Pratique des Hautes Etudes*, Sorbonne. He was the author of *Megalopolis.*

Guyot, Arnold (1807–1884), Swiss scholar, who was a disciple of Carl Ritter. He was professor of Physical Geography and Geology at the college of New Jersey (Princeton) from 1854 to 1884.

Hägerstrand, Torsten (1916–2004), Swedish geographer who developed mathematical and theoretical geography at the Lund University (Sweden).

Haggett, P. (1933–), A leading British geographer known for his mathematical techniques and book *Models in Human Geography.*

Halley, Edmond (1656–1742), A British astronomer who wrote a treatise on the winds (1686) and was the first to show magnetic variation on a map by connecting points of equal variation with lines.

Hartshorne, Richard (1935–1992), American geographer, who obtained his Ph.D. from the Chicago University. He taught at Minnesota (1924–1940) and Wisconsin (1940–1970). His book *Nature of Geography* published in 1939 is a masterpiece in Geography.

Harvey, David (1935–), A British geographer, who taught at Johns Hopkins since 1969. He authored the book *Explanation in Geography.*

Haushofer, Karl (1869–1946), A German geopolitician of Nazi period.

Hecataeus (550–475 BC), Greek scholar who is considered as the Father of Geography. He was the first Greek prose writer. He used Europe, Asia and Libya as regional divisions of his world geography.

Henry, Prince (1394–1460), He was the second son of the Portuguese king who established the Geographical Institute at Sagres.

Herbertson, A.J. (1865–1915), British geographer. He developed a system of world regions.

Herodotus (484–425 BC), Greek historian, ethnographer, and historical geographer. He is known as the Father of History.

Hettner, Alfred (1859–1941), German geographer. He taught at Leipzig, Tubingen and Heidelberg Universities (1899–1928).

Hipparchus (190–120 BC), He was the first in the western world to introduce the use of grid of latitude and longitude as a basis for locating places on the face of the earth. He developed the stereographic and orthographic projections. He also used a circle divided into 360°.

Hippocrates (460–370 BC), Greek physician who produced the first medical geography. He developed ideas of environmental influence on human behaviour.

Homer (8th century BC), The Greek poet described by Strabo as the first geographer.

Hsuan-Tsang (602–664), Chinese Buddhist monk who crossed Tibet to reach India.

Humboldt, Alexander von (1769–1859), The great German geographer who brought the classical period of universal scholars to an end as he foreshadowed the modern period. He extensively travelled in South and North America, Europe, Russia and Central Asia.

Huntington, Ellsworth (1876–1947), American geographer, who taught at Yale (1907–1915). He was an environmental determinist, who authored the book *Principles of Human Geography*.

Hutton, James (1726–1797), Scottish geologist who was the first European fluvialist and developed the concept of uniformitarianism.

Ibn-Battuta (1304–1368), He was one of the most important Arab travellers who visited Asia and Africa. He was appointed as Ambassador by the ruler of Delhi (Mohammad Bin Tughlaq) to China.

Ibn-Khaldun (1332–1406), Arab historian whose book *Muqaddimah* (Introduction to Universal History) was published in 1377. He was a determinist who said history should be read geographically and geography should be studied historically.

Isard, Walter (1904–), American economist; Ph.D. Harvard; founded the Regional Science Association in 1954. He was the chairman, Department of Regional Science, University of Pennsylvania.

James, Preston E. (1899–1986), American geographer, Ph.D. from the Clark University. He taught at Michigan (1923–1945) and Syracause (1945–1970). He authored the book *All Possible World: A History of Geographical Thought*.

Jefferson, Mark (1863–1949), American geographer, studied with W.M. Davis at Harvard. He taught at Michigan State Normal College (Ypsilanti) from 1901 to 1939.

Kant, Immanuel (1724–1804), German philosopher and geographer who rejected the doctrine of final cause, lectured on physical geography at Konigsberg (1755–1796). He provided an early statement of geography as chorology.

Kepler, Johannes (1571–1630), German astronomer who rejected Ptolemy's concept of circular orbits in favour of elliptical orbits.

Khan, I.R. (1897–1960), Started the first Indian Department of Geography at the Aligarh Muslim University, Aligarh in 1931.

Koppen, Wladimir (1846–1940), German meteorologist and climatologist who delineated the climatic regions of the world on a scientific basis.

Kropotkin, Prince Peter (1842–1921), Russian revolutionary whose lifelong interest was in physical and economic geography. He worked at the Royal Geographical Society, London.

Lenin, N. (1870–1924), The founder of Bolshevik Communism, he was an important contributor to the geographical ideas and concepts.

Losch, August (1906–1945), A German geographer who formulated theoretical models to explain Central Place Hierarchies.

Mackinder, Halford J. (1861–1947), A British geographer who started the Department of Geography at the Oxford University. He presented the concept of 'Heartland' in political geography. This concept affected the world politics up to the Second World War significantly.

Magellan, Ferdinand (1480–1521), Portuguese captain and explorer who was the first to sail around the world for Spain. He was killed at Philippines by the native people.

Malthus, Thomas Robert (1760–1834), An English economist who developed the first theory of population growth in 1798.

Marinus of Tyre (100–150). Teacher of Ptolemy, he accepted smaller circumference of the earth.

Marx, Karl (1818–1883), A political thinker who wrote *Das Capital* and advocated communism.

Mercator, Gerardus (1512–1594), He devised the Mercator Projection for use in navigation.

Monkhouse, F.J.K. (1914–), British geographer who was an expert in cartography.

Peary, Robert E. (1856–1920), He was the first to reach the North Pole in 1909.

Penck, Albrecht (1858–1945), Professor of Physical Geography at Vienna (1885–1906).

Penck, Walther (1888–1923), Son of Albrecht Penck; was a leading German geomorphologist. He presented a major challenge to W.M. Davis' concept of Normal Cycle of Erosion.

Peschel, Oscar (1826–1857), A German geographer, journalist, editor. His main contribution was historical geography.

Plato (428–348 BC), Greek philosopher. He believed the world had been created in perfection and was in process of deteriorating. He stressed deductive method, and accepted idea of a round earth. He was the founder of the Academy at Athens.

Polo, Marco (1254–1323), A traveller from Venice who visited China in 1271–1295. His book describes the many places he visited in Central and Eastern Asia.

Posidonius (135–50 BC), Greek geographer who estimated the circumference of the earth to be smaller than that of Eratosthenes. He believed the highest temperatures are along the tropics, not along the Equator. He was one of the earliest philosophers to face the apparent dichotomy between factual descriptions of unique phenomena and the formulation of theoretical explanations.

Powell, John Wesley (1834–1902), American explorer and pioneer geographer. He was the first to explore the Grand Canyon. He was the leader of US Geographical and Geological Survey of the West, and the second Director of US Geological Survey.

Ptolemy (90–168), He summarized the Greek knowledge of astronomy; gazetteer of places located by latitude and longitude. In his world map he showed the Indian Ocean as an enclosed Ocean, bounded by *Terra-Incognita*. He accepted smaller earth circumference of Marinus.

Pythagoras (570–495 BC), A Greek mathematician, who believed in the concept of a round earth.

Ratzel, Friedrich (1844–1904), Got his training in Zoology and Geography at Munich. He travelled as roving reporter, 1871–1875, including visit to America. He taught geography in Munich (1875–1886), at Leipzig (1886–1904).

Reclus, Elisée (1830–1905), Student of Carl Ritter. He was banished for revolutionary activity in France in 1851 and again in 1871. He was a professor at the University of Brussels (1898–1905).

Ricardo, David (1772–1823), British economist who developed the theory of rent.

Richthofen, Ferdinand von (1833–1905), Trained as a geologist; after field studies in China and California, taught geography at Bonn (1877–1883).

Ritter, Carl (1779–1859), the great German teacher of geography who, with Humboldt, brought the classical period in geographical thought to an end. His book *Erdkunde* is highly appreciated as a great work in geography.

Rousseau, Jean-Jacques (1712–1778), A follower of Montesquieu in relating politics to climate; suggested new methods of teaching to avoid rote learning.

Salisbury, Rollin D. (1858–1922), American geologist ad geographer; Chairman of the Department of Geography at Chicago (1903–1919).

Sauer, Carl O. (1889–1975), American geographer; Ph.D. Chicago, taught at Michigan (1915–1923) and at the University of California, Berkeley (1923–1957). He developed the School of Cultural Geography.

Schaefer, Fred Kurt (1904–1953), German economist who taught geography at Iowa, 1946–1953. He pleaded against ideographic geography. He was a strong supporter of nomothetic geography.

Semple, Ellen Churchill (1863–1932), American geographer with M.A. in history from Vassar; introduced Ratzel's anthropogeography to America. She was a staunch environmental determinist who stressed that the history of people and lifestyle are controlled by the topographical features.

Smith, Adam (1723–1790), Scottish political economist.

Soja, E.W. (1940–), American geographer, who obtained his Ph.D. from the Syracuse and taught at Northwestern University.

Spate, O.H.K. (1911–2000), British geographer who taught at the Australian National University.

Spencer, Herbert (1820–1903), British philosopher who applied Darwin's biological concepts to human societies.

Stamp, L.D. (1898–1966), British geographer who directed the British Land Use Survey. His book *Land of Britain its Use and Misuse* is the pioneer work in the field of land use studies.

Strabo (c. 64 BC–20 AD), Summarized Greek Geography in his books written for use of Roman administrators.

Strahler, A.N. (1918–2002), American geomorphologist who was a Professor of Geography at the University of Columbia.

Suess, Eduard (1831–1914), Professor of Geology at Vienna (1857–1901).

Taylor, Griffith (1880–1963), British geographer who introduced modern geography in Australia.

Thales of Miletus (624–548 BC), Greek merchant and scholar; originator of several basic theorems of geometry.

Ullman, Edward L. (1912–), American geographer; Ph.D. Chicago; Professor at the University of Washington; specialist on transportation and urban geography.

Vidal de la Blache, Paul (1845–1918), French geographer who was the first Professor of Geography in France. A staunch believer in possibilism, he was a committed Human Geographer.

Voeikov, Aleksndr Ivanovich (1842–1916), Russian geographer and climatologist. He studied the effects of human action on the physical and biological features of the earth.

Wallace, Alfred Russel (1823–1913), British naturalist; coauthor with Charles Darwin of the theory of selection.

Weaver, John C. (1915–), American geographer; Ph.D. 1942, President, University of Missouri (1966–1971), President, University of Wisconsin, 1971–1976.

Whittlesey, Dewent S. (1890–1956), Amercian geographer who obtained his Ph.D. from the Chicago University; taught at Chicago from 1920 to 1928, and at Harvard from 1928 to 1956.

Wooldridge, Sidney W. (1900–1963), British geomorphologist who was appointed Professor of Geography in Kings College, London.

Wright, John K. (1891–1969), American geographer, specialist in the history of geography and historical geography at the American Geographical Society as librarian (1920–1938).

Yoshida, Togo (1864–1918), Japanese historical geographer.

Zelinski, Wilbur (1921–), American geographer, Ph.D. from California (Berkeley), teaching at Pennsylvania State University, specialist on Population Geography.

Glossary

Activity Allocation Models: Planning models that are used to decide where activities will be located in a region.

Alonso Model: A model developed by William Alonso (1964) to account for intra-urban variations in land values, land use and land use intensity. It builds on the von Thunen model of agricultural land use pattern.

Anarchism: A political philosophy that advocates the removal of state and its replacement by voluntary groups of individuals who can sustain social order without any external authority.

Applied Geography (Stamp, 1947): The application of geographical knowledge and skills for the solution or resolution of problems of the society.

Areal Differentiation (Hartshorne, 1939): The study of the variation of the physical and human phenomena as they relate to other spatially proximate and causally linked phenomena.

Behavioural Environment (Gregory, 1978): The perceptual or subjective environment in which the facts of the phenomenal world are organized into conceptual patterns and given meanings or values by individuals within particular cultural contexts.

Behavioural Geography: A psychological turn in human geography which emphasized the role of cognitive (subjective) and decision-making variables as mediating the relationship between environment and spatial behaviour.

Berkeley School: The collective term applied to the group of geographers influenced by Carl O. Sauer during his long years in the Department of Geography at the University of California, Berkeley.

Cartogram (Kidron and Segal, 1984): A specific map projection transforming topographic space according to statistical factors, so that the largest mapping units relate to the greatest statistical values, e.g. census boundaries relative to size of population.

Cartography: The art and science of map-making.

Central Business District (CBD): The nucleus of an urban area, containing the main concentration of commercial land uses (shops, offices and warehouses). This concentration is associated as both cause and effect with both the most accessible point in the urban area and its peak land value. The CBD contains the densest concentration of land uses and the tallest non-residential buildings within the urban area, and is spatially structured internally, with different uses and category of use (e.g. clothing shops). Most CBDs are relative if not in absolute decline, as their characteristic uses are increasingly decentralized to suburban and ecurban locations, as with the growth of planned shopping centres and office parks close to major highway intersections.

Central Place Theory (Christaller, 1933): A theoretical account of the size and distribution of settlements within an urban system in which marketing is the predominant urban function.

Chicago School: The social scientists and geographers of the University of Chicago who concentrated on the study of American cities.

Chorology (or Chorography): The study of areal differentiation of the earth's surface (regional geography).

Christaller Model: A prediction of the spatial arrangement of urban places as service centres, combining nested hierarchies of settlements and market area.

Command Economy: An economy in which the means of production are owned and controlled by the state, and in which central planning of the structure and quantity of outputs prevails.

Critical Rationalism (Popper, 1976): A philosophy of science developed by Karl Popper, originally as a critical response to the logical positivism of the Vienna Circle.

Compage: The concept of compage was introduced by D. Whittlesey in 1956. All of the features of the physical, biotic and societal environments that are functionally associated with the human occupancy of the earth. It means a way of joining or connecting matter. Compage applies a highly diversified although unitary complex.

Concentric Zonal Model (Burgess, 1924, 1927): A model of urban spatial organization developed by Burgess Chicago School.

Conurbation: A term coined by Patrick Geddes to describe a built-up area by the coalescence of once separate urban settlements, initially through Ribbon Development along the main inter-urban routes. The term has been replaced by concepts such as daily urban system, megalopolis, metropolitan area and metropolitan labour area.

Core-Periphery Model (Friedmann, 1966): A model of the spatial organization of human activity based upon the unequal distribution of power in economy, society and polity. The core dominates (although in turn may be dominated from outside), while the periphery is dependent largely on the core.

Cultural Landscape: Landscape developed by man.

Cultural Area: A concept derived from cultural ecology, referring to the geographical region over which a degree of homogeneity in cultural traits may be identified.

Cultural Ecology (Steward, 1955): An approach to the study of relations between the cultural group and its natural environment.

Cultural Hearth (Sauer and Berkeley School, 1952): The area of origin of a cultural group involved in the creation of a particular cultural landscape.

Darwinism (Darwin, 1859): Though Darwinism refers to the version of evolutionary theory, it is difficult to provide a more precise definition of this concept. Darwinism is a group or system of ideas more related by family resemblance than by genetic identity. Darwin argued that a struggle for existence must take place; it followed that those who survived were better adapted to their environments than competitors.

Demographic Transition: (Thompson, 1929 and Nottestien, 1945): A general model describing the evolution of levels of fertility and mortality over time. The model suggests four highly stylized phases in the process of population.

Dialectic (Hegel, 1770–1813 and Marx, 1818–1883): The perpetual resolution of binary opposition.

Diffusion Model (Hägerstrand et al.): The spread of a phenomenon over space and through time. There is a long and distinguished tradition of diffusion studies in cultural geography. According to Sauer (1941), it was Ratzel who founded the study of diffusion of cultural traits, presented in the forgotten second volume of his Anthropogeography. In the opinion of Sauer, 'the filling of the space of the earth' was a 'general problem of social science'. Hägerstrand made a formal study of innovation diffusion. Hägerstrand's *Innovation Diffusion as a Spatial Process* was published in 1968.

Distance Decay (Nystuen, 1968): The attenuation of a pattern or process with distance.

Ecological Fallacy (Robinson, 1950): The problem of inferring characteristics of individual from aggregate data referring to a population.

Ecology (Haeckel, 1866): The study of interrelations between organism and their biotic and abiotic environments.

Ecosystem (Tansley, 1935): The concept of ecosystem means that organism and environment are one interacting unity.

Empiricism: A philosophy of science which accords a special privilege to empirical observations over theoretical statements. Specifically, it assumes that observational statements are the only ones which make direct reference to phenomena in the real world.

Environmental Determinism: The doctrine that human activities are controlled by the environment.

Environmentalism: Environmentalism flourished in the pre-Darwinian period among those such as Henry Buckle who sought for a historicist's history that subjected human activities to natural law, among regional sociologists such as La Play.

Epistemology: The study of knowledge and the justification of belief. An epistemology is a theory which seeks to determine the correspondence between realm of knowledge, e.g. concepts and propositions, and a realm of objects, e.g. experiences and things.

Ethnography: Based on first-hand observation in the field, ethnography employs participation, observation and other qualitative methods to convey the inner life and texture of a particular social group or neighbourhood.

Ethnomethodology: Procedures to discover the general and universal methods by which people make sense of, and give order to, the world. It is based on participation, observation and discussion method to ascertain the geographical reality.

Exceptionalism (regional approach): The belief that geography and history are methodologically distinct from other fields of enquiry, because they are peculiarly concerned with the study of unique and the particular. The idea is closely associated with Kantianism, but in geography, the term is usually identified with Schaefer (1953). Schaefer believed in nomothetic (systematic) geography and not in idiographic (regional) geography.

Footloose Industry: An industry which can locate virtually anywhere because it has no strong material orientation or market orientation in the locational requirements and very wide spatial margins; transport usually involves only a small proportion of the total cost structure.

Functionalism: A perspective from which the world is seen as a set of differentiated and interdependent systems, the collective actions and interactions of which are instances of repeatable and predictable regularities can be assumed to be related, and which explains these form function relations in terms of their role in maintaining the continuity or integrity of the system(s).

General System Theory (Bertalanffy, 1968): An attempt to develop general statements about the common properties of superficially different systems, usually identified with the work of Bertalanffy.

Genre de vie (Vidal de la Blache, 1911): Lifestyle – a form of livelihood functionally characteristic of a human group; for example, transhumants and peasant agriculturists. Along with milieu and circulation, *genre de vie* was a pivotal concept in Vidalian geography (Possibilism). The French geographers produced a number of regional monographs which sought to identify localized cultural landscape to which the interaction of the three gave rise to.

Geocentric Concept: The concept advocated by Ptolemy stating that the earth lies in the centre of solar system and all the planets including the sun (star) and the moon (satellite) revolve around it.

Geography: Geography can be formally defined as the study of earth's surface as the space within which the human population lives (Haggett, 1990), or simply as the study of earth as the home of people (Tuan, 1991). The word is derived from the Greek geo, the earth, and graphein, to write.

Geopiety (Wright, 1966): A term initially coined by Wright to denote the sense of thoughtful piety (sense of human territoriality), aroused by human awareness of the natural world and geographical space, and thus closely connected to topophilia.

Growth Pole Model (Perroux, 1955): A dynamic and highly integrated set of industries organized around a propulsive leading sector or industry (industrie motrice).

Heartland (Mackinder, 1904): A geopolitical concept developed by Mackinder. For Mackinder, the one who controls heartland Siberia, east of Urals, rules the World Island (Asia, Europe and Africa), and rules over the world.

Holism: A belief that the whole organism is greater than simply the sum of its parts. In biology, this implies that there are characteristics of an organism which are not a function of their individual components, which produces laws of the whole (as studies in systematic biology) as well as laws of the parts. In geography, some approaches to the study of regions have adopted a holistic, organismic analogy, and the concept of lebensraum, central to the 1930s German School of Geography, was similarly based on such an analogy.

Human Agency: The capabilities of human beings. Human agency is a central concern of humanistic geography in particular and 'humanism' in the social sciences more generally.

Human Geography: That part of the discipline of geography concerned with the spatial differentiation and organization of human activity and with human use of the physical environment.

Humanistic Geography: An approach to human geography distinguished by the central and active role that it gives to human awareness and human agency, human consciousness and human creativity. It is 'an expansive view of what the human person is and can do' (Tuan, 1916).

Human Ecology (Barrow, 1923): It was in 1923 when H.H. Barrow (University of Chicago) defined geography as 'the science of human ecology'. For geography it is 'the relationship existing between natural environments and the distribution and activities of man', with a focus on human 'adjustment', to those environments. Barrow opined that geography is a nomothetic (systematic) science. The focus of 'human ecology' is on human adjustment to the natural environment.

Iconography (Gottmann, 1952): The description and interpretation of visual image in order to disclose their symbolic meanings. Conventionally applied to religious icons and painted images, 'iconography' was initially introduced into geography by Jean Gottmann (1952).

Idiographic (Windelband, W. and N. Rickert): This is a concept concerned with the unique and the particular (regional geography).

Industrial Location Theory (Weber, A.): A theory about the location of industries which provided the base for industrial geography.

Inner City (Chicago School): An ill-defined area close to the Central Business District (CBD).

Intervening Opportunities: A concept developed by American sociologist, S.A. Stouffer (1940) to explain the pattern of human migration, but since then applied in studies of commodity flow, passenger trips, traffic movements, etc.

Kantianism: A philosophy developed by Immanuel Kant (1724–1804). Kant considered that knowledge could be classified in two ways, either logically or physically.

Lamarckianism: A non-Darwinian theory of evolutionary change originating with the French naturalist Jean Baptiste de Lamarck. As a doctrine of organic progression, Lamarckianism in the pre-Darwinian period differed substantially from its post-Darwinian (neo-Lamarckian successor) period. Lamarck himself did not conceive of evolution as a system of common descent, but rather of 'separate lines progressing in parallel along the same hierarchy'.

Landscape: A polysemic term referring to the appearance of an area, the assemblage of objects used to produce that appearance and the area itself.

Landschaft: Literally and most generally landscape, but the German term is more particularly associated with the continental European School of Landschaftsgeograhie, a tradition which can be traced back to the end of the 19th century when German geographers started to define the subject as 'landscape science'.

Law of Sea: Prior to 1958, most of the law of the sea was derived from customary law, but subsequently three United Nations conferences in a series of convention, have had some success in bringing the oceans within a single body of international law. The most recent is UNCLOS III, agreed on 17 December, 1982 at Montego Bay, Jamaica, and signed by the majority of maritme states. The convention defines seven maritime jurisdictional zones as under:

(i) *Internal water*: All waters landward of the baseline from which the territorial sea is measured, such as rivers, lakes, bays, ports and any waters landward of the low tide line.

(ii) *The territorial sea*: States exercise total sovereignty over these waters, except the rights of innocent passage. They extend out for 12 nautical miles from the baseline.

(iii) *The contiguous zone*: An area 12 nautical miles beyond the limit of the territorial sea, within which states are free to apply customs and other national regulations.

(iv) *The continental shelf*: As area extending 200 nautical miles from the baseline, within which states may claim virtually exclusive rights to the seabed resources.

(v) *Fishing*: Most states now claim excessive, exclusive fishing rights upto 200 nautical miles out to sea from their coasts.

(vi) *Exclusive economic zone*: Synonymous with the redefined continental shelf described in (iv).

(vii) *The high sea*: It includes all waters other than those defined in (i–iv).

Lebensraum (Ratzel): Literally, 'living space' or 'the geographical area within which living organism develops'. In his book on Political Geography, Ratzel equated a nation with a living organism, and argued that a country's search for territorial expansion was similar to a growing organism's search for space.

Limits of Growth (Meadows et al., 1972): A phrase that gained currency as the title of a report issued in 1972 by the Club of Rome on global resource and pollution trends.

Locale (Giddens, 1979): A setting or context for social interaction.

Location Theory: A body of theories which seek to account for the location of economic activities, e.g. von Thunen model of land use (1926), Weber's model or industrial location (1909), Christaller's Central Place Theory (1933), August Losch's Theory of Central Places (1944), etc.

Locational Analysis: An approach to human geography which focuses on the spatial arrangement of phenomena. Its usual methodology is that of spatial science.

Locational Triangle: A simple graphic model devised by Alfred Weber for the analysis of industrial location.

Mega City: A large city (megalopolis) having a population of eight or more than eight millions (e.g. Tokyo, Mexico, New York, Mumbai, Shanghai, etc.).

Megalopolis (Gottmann, 1946): A Greek word (combining the terms for 'great' and 'city' adapted by Jean Gottmann to describe the urban complex, north-eastern seaboard of the USA.

Mental Map (Lynch, 1960): A term referring to the psychological representation of space as shown by simple paper and pencil tests. A psychological turn in human geography in the late 1960s directed attention to the central role of perception as a mediation between the environment and human action.

Mobility Transition Model (Zelinsky, 1971): Mobility is a general term which includes all types of territorial movements, including migration. The social mobility is often designated as circulation, which covers a great variety of movements, usually short-term, repetitive or cyclical in character, but all having in common the lack of any declared intention of a permanent or long-lasting change in residence.

Model: An idealized and structured representation of the real earth surface.

Morphology: The science of forms (landforms).

Multiple Nuclei Model (Harris and Ullman, 1945): A model of internal structure of city having more than one CBD.

Naturalism: The thesis that there is (or can be) an essential unity of method between the natural and the social sciences.

Nomothetic: Concerned with the universal/systematic or the general. The term derived from neo-Kantian epistemology, and most notably from an 1894 address by the philosopher Windbelband, who used it to signify one of the two possible goals of concept formation.

Ontology (Bhaskar, 1978): Theories, sometimes called meta-theories, which seek to answer the question of what the world must be like for.

Paradigm: The working assumption's procedures and findings routinely accepted by a group of scholars, which together define a stable pattern of scientific activity, which in turn defines the community that shares in it.

Pays (Vidal de la Blache, 1927): A micro-region having homogeneity in physical and cultural phenomena, used by Vidal de la Blache in his study of the Massif of France.

Phenomenal Environment (Kirk, 1963): The physical and cultural environment that lies outside individual's perceptions.

Place: A portion of geographical space occupied by a person or thing.

Positivism (Comte, 1820): A philosophy of science originally proposed by Auguste Comte in 1820s and 1830s. The primary purpose of positivism was to distinguish science from metaphysics.

Possibilism (Febvre, 1952): The view that the physical environment provides the opportunity for a range of possible human responses and that people have considerable discretion to choose between them.

Postmodernism: A recent movement in philosophy, arts and social sciences characterized by skepticism towards the grand claims and Grand Theory of the modern era and their privileged vantage point, stressing in its place an openness to a range of voices in social enquiry, artistic experimentation and political empowerment.

Pragmatism: A philosophical perspective, which is centrally concerned with the construction of meaning through the practical activity of human beings.

Principle of Activity (Brunhes, 1920): The principle stressing that everything is changing with the change in time.

Probabilism (Spate, 1957): The view that although the physical environment does not uniquely determine human actions, it does nevertheless make some response more likely than others.

Production of Space: The social production of the 'mental' and 'material' space within which social life takes place.

Qualitative Methods: A set of tools developed to pursue the epistemological mandate of the philosophies of meaning (see epistemology). They are a product of the advent of humanistic geography – a recognition of what Cloke, P. et al. call the 'peopling of human geography'.

Quantitative Methods: The use of mathematical and statistical techniques, theorems and proofs in understanding geographical systems.

Quantitative Revolution: The radical transformation of spirit and purpose (Bunon, 1963) which Anglo-American geography underwent in the 1950s and 1960s, replacing an earlier ideographic concern and areal differentiation by a nomothetic search for models of spatial structure.

Radicalism (Peet, 1977): A term introduced in the 1970s as a reaction to spatial science and positivism. The radicalists stressed on three political issues: the Vietnam war, civil rights of American blacks, and the pervasive poverty and inequality suffered by residents of urban ghettos. Their approach was close to Marxism and *Antipode* was their journal.

Rank-Size Rule (Zipf, 1949): An empirical regularity in the city-size distributions of countries and regions. In its most general form, the cities are rank-ordered in terms of their populations from the largest (l) to the smallest (n).

Region (Whittlesey, 1929): A differentiated segment of earth surface.

Regional Geography: The study of the geography of regions.

Resource: A concept employed to denote sources of human satisfaction, wealth or strength.

Rural Urban Fringe (Pryor, 1980): A zone of transition between the continuously built-up urban and suburban areas of the central city and the rural hinterland. This area displays changes in mix of land use, social and demographic characteristics, and is an arena in which issues of the location of large-scale urban amenities (such as airport and sewage works), the problems of 'fringe' agriculture, the acquisition of land banks for development, and the social integration of commuters have all been prominent. It is also characterized by green belt, floriculture, cultivation of vegetables, etc.

Scientific Determinism: The doctrine that human activities are controlled by the environment. The post-Darwinian determinism is termed as scientific determinism.

Sectoral Model: A model of intra-urban land use patterns developed by Homer Hoyt (1939) using housing (including rental and value) data.

Sense of Place: The character intrinsic to a place itself. The attachments that people themselves have to a place. These are two distinct but interlocking perspectives.

Sequent Occupance (Whittlesey, 1929): The chronological approach stressing that 'each generation of human occupance is linked to its forbear and to its offspring'. The view of geography as a succession of stages of human occupance ... which establishes the genetics of each stage in terms of its predecessors. Whittlesey's scheme owed much to human ecology, but although he knew that the analogy between sequent occupance in chronology and plant succession in botany will be apparent to all, he insisted that his own conception was more intricate.

Social Darwinism: The application of Darwinism evolution to socio-economic and political affairs.

Social Geography: The study of social relations in space and the spatial structures that underpin those relations.

Social Physics: An approach which suggests that aggregate human interactions over space can be explained and predicted using physical theories and laws.

Social Reproduction: Reproduction both of social relations within which, and the material means by which, social life premised.

Social Space (Buttimer, 1969): Space as it is perceived and used by social groups. Space as perceived and used by individuals is usually termed personal space.

Space Economy: The spatial structure of an economy. The term is used to describe economic landscapes of various scales, and geographers typically distinguish between the regional, national and international space economy.

Spatial Analysis (Unwin, 1991): The quantitative procedures employed in locational analysis. The term is sometimes used as a synonym for that portion of the discipline.

Spatial Interaction (Ullman, 1980): A term coined by Ullman to indicate interdependence between geographical areas.

Spatial Science (Berry and Marble): It was formulated during the quantitative revolution and is closely associated with the philosophy of positivism. This is the presentation of human geography as that component of the social sciences which focuses on the role of space as a fundamental variable influencing both society's organization and operation of its individual members.

Stages of Growth (Rostow, 1971): A five-stage sequence of economic and social development, postulated by American economic historian, W.W. Rostow in 1971, through which, he argued, all societies may pass.

System (Chapman, 1977): A group of elements organized in such a way that every element is to some degree interdependent (directly or indirectly) with every other element.

System Analysis: A mathematical approach to the modelling of systems which uses techniques developed in control engineering to investigate the behaviour or systems in response to external stimuli or inputs.

Take for Granted World (Ley, 1977): Usually, a synonym for life world. The term gained currency in contemporary humanistic geography.

Teleology: A theory that events can only be accounted for as stages in the movement towards a pre-oriented end. The end may be determined by those involved in the event, as with various forms of planning, or it may be externally defined, as in many religions.

Terra-Incognita (unknown land): The concept of *Terra-Incognita* was propounded by Ptolemy. He opined that there should be a huge land in the Southern Hemisphere called the *Terra-Incognita*, which was not discovered till the period of Ptolemy. During the period of 'Great Age of Discovery' several explorers attempted to discover this 'unknown land'.

Territoriality (Sack, 1986): The spatial organization of persons and social groups through the demarcation of boundaries.

Thematic Map: A map that depicts statistical variations of objects in space.

Time Geography: An approach in contextual theory originally developed by Hägerstrand and his associates in the University of Lund, which conceives of time and space as providers of 'room' for collateral processes.

Topophilia (literally, the love of place): The term was coined by the French phenomenologist Gaston Bachelard in La Poetique de Pespace (1958). In geography, it was introduced by Yi-Fu Tuan in 1961.

Welfare Geography (Smith, 1979): An approach to human geography emphasizing the questions of inequality and social justice. It developed as a reaction to quantitative models.

World Systems Analysis (Wallerstein, 1974): A materialistic approach to the study of social change.

Bibliography

Abler, R., Adams, J. and Gould, P.R., 1971, *Spatial Organisation: The Geographer's View of the World*, Prentice Hall, Englewood Cliffs.

Ackerman, E.A., 1965, *The Science of Geography*, Washington.

—, 1963, Where is Research Frontier? *AAAG* 53, pp. 429–440.

—, 1958, *Geography as a Fundamental Research Discipline*, University of Chicago, Department of Geography, Research Paper No. 53.

Acton, H.B., 1967, 'Idealism' in P. Edwards (ed.), *The Encyclopedia of Philosophy*, New York. vol. iv, p. 110.

Adams, J. (ed.) 1976, *Urban Policy-Making and Metropolitan Dynamics: A Comparative Geographical Analysis*, Cambridge MA, Ballinger.

Agnew, J. et al. 1996, *Human Geography*, Oxford, Blackwell.

Ahmad, A., 1983, 'Nature of Dichotomy in Geography: General and Particular', *The Indian Geographical Journal*, Vol. 51, No. 2, pp. 97–107.

—, 1978, Dualism and Dichotomies in Geography, *A Discussion Paper, CSRD, JNU*, New Delhi.

Ahmad, N., 1947, *Muslim Contributions in Geography*, Lahore.

Ali, S.M., 1959, *Arab Geography*, Aligarh:

—, *The Geography of Puranas.*

—, 1976, *Arab Geographical Thought*, Aligarh.

Althuser, L., 1978, 'Marxism, Leninism and Class Struggle' *Theoretical Review*, Vol. I, pp. 17–20.

Al-Masudi, *Muruj*, Vol. I.

Alvi, S.M.Z., 1960, Al-Masudi's Conception of the Relationship Between Man and Environment, *Al-Masudi-Millenary Commemoration Volume*, Aligarh.

Amedee, D. and Golledge, R.G., 1975, *An Introduction to the Scientific Reasoning in Geography*, New York.

Anuchin, V.A., 1956, *Regional Study of Transcarpathian.*

—, 1973, 'Theory of Geography' in *Directions in Geography*, ed. R.J. Chorley & London, Methuen pp. 43–63.

—, 1981, The Relationship between Society and Nature in the Geographical Environment and Philosophical Aspects of Geography, In *Soviet Geography Today*, Moscow.

Aristotle, *Meteorologica*, Vol. I.

Atwood, W.W., 1935, 'The Increasing Significance of Geographic Conditions in the Growth of Nation States' *AAAG*, 25, pp. 1–16.

Aune, B., 1970, *Relationalism, Empiricism and Pragmatism: An Introduction*, Random House, New York.

Baker, J.N.L., 1955, *Geography and Its History, Advancement of Science 12*, pp. 188–198.

—, 1963, *The History of Geography*, New York, pp. 130–157.

Barbour, I., (ed), 1973, *'Western Man and Environmental Ethics: Attitudes Towards Nature and Technology'* Addison-Wesley, Massachussets.

Bagrow, L., 1945, 'The Origin of Ptolemy's Geographia', *Geografiska Annaler*, 1945, pp. 318–387.

—, and Skelton, R.A., 1964, *History of Cartography*, Cambridge.

Bartels, D., 1973, 'Between Theory and Metatheory' in Richard Chorley (ed) *Directions in Geography*, London.

Barrow, H.H., 1923, *Geography as Human Ecology, AAAG* 13, pp. 1–14.

Ballabon, M.G., 1957, 'Putting the 'Economic' in Economic Geography' *Economic Geography*, Vol. 43.

Baulig, H., 1950, 'William Morris Davis: Master of Methods' *AAAG*, 40, pp. 188–195.

Beazley, C.R., 1949, *The Dawn of Modern Geography*, Vol. III, New York.

—, 1895, *Prince Henery the Navigator*, New York.

Beck, H., 1959–61, *Alexander von Humboldt*, Vol. I, (1959).

Berry, B.J.L, 1961, 'A Method of Deriving, Multifactor Uniform Region' *Przeglad Geogrficzny*, 20. Trans.

—, 1973, 'A Paradigm for Modern Geography' in *Directions in Geography* ed. R.J. Chorley, London, pp. 3–21.

—, and Marble, D.F., 1968, *Spatial Analysis.*

—, 1973, 'A Paradigm from Modern Geography' in *Directions in Geography*, London.

Bertalanffy, L., 1968, *General System Theory: Foundations, Development, Applications*, New York.

Bird, J.H., 1975, 'Methodology and Philosophy', *Progress in Human Geography*, 2 pp. 133–140.

Blablock, H.M., and Blablock, A., 1959, 'Towards Clarification in System Analysis in Special Series', *Philosophy Sciences*, Vol. 26.

Blanshard, B., 1958, 'The Case of Determinism' in Sidney Hook (ed.) *Determinism and Freedom in the Age of Modern Science*, New York.

Bogekomp, H., 1863, 'An Account of Ritter's Geographical Labour' in W.L. Gage, Trans. *Geographical Studies by the Late Prof. Ritter*, Berlin, pp. 33–51.

Boulding, K.E., 1956, 'General System Theory: The Skeleton of a Science', *Management Science*, 2, pp. 197–208.

Bowman, I., 1916, *The Andes of Southern Peru*, New York.

—, 1934, *Geography in Relation to Social Science*, New York.

—, 1911, *Forest Physiography of the United States and Principles of Soil in Relation to Forestry*, New York.

Braithwaite, R.B., 1953, *Scientific Explanation*, London.

Brigham, A.P., 1903, *Geographic Influence in American History*, Boston.

Broek, J.O.M., 1965, *Geography: Its Scope and Spirit*, Columbus, Ohio.

——, 1968, *Explanation in Social Science*, Chicago.

Brown, L.A., 1960, *Map Making: The Art that Became a Science*, Boston.

——, 1969, 'Disfusion Research in Geography: A Perspective' *Progress in Geography*, 1, pp. 121–157.

Brunhes, J., 1920, *Human Geography*, Chicago, tras. by T.C. Le Compte and edit. by Isaiah Bowman.

Bryan, K., 1935, 'William Morris Davis: Leader in Geomorphology and Geography' *AAAG*, 25, pp. 23–31.

Buch-Hanson, ed. Nielson, B., 1977, 'Marxist Geography and the Concept of Territorial Structure', *Antipode*, 9, pp. 1–12.

Bunbury, E.H., 1883, *A History of Ancient Geography*, Vol. II, London.

Bunge, W., 1962, *Theoretical Geography*, Lund, Sweden.

Bunting, T.E. and Guelke, L., 1979, 'Behavioural and Perception Geography: A Critical Appraisal' *Ann. Ass. Am. Geog.*, 63, pp. 448–462.

Burton, I., 1963, 'The Quantitative Revolution and Theoretical Geography', *The Canadian Geographer*, Vol. VII, pp. 151–162.

Bultimer, A., 1970, *Society and Milieu in the French Geographic Tradition*, Chicago.

——, 1978, Charisma and Context: 'The Challenge of la Geographic Humaine' in *Humanistic Geography: Prospects and Problems*, ed. D. Ley and M.S. Samuel, Chicago, pp. 58–76.

Cahanman, W.J., 1948, 'Outline of Theory of Area Study', *Annals*, Vol. 38.

Carter, G.F., 1950, *Isaiah Bowman: 1878–1950, AAAG*, 40, pp. 335–350.

Castor, de J., 1952, *The Geography of Hunger*, Boston.

Chapman, G.P., 1977, *Human and Environmental Systems: A Geographer's Appraisal*, London.

——, 1966, 'The Status of Geography', *The Canadian Geographer*, Vol. x 3, 1.

Chappel, J.E. Jr., 1967, 'Marxism and Environmentalism' *AAAG*, 57, pp. 203–206.

Chisholm, M., 1962, *Rural Settlement and Land-Use*, London.

——, 1975, *Human Geography: Evolution or Revolution*? Pelican.

Chojnicki, Z., 1970, 'Prediction in Economic Geography', *Eco. Geog.*, 46, pp. 213–22.

Chorley, R.J., 1973a 'Geography as Human Ecology' in *Directions in Geography*, ed. R.J. Chorley, London.

——, Haggett, P. (eds) 1965, *Frontiers in Geographical Teaching*, London, Methuen.

——, 1967, *Models in Human Geography*, London.

Christaller, W., 1933, Eng. Trans, 1966, C.W. Baskin, *Central Places in Southern Germany*, Englewood Cliff.

Claval, P., 1980, 'Epistemology and History of Geographical Thought', *Progress in Human Geography*, 4, London, pp. 371–847.

Clark, A.H., 1949, *The Invasion of New Zealand by People, Plants and Animals-The South Island*, Rutgur Univ. Press.

——, 1954, 'Historical Geography' in P.E. James and C.F. Jones, ed. *American Geography: Inventory and Prospects*, pp. 70–105.

Coates, B.E., Johnston, R.J. and Knox, P.L., 1977, *Geography and Inequality*, London.

——, 'Changing Currents in Geographical Thought', *AAAG*, 26, pp. 1–37.

Colby, C.C., 1933, 'Ellen Churchill Semple', *AAAG*, 23, pp. 229–240.

Commission on History of Geographical Thought, IGU, 1972, *Geographical Thought: A Century of International Congress.*

Cole, J.P. and King, C.A.M., 1968, *Quantitative Geography: Techniques and Theories in Geography*, Wiley & Sons.

Cresswell, Tim, 2013, *Geographic Thought: A Critical Introduction,* Oxford, Wiley-Blackwell.

Crone, G.R., 1950, *Maps and Their Makers: An Introduction to the History of Cartography*, New York.

Collingwood, R.G., 1956, *The Idea of History*, Oxford Univ. Press, New York.

Darwin, C.R., 1859, *On the Origin of Species by Means of Natural Selection or Preservation of Favoured Races in Struggle for Life*, London.

Davies, W.K., 1972, *Conceptual Revolution in Geography*, London.

Davis. W.M., 1899, 'The Geographical Cycle', *Geographical Journal*, 14, pp. 481–504.

Dear, M., 1986, 'Postmodernism and Planning', *Env. Plann.* D. Society and Space 4, pp. 367–84.

Debenham, F., 1960, *Discovery and Exploration in Atlas*, New York.

De Geer, S., 1923, 'On the Definition, Method and Classification of Geography', *Geografiska Annaler*, pp. 1–37.

Demangeon, A., 1905, *La Picardie et les regions voisines, Artois, Cambresis Beauvaises*, Paris.

Dickinson, R.E., 1969, *The Makers of Modern Geography*, London.

—, 1976, Regional Concept: The Anglo-American Leaders, Routledge & Kegan Paul.

—, and Howarth, O.J.R., 1933, *The Making of Geography*, Oxford.

East, W.G., 1935, *An Historical Geography of Europe*, London.

—, 1951, 'Historical Geography' in S.W. Wooldridge ed. *The Spirit and Purpose of Geography*, pp. 80–102.

Einstein, A. and Infeld, L., *The Evolution of Physics*, Cambridge.

Eliot-Hurst, M., 1980 'Geography, Social Science and Society: Towards a De-definition', *Australian Geographical Studies*, 8.

Em. de Martonne, 'Albert-Demangeon', *Annals de Geographie*, 65.

Encyclopaedia Britannica, 1971, Vol. 3, London, p. 894.

English, P.W., 1968, 'Landscape, Ecosystem and Environment Perception: Concepts in Cultural Geography', *The Journal of Geography*, Vol. 67.

Entrikin, J.N., 1977, *Contemporary Humanism in Geography, AAAG*, 66, pp. 615–632.

Eyre, S.R. and Jones, G.J.R., 1966, *Geography as Human Ecology, Methodology by Example*, London.

Fawcett, C.B., 1919, *The Provinces of England*, London.

Febvre, L.A., 1932, *Geographical Introduction to History*, London.

—, 1952, *Le Terra et i' Evolution Humaine Introduction Geographique a l' Historie*, Paris.

Fenneman, N.M., 1919, 'The Circumference of Geography', *AAAG*, 9, pp. 3–11.

Fischer, E., Campbell, R.D. and Miller, E.S., 1969, *A Question of Place: The Development of Geographical Thought.*

—, and Marke, F., 1971, *The Essential Marx*, New York.

Floyd, B., 1971, 'The Quantitative and Model Building Revolution in Geography' *Geography*, Vol. LV, p. 40.

Fleure, H.J., 1917, 'Regions Huamaines', *Annals of Geographie*, 26, pp. 161–74.

—–, 1919, 'Human Regions', *Scottish Geographical Magazine*, 35, pp. 94–105.

Folke, S., 1972, 'Why a Radical Geographer Must be a Marxist', *Antipode* 4, 13–18.

Foote, D.C. etc. 1968, 'An Approach to Systems Analysis in Cultural Geography', *Prof. Geogr.* Vol. 20, pp. 86–91.

Forbiger, *Geographie*, Vol. I.

Forde, D., *Habitat, Economy and Society*, London.

Forer, P., 1978, 'A Place for Plastic Space', *Human Geography*, Vol. II, pp. 203–67.

Fraizer, J.W., 1983, 'Pragmatism: Geography and the Real World, in *Themes in Geographic Thought* ed, Harvey, London.

Freeman, N.M., 1919, 'The Circumference of Geography', *AAAG*, 9, 3–11.

Freeman, T.W., 1961, *A Hundred Years of Geography*, London.

French, R.A., 1968, 'Historical Geography in the U.S.S.R.', *Soviet Geography Reviews*, Transaction, 9, 551–553.

Fuchs. R.J. and Demke, G.J. eds. 1977, *Theoretical Problems of Geography*, ed. V.A. Anuchin, Columbus, Ohio State Press.

Gage, W.L., Trans. 1863, *Geographical Studies by late Prof. Carl Ritter of Berlin*, Boston.

George, N., 1966, *Existentialism Versus Marxism*, New York.

Gerasimov, I.P., 1962, *Soviet Geography*.

—–, 'The Past and Future of Geography' *Soviet Geography*, 7. p. 3.

—–, 1968b, 'Fifty Years of Development of Soviet Geographic Thought'.

—–, 1981, *Soviet Geography*, 9, pp. 238–253.

Gibson, E.M.W., 1983, 'Realism' in *Geographic Themes*, London.

Gilbert, E.W., 1960, 'The Idea of Region', *Geography*, 45, pp. 157–175.

—–, 1933, *The Exploration of Western America, in 1800–1850: A Historical Geography*, Cambridge.

Glacken, C.J., 1956, 'The Changing Ideas of the Hobitable World' in W.L. Thomas ed. *Man's Role in Changing the Face of the Earth*, Chicago.

Godolier, M., 1977, *Perspective in Marxist Anthropology*, Cambridge.

Goldman, R.R., 1973, 'Environmental Limits, Their Perception and Prescription', *International Journal of Environmental Studies*, 5, 193–204.

Goldsmith, W.R., 1966, *Comparative Functionalism: An Essay in Anthropological Theory*, University of California Press.

Golledge, R.G., 1979, 'The Development of Geographical Analysis', *AAAG*, 69, 151–154.

—–, and Amedee, D., 1968, 'On Laws in Geography', *AAAG*, 58, pp. 160–174.

Goode, J., 1973, *Explanations in Social Theory*, New York.

Gottmann, J., 1946, 'French Geography in War Time', *Geographical Review*, 36, pp. 80–91.

Gould, P. and White, R., *Mental Maps*, Baltimore.

Haggett, P., 1966, *Locational Analysis in Human Geography*, New York.

Haeckel, E., 1876, *The History of Creation*, New York, Vol. II.

—–, 1979, *Geographical Synthesis*, New York.

Hale, J.R., 1966, *Age of Exploration*, New York.

Hakluyt, R., 1965, *Hakluytis Voyages*, ed. I.R. Blacker, New York.

Harrison, Church, R.J., 1951, The French School of Geography, in G. Taylor ed. *Geography in the Twentieth Century*, New York.

Hartake, W., 1956, 'Die Sozialbrache als Phanomen der Geographisches Differnzierung der Landschaft, *Erdkunde*, No. 4.

Hartshorne, R., 1950, 'The Functional Approach in Political Geography,' *AAAG*, 40.

—, 1955, 'Exceptionalism in Geography: Re-examined', *AAAG*, 45, 205–44.

—, 1958, 'The Concept of Geography as a Science of Space', from Kant and von Humboldt to Hattner, *AAAG*, 48, pp. 97–108.

—, 1959, *Perspectives in the Nature of Geography*, Chicago.

—, 1961, *The Nature of Geography*, Lancaster.

Harvey, D., 1969, *Explanation in Geography*, London, Edward Arnold.

—, 1973, *Social Justice and the City*, London, Edward Arnold.

Harvey, D., 1989, *The Condition of Postmodernity*, Oxford: Blackwell.

Harvey, M.E., and Holly, B.P. 1983, *Themes in Geographic Thought*, London.

Hawking, S. Einstein Memorial Lecture, New Delhi, *Times of India*, 19 January 2001.

Heinemann, F.H., 1958, *Existentialism and the Modern Predicament*, New York.

Heidel, W.A., 1937, *The Frame of Ancient Greek Maps*, New York.

Hempal, C.G., 1959, The Logic of Functional Analysis. in Cross ed. *Symposium on Sociological Theories.*

Herbertson, A.J., 1905, 'The Major Natural Regions: An Essay in Systematic Geography', *Geographical Journal* 25: 300–312.

Hesse, M., 1966, *Models and Analogues in Science*, Notre Dame, 111.

Hettner, A., 1935, *Vergleichende Landerkunde*, Vol. IV, Berlin.

Hicks, D., 1938, *Critical Realism*, London.

Hill, M.R., 1983, 'Positivism' in *Themes in Geographic Thought*, London.

Hirsch, F., 1976, *Social Limits of Growth*, Cambridge.

Hofstadter, R., 1955, *Social Darwinism in American Thought*, Boston.

Holt-Jensen, Arild, 2009, *Geography 4e,* London, Sage.

Hooson, D.J.M., 1959, 'Some Recent Developments in the Content and Theory of Soviet Geography' *AAAG*, 49, pp. 73–82.

Houston, J.M., 1953, *A Social Geography of Europe*, London.

Hudson, J., 1969, 'A Locational Theory of Rural Settlement' *AAAG*, 52.

Humboldt, A.V., *Kosmos* (English Translation)

Huntington, E., 1907, *The Pulse of Asia*, New York.

—, 1923, *Earth and Sun*, Yale University Press.

—, 1924, *Civilization and Climate*, Yale University Press.

—, 1945, *Main Springs of Civilisation*, New York.

—, 1920, *Principles of Human Geography*, New York.

Husain, M., 2004, *Human Geography*, Jaipur, Rawat Publications

—, 2012, *World Geography*, Jaipur, Rawat Publications.

—, 2014, *Geography of India,* New Delhi, McGraw Hill.

—, 2014, *Ecology and Environment,* 2nd ed. N. Delhi Access Publishing India.

Huxley, T.H., 1977, *Physiography*, London.

Ibn-Battuta, *The Travels of Ibn-Battuta, A.D. 1325–1354*, Cambridge.

Ibn-Khaldun, *The Muqaddimah*. Trans. F. Rosenthal 1958, New York.

Ihde, D., 1977, *Experimental Phenomenology*, New York.

Isachenke, A.G., 1968, 'Fifty Years of Soviet Landscape Science', *Soviet Geography*, 9, pp. 402–407.

Isard, W., 1956, 'Regional Science: The Concept of Region and Regional Structure Papers and Proceedings, *Regional Science Association*.

Jager, H., 1972, 'Historical Geography in Germany, Austria and Switzerland', *Progress in Historical Geography*, pp. 45–62. New York.

Jafferson, 1928, 'The Civilising Rails', *Economic Geography*, Vol. IV. pp. 217–231.

—, 1939, 'The Laws of Primate City', *Geographical Review*, 29, pp. 226–232.

James, P.E., 1972, *All Possible Worlds: A History of Geographical Ideas*, New York.

Jarrel, J.L., 1954, 'Pragmatism' in Jarrett ed. *Contemporary Philosophy*, Chicago.

Jawad, A., 1992, *Development and Social Order*, Anmol Publications, New Delhi.

Joerg, W.L.G., 1922, 'Recent Geographical Changes in Europe', *Geographical Review*, 12. pp. 431–484.

Johnson, D.W., 1909, Twenty-six Davis Papers, *Geographical Essays*.

Johnston, R.J., 1976, *Geography and Geographers: Anglo-American Human Geography Science*, 1945, New York.

—, 1994, *The Dictionary of Human Geography*, 3rd ed., Oxford, Blackwell.

—, 1978, 'Paradigms and Revolution or Evolution', *Progress in Human Geography*, 2, pp. 189–206.

Jones, E. 1964, *Human Geography*, Chatte and Windus.

Kalesnik, S.V., 1964, 'General Geographic Regularities of the Earth', *AAAG*, 54, pp. 160–164.

Karl, A.S., 1959, 'Carl Ritter, 1779–1859', *Scottish Geographical Magazine*, Vol. 75.

Kazakova, O.N., 1966, 'The Development of Theory of Geography and Landscape in East West Germany', *Soviet Geography*, 7, pp. 40–70.

Keat, R. and Urry, J., 1975, *Social Theory as Science*, London.

Kellner, L., 1963, *Alexander von Humboldt*, London.

Kimble, G.H.T., 1938, Geography in the Middle Ages, London.

—, 1945, 'The Craft of Geographer', *The Canadian Geographical Journal*.

—, 1951, 'The Inadequacy of Regional Concept', In L.D. Stamp, and S.W., Wooldridge, eds. *London Essays in Geography*, pp. 151–174.

King, L.W., 1969, *Statistical Analysis in Geography*, Englewood Cliff.

Kirk, W., 1951, 'Historical Geography and the Concept of Behavioural Environment', *Indian Geographical Journal*, Silver Jubilee Volume, 152–160.

—, 1963, Problems of Geography, *Geography*, 48, pp. 357–371.

Klish, J., and Valach, M., 1967, *Cybernetic Modelling*, London.

Knox, P.L., 1975, *Social Well-being: A Spatial Perspective*, London.

Kuhn, T.S., 1970, *The Structure of Scientific Revolution*, Chicago.

Lamb, H.H., 1972, *Climate: Present, Past, and Future*, Vol. 2. London.

Lazle, E., 1972, *Introduction to System Philosophy*, New York.

Leighly, J., 1976, 'Carl Ortwin Sauer, 1889–1975', *AAAG*, 66, pp. 337–348.

Leiss, W., 1974, *Domination of Nature*, New York.

Lewthwaite, G.R., 1966, 'Environmentalism and Determinism: A Search for Clarification', *AAAG*, Vol. 56.

Ley, D., and Samuels, M.S. eds. 1978, *Humanistic Geography: Prospect and Problems*, Chicago.

Lukermann, F., 1958, Towards a more Geographic Economic Geography, *Professional Geographers* 10 (1), pp. 2–10.

Lyell, S.C., *Principles of Geology*, New York.

Macdonel, J.R., 1964, 'Current Trends in French Geography', *Professional Geographers* 17, pp. 15–18.

Mackinder, H.J., 1887, 'On the Scope and Methods of Geography, *Proc. Royal Geographical Society*, 9 pp. 140–160.

—, 1902, *Britain and British Seas*, New York.

—, 1904, 'The Geographical Pivot of History', *Geographical Journal*, Vol. 23.

—, 1919, *Democratic Ideals of Reality*, New York.

Martin, A.F., 1951, 'The Necessity of Determinism' *Transaction*, I.B.G. 17, pp. 1–12.

Martin, G.J., 1968, Mark Jafferson, *Geographers*, E. Michigan University.

Marx, K., 1967, *Capital*, Vol. III, New York.

—, and Engles, F., 1975, *The German Ideology in Marx and Engels*, Collected Works, Vol. V, New York.

Masterman, M., 1970, The Nature of a Paradigm, in Imre Lakates ed., *Criticism and Growth of Knowledge*, Cambridge, pp. 59–89.

Meadows, D.H., and Meadows, D.L., 1972, *Limits of Growth*, London.

Mesrovic, M., and Pestel, E., 1974, *Mankind at the Turning Point*, New York.

Mill, H.R., 1891, *The Realm of Nature*, London.

Minshull, R., 1970, *The Changing Nature of Geography*, London.

Montefiere, A.C. and William W.M., 1955, 'Determinism and Possibilism', *Geographical Studies*, 2, pp. 1–11.

Morgan, W.B., and Moss, R.P., 'Geography and Ecology', *AAAG*, 55.

Morison, S.E., 1971, *The Europeans Discovery of America: The North Voyages*, New York.

Morril, R.J., 1969, 'Geography and the Transformation of Society', Part I, *Antipode*, pp. 6–9.

Myrdal, G., 1953, 'The Relation Between Social Theory and Social Policy', *Br. J. Social*, 23, pp. 210–42.

Natason, M., 1963, *Philosophy of Social Science*, New York.

Neef, E., ed. 1956, *Das Gesichider Erde*, 2nd ed. 1962, Leipzig.

Negel, E., 1961, *The Structure of Science*, London.

Nergdard, A., 1972, *Korologiska*, Oslo.

Nick, M., 1945, *Die Entdeckung von Europe durch die Griechen, Bansal, Benno-Schwal.*

Nikitin, N.P., 1966, 'A History of Economic Geography in Pre-Revolutionary Russia', *Soviet Geography*, 7, pp. 3–37.

Obituary Article, by Cholley, V., *Annales de Geographie*, 65.

Odum, E.P., 1975, *Ecology*, Rienhart and Winston, New York.

Oliveira Martin, J.P., *The Golden Age of Prince Henery the Navigator*, Trans. JJ. Abraham, London.

Olsson, G., 1965, *Distance of Human Interaction: A Review of Biogeography.*

Olsson, G., 1980, *Birds in Egg: Eggs in Bird*, London, Pion.

Orr, L.B., 1950, *The Food Problem, Scientific American*, Vol. 183.

Paschel, O., 1865, *Geschichte der Erdkunde bis auf A.V. Humboldt and Carl Ritter*, Munich.

—, 1879, Psysische Erdkunde, ed. G. *Leipoldt*, Leipzig.

Passmore, J., 1966, *A Hundred Years of Philosophy*, New York.

Peet, R., 1977, The Development of Radical Geography in U.S., in R. Peet ed *Radical Geography*, Chicago, pp. 6–30.

—, 1978, 'The Geography of Human Liberation', *Antipode*, 10, pp. 119–34.

—, 1988, Modern Geographical Thought, Jaipur, Rawat Publications.

Peet, R.J. and Lyons, J.M., Marxism: Dialectical Materialism, Social Formation and ... in *Geographic Themes Thought*, London, p. 190.

Penrose, B., 1952, *Travel and Discovery in Renaissance*, 1420–1620, Cambridge.

Perpillou, A.V., 1966, *Human Geography*, Longmans.

Platt, R.S., 1948, 'Determinism in Geography', *AAAG*, 38, pp. 126–132.

Pred, A., 1967, *Behaviour and Location: Foundation for Geographic and Dynamics Location Theory*, Pt. I (Lund: Gleerup).

Popper, K.R., 1959, *The Logic of Scientific Discovery*, London.

Ratzel, F., 1982–91, *Anthropo Geographie*, Vol. I.

—, 1897, *Politische Geographie* Oder *die Geographie der Staanten, des Verkers und der Krieges*, 2nd ed.

Ravetz, J.R., 1973, *Scientific Knowledge, and its Social Problems*, Oxford.

Raza, M., 1979, *A Survey of Research in Geography*, ICSSR.

Rechis, E., 1866–67, *Le Terre*, Paris.

Relph, E., 1976, 'The Phenomenological Foundations of Geography', *Discussion Paper 21*, Dept. of Geography, Toronto.

Rink, F.T., 1802, *Emmanuel Kant's Physische Geographer Kongsberg.*

Ritter, C., 1822–59, *Die Erdkunde.*

—, *Comparative Geography.*

Rose, G., 1993, *Feminism and Geography*, Cambridge, Polity Press.

Rostow, W.W., 1960, *The Stages of Economic Growth*, London.

Russel, B., 1959, *Wisdom of the West*, Doubleday, Garden City.

Said, E., 1979, *Orientalism*, New York, Vintage.

Samuels, M.S., 1978, 'Existentialism and Human Geography' in D. Ley and M.S. Samuels eds. *Humanistic Geography*, Chicago.

Sartre, J.P., 1966, *Being and Nothingness*, New York.

Sauer, C.O., 1925, 'The Morphology of Landscape', in *Geography* 2, pp. 19–35.

—, 1971, 'The Formative Years of Ratzel in the U.S.A.', *AAAG*, 61.

—, 1941, 'Foreword to the Historical Geography', *AAAG* 31, pp. 1–24.

Schaefe, F., 1953, 'Exceptionalism in Geography', *AAAG*, Vol. 43.

Schultz, J.H., ed. 1959, *Alexander von Humboldt*, Berlin.

Semple, E.C., 1911, *Influences of Geographic Environment on the Basis of Ratzel's Systems of Anthropogeography*, New York.

Shafi, M., 1960, 'Al-Masudi as a Geographer' in *Al-Masudi Millenary Commemoration Volume*, Aligarh.

Simmons, I.G., 1997, '*Changing the Face of the Earth: Culture, Environment, History*', Oxford, Blackwell.

Simon, M., 1960, 'What is a Geographical Factor?', *Geography*, Vol. 51.

Sinnhuber, K.A., 1959, 'Carl Ritter, 1779–1859', *Scottish Geographical Magazine 75*, pp. 152–163.

Smith, D.M., 1947, *Human Geography*, London.

Soja, E., 1989, *Postmodern Geographies: The Reassertion of Space in Critical Social Theory*, London, Verso.

Spate, O.H.K., 1957, 'How Determinism is Possibilism', *Geographical Studies*, London.

—, 1958, 'The End of an Old Song?', The Determinism and Possibilism in G. Taylor ed. *Geography in the 20th Century*, London.

Sarton, G., 1925, *A History of Science: Ancient Science Through the Golden Age of Greece*, New York.

Stamp, L.D., 1947, *The Land of Britain: Its Use and Misuse*, London.

Stewart, J.Q., 1947, 'Empirical Mathematical Rules Concerning the Distribution and Equilibrium of Population', *Geographical Review 37*, pp. 461–85.

Stewart, J.Q., 1965, 'The Development of Social Physics', *American Journal of Physics*, Vol. 18.

Stoddart, D.K., 1965, 'Organisation and Ecosystem as Geographical Models' in Chorley, R.J. ed. Haggett, *Frontier of Geographical Teachings*, London.

—, 1966, 'Darwin's Impact on Geography', *AAAG*, 56, pp. 683–698.

—, 1975, 'The Victorian Science-Huxley's Physiography and Its Impact on Geography', *Transaction*, IBG, 66, pp. 17–40.

Strabo, *The Geography of Strabo*, Trans. H.L. Jones, 1917, New York.

Suskin, Y.G., 1962, 'A History of Soviet Economic Geography', *Soviet Geography*, 7, p. 103.

Sykes, P., 1961, *A History of Explorations from the Earliest Times to the Present Day*, New York.

Taaffe, E., 1974, 'The Spatial View in Context', *AAAG*, 64, pp. 1–16.

—, 1970, *Geography*, Englewood Cliff.

Tatham, G., 1959, 'Geography in the 19th Century' in G. Taylor ed. *Geography in the 20th Century*, London.

Taylor, G., *Geography in the Twentieth Century*, London.

Taylor, P.J., 1976, 'An Introduction of Quantification: Debate in British Geography', *Transaction*, I.B.G. New Series.

Teeph, 1951, 'History of Plants', *Geog. in the 20th Century* by G. Taylor, London.

Terra, H. de, 1955, *The Life and Times of Alexander von Humboldt 1769–1859* New York.

The Encyclopaedia Americana, 1960, Vol. 8, pp. 485–86.

Thomas, W.L., ed. 1956, *Man's Role in Changing the Face of the Earth*, Chicago.

Thomson, J.O., 1965, *A History of Ancient Geography*, New York.

Tooley, R.V., 1949, *Maps and Map Makers*, New York.

Tower, W.S., 1910, 'Scientific Geography', *Bulletin of American Geographical Society*, 42, pp. 801–825.

Tozer, H.F., 1897, *A History of Ancient Geography*, Cambridge.

Thunen von, J.A., 1826, Eng. trans. C.M. Wartenburg in 1966.

Trell, C., 1949, 'Geographical Science in Germany During the Period 1933–1945', *AAAG* 39, pp. 100–137.

—, 1959, Die Geographiche Landschaft and Enforschung, *Stadium Generale*, Vol. 3.

Tuan, Yi Fu, 1971, 'Geography, Phenomenology and the Study of Human Nature', *Canadian Geographer*, 15, pp. 181–192.

—, 1976, 'Humanistic Geography', *AAAG* 66, pp. 266–276.

Ullman, E.L., 1941, 'A Theory of Location for Cities', *American Journal of Sociology*, 46, pp. 853–864.

Ullman, E.J., 1953, 'Human Geography and Area Research', *AAAG* 43, pp. 54–66.

Unstead, J.F., 1916, 'A Synthetic Method of Determining Geographical Region' *Geographical Journal*, 48, pp. 230–249.

Unstead, J.F. et al., 1937, 'Classification of Regions of the World', *Geography*, 22, pp. 253–282.

Van Valkerberg, 1951, 'The German School of Geography' in G. Taylor ed. *Geography in the Twentieth Century*, New York.

Van Valkerberg and Held, C.C., 1952, *Europe*, New York.

Varenius, B., 1650, *Geographic Generalis*, Amsterdam.

Vidal de la Blache, 1926, *Principles of Human Geography*, Paris.

—, 1927,*Geography Universale*, Vol. XV, Paris.

Voeikov, A., 1901, 'De i' influence de I' homme suri 'Terre' *Annales de geographie* 10, 97, 114, 193, 215.

Wanklyn, H., 1961, 'F. Ratzel: A Biographical Memoirs and Bibliography', Cambridge.

Weber, A., 1909, Trans. C. Friederich 1939, as *'Alfred Weber's Theory of Location of Industries'*, Chicago.

Weaver, J.C., 1954, 'Crop Combination Regions in Middle West', *Geographical Review*, 44, pp. 175–200.

White, R., and Gould, P., 1974, *Mental Maps*, Penguin.

Whittlesey, D., 1929, 'Sequent Occupance', *AAAG*, 19, pp. 162–165.

Wiener, N., 1961, *Cybernetic*, Cambridge.

Wilbank, T., and Symanski, R., 1968, 'What is Systematic Analysis', *Professional Geographer*, 20, pp. 86–91.

Wise, M.J., 1977, 'On Progress and Geography' in *Progress in Human Geography*, pp. 1–11.

Wooldridge, S.W., and East, W.G., 1951, *The Spirit and Purpose of Geography*, London.

Wright, J.K., 1925, *The Geographical Lore at the Time of Crusades*, New York.

—, 'The History of Geography: A Point of View', *AAAG*, 15, pp. 200–210.

Wrightly, G.M., 1951, Isaiah Bowman, *Geographical Review*, 14: 7–65.

Yeates, M., 1968, *Introduction to Quantitative Analysis in Economic Geography*, New York.

Zebler, L., 1958, 'Decision-Making in Regional Construction' *AAAG*, 48: 140–148.

Zvonkovo, T.V. and Sauskin, Y.G., 1968, 'Problems of Long Term Geographical Predictions', *Soviet Geography*, 9, pp. 755–765.

Index